DEVELOPMENT IN MAMMALS
VOLUME 1

DEVELOPMENT IN MAMMALS

VOLUME 1

Editor
Martin H. Johnson
Lecturer in Anatomy, University of Cambridge, Director of Studies & Fellow, Christs' College, Cambridge.

1977

NORTH-HOLLAND PUBLISHING COMPANY
AMSTERDAM · NEW YORK · OXFORD

ISBN North-Holland for the series: 0 7204 0632 3
ISBN North-Holland for this volume: 0 7204 0631 5

Published by:

Elsevier/North-Holland Biomedical Press
335 Jan van Galenstraat, P.O.Box 211,
Amsterdam, The Netherlands.

Sole distributors for the U.S.A. and Canada:
Elsevier/North-Holland Inc.
52 Vanderbilt Avenue
New York, N.Y. 10017.

Printed in the Netherlands

PREFACE

This is the first volume in a new series which will be published once or
twice a year. The first two volumes will concentrate on consideration of
peri-implantation embryogenesis and the maternal influences on the embryo.
Contributions to subsequent volumes, both by invitation and submission, may
take the form of critical responses, informed speculation, or new data placed
in the context of a review. No rigid format of length or style will be
applied, and publication will be rapid. It is hoped particularly that younger
scientists will contribute. Any area of mammalian development may be discussed.
With this approach the series will hopefully be provocative and informative.

Attendant in the first appearance of a new series are many practical
problems. The editor wishes to express his particular thanks to Shirley French,
Mitzi Gooding, Debbie Eager, Raith Overhill, Chris Burton, Tim Crane, Freddie
Algate, Don Manning, Judy Winton-Thomas and Debby Hickman for their help and
advice in making the production of this volume possible.

Martin H. Johnson

Cambridge

20th November, 1976.

VI

Acknowledgement

Several illustrations and diagrams in this volume have been obtained from other publications. Some of the original figures have been slightly modified. In all cases reference is made to the original publications in the figure caption. The full sources can be found in the reference lists at the end of each chapter. The permission for the reproduction of this material is gratefully acknowledged.

CONTENTS

DEVELOPMENT IN MAMMALS

Martin H. Johnson

Department of Anatomy

Downing Street,

Cambridge CB2 3DY, U.K.

Mammalian embryonic development comprises three general categories of process i) growth, an increase in size, ii) differentiation, the acquisition of phenotypic variation amongst cells of identical genotype, and iii) morphogenesis, the attainment of the correct spatial and temporal relationships between phenotypically distinct cells. The blue-print required for the regulation of this complex process, achieved nonetheless with such regular success, resides within a single cell - the newly fertilized egg. Studies to elucidate the nature of the blue-print and of the mechanisms by which it is decoded in an orderly and systematic manner have followed two main paths. An indirect approach to the problem has made use of model systems such as teratomas, slime-moulds, plants or free-living larvae, and the results then extrapolated to the mammalian embryo. There have been distinct advantages in the use of this approach. Experimental access to material is easier, and the greater quantity of biological material available has facilitated biochemical and molecular analysis. Many of the conceptual and experimental approaches derived from work on these systems have strongly influenced the studies of those working more directly on mammalian development. Although the validity of simply extrapolating conclusions drawn from model systems to mammalian embryos is questionable, there is at present no compelling reason to believe that the development of the mammalian embryo will prove to operate through mechanisms fundamentally different from those observed and studied in other eukaryotes. In one regard, however, the mammalian embryo is distinctive, since it develops not independently but in prolonged parabiotic liason with a genetically alien individual. The biological advantage conferred by internal fertilization and viviparity has opened the possibility of maternal regulation or control. The cellular and molecular conversations which occur between mother and embryo are critical to normal development as contributions to this first volume show.

The fact of viviparity while beneficial to the embryo has hindered the scientist in the direct study of mammalian embryos themselves. Success in this approach has hinged on the solution to three major technical problems - the successful production and in vitro culture of large numbers of embryos, the development of manipulative techniques for handling the embryos, and the

application of microanalytic techniques to the relatively small quantities of
material available.

The solution to the first of these problems is by no means complete for all
species, but has been achieved for some. The techniques of super-ovulation
originally developed in mice (Edwards & Gates, 1959; Gates, 1971) are now widely
applied to many species including man. Systematic analysis of the culture
conditions required for in vitro fertilization (Chang, 1959; Austin, 1961;
Whittingham, 1968) and for preimplantation development (Brinster, 1968; Biggers,
Whitten & Whittingham, 1971) encourage confidence that, by all available
criteria, a 'normal' environment can be created artificially for several species.
More recently, the achievement of in vitro culture of post-implantation stages
giving growth rates as good as those observed in utero has facilitated study
of whole embryos undergoing organogenesis (New, Coppola & Cockcroft, 1976).

Major experimental manipulation of cultured cleavage stage embryos without
loss of subsequent viability came with the production of aggregation chimaeras
(Tarkowski, 1961; Mintz, 1962). This approach yielded useful information, but
it rapidly became clear that if substantial advances in our understanding of cell
commitment during this early phase of development were to occur, it would be
necessary to surgically dissect the differentiating components of the early
peri-implantation embryo and recombine the components in different spatial and
proportional arrays. Gardner (1968) pioneered the elegant techniques of
microsurgical dissection of embryos that has led to the major breakthrough in
this type of analysis. This approach, supplemented with enzymic (Skreb,
Svajger & Levak-Svajger, 1976) and immunological (Solter & Knowles, 1975;
Handyside & Barton, 1976) approaches to 'embryo surgery', has revolutionised the
study of early mammalian development. The ability to isolate individual or
groups of cells from the preimplantation embryo, to recombine these with complementary cells taken from a genetically distinct embryo, and then to follow
the developmental fate of each cell population has permitted assessment of cell
origin, commitment, potentiality and interaction during the primary differen-
tiative and morphogenetic events of early development (Gardner & Papaioannou,
1975).

An essential feature of these experiments has been the identification of a
reliable and stable cell marker, that enables the fate of the variously
recombined cells to be traced. Physical markers such as dyes or radiolabels
are diluted out with division and may adversely influence cell activity.
Genetic markers should ideally be detectable in all cells at all stages of
development of intact or serially-sectioned embryos. Such a marker has proved
elusive, since use of chromosomal anomalies or isozymes has hitherto required
tissue homogenisation with attendent loss of detailed spatial information.
Despite their limitations, however, these approaches have yielded a great deal

of information (see Gardner & Papaioannou, 1975). The only marker fulfilling all the required criteria has come through the use of interspecific chimaeras between the rat and mouse, in which antigenic distinction of the different cell populations by immuno-fluorescence can be undertaken (Gardner & Johnson, 1973, 1975). With the reservation that the cell interactions occurring in inter-specific chimaeric embryos might disturb the normality of development, the results obtained by this approach have complemented those from use of intra-specific chimaeras, to give an otherwise unattainable level of detail about early developmental interactions.

Surgical manipulations of intact post-implantation stage embryos has only recently become a feasible proposition, although use of isolated foetal tissues to study cell sorting and interaction in development has long been established and has the added advantage of increased quantities of material, albeit often of greater cellular heterogeneity (Saxen & Korkinen-Jaaskelainen, 1975; Grobstein, 1957; Rutter et al., 1968; Dienstman & Holtzer, 1975).

The third major technical requirement for productive analysis of mammalian embryos is the development of micromethods of sufficient sensitivity to probe the molecular behaviour of the few cells of the preimplantation embryos or of small but pure populations of cells taken from later stage embryos. It will be evident from the contributions in this and ensuing volumes that such methods are now not only available but also being applied. Of particular importance are the development of sensitive techniques for detection of protein and RNA synthesis and for analysing the distribution of ions in embryonic tissues. These tech-niques make possible a powerful dual analysis of cellular behaviour and its molecular correlates by which descriptive embryology can be reinterpreted in terms of dynamic molecular and cellular interactions.

It is this new phase in the study of mammalian development, its controversies and its successes, that this series is attempting to reflect.

REFERENCES

AUSTIN, C.R. (1961) Fertilization of mammalian eggs in vitro. Int. Rev. Cytol. 12, 337-356.

BIGGERS, J.D., WHITTEN, W.K. & WHITTINGHAM, D.G. (1971) The culture of mouse embryos in vitro. In: J.C. Daniel, Jr. (Ed.). Methods in Mammalian Embryology. W.H. Freeman & Co., San Francisco, 86-116.

BRINSTER, R.L. (1968) Mammalian embryo culture. In: E.S.E. Hafez and R.J. Blandau (Eds.), The Mammalian Oviduct. University of Chicago Press, Chicago, 419-444.

CHANG, M.C. (1959) Fertilization of rabbit ova in vitro. Nature, 184, 466-467.

DIENSTMAN, S.R. & HOLTZER, H. (1975) Myogenesis: a cell lineage interpretation. In: J. Reinert and H. Holtzer (Eds.), Cell Cycle and Cell Differentiation. Springer-Verlag, Berlin, 1-26.

4

EDWARDS, R.G. & GATES, A.H. (1959) Timing of the stages of the maturation divisions, ovulation, fertilization and the first cleavage of eggs of adult mice treated with gonadotrophin. J. Endocr. 18, 292-304.

GARDNER, R.L. (1968) Mouse chimaeras obtained by the injection of cells into the blastocyst. Nature 220, 596-597.

GARDNER, R.L. & JOHNSON, M.H. (1973) Investigation of early mammalian development using interspecific chimaeras between rat and mouse. Nature New Biology 246, 86-89.

GARDNER, R.L. & JOHNSON, M.H. (1975) Investigation of cellular interaction and deployment in the early mammalian embryo using interspecific chimaeras between the rat and mouse. In Ciba Foundation Symposium on Pattern Formation. 183-200.

GARDNER, R.L. & PAPAIOANNOU, V.E. (1975) Differentiation in the trophectoderm and inner cell mass. In: M. Balls and A.E. Wild (Eds.) The Early Development of Mammals. Cambridge University Press, Cambridge, 107-132.

GATES, A.H. (1971) Maximizing yield and developmental uniformity of eggs. In: J.C. Daniel, Jr. (Ed.). Methods in Mammalian Embryology. W.H. Freeman & Co., San Francisco, 64-75.

GROBSTEIN, C. (1957) Some transmission characteristics of the tubule-inducing influence on mouse metanephrogenic mesenchyme. Exp. Cell Res. 13, 575-587.

HANDYSIDE, A.H. & BARTON, S.C. (1976) Evaluation of the technique of immuno-surgery for the isolation of inner cell masses from mouse blastocysts. J. Embryol. exp. Morph. (in press).

MINTZ, B. (1962) Formation of genotypically mosaic mouse embryos. Am. Zool. 2, 432.

NEW, D.A.T., COPPOLA, P.A.T. & COCKCROFT, D.L. (1976) Comparison of growth in vitro and in vivo of post-implantation rat embryos. J. Embryol. exp. Morph. 36, 133-144.

RUTTER, W.J., KEMP, J.D., BRADSHAW, W.S., CLARK, W.R., RONZIO, R.A. & SANDERS, T.G. (1968) Regulation of specific protein synthesis in cytodifferentiation. J. Cell Physiol. Suppl. 1. 72, 1-18.

SAXEN, L. & KARKINEN-JAASKELAINEN, M. (1975) Inductive interactions in morphogenesis. In: M. Balls and A.E. Wild (Eds.) The Early Development of Mammals. Cambridge University Press, Cambridge, 319-334.

SKREB, N., SVAJGER, A. & LEVAK-SVAJGER, B. (1976) Developmental potentialities of the germ layers in mammals. Ciba Foundation Symposium on Early Mammalian Development. 27-39.

SOLTER, D. & KNOWLES, B.B. (1975) Immunosurgery of mouse blastocysts. Proc. Natl. Acad. Sci. USA 72, 5099-5102.

TARKOWSKI, A.K. (1961) Mouse chimaeras developed from fused eggs. Nature 190, 857-860.

WHITTINGHAM, D.G. (1968) Fertilization of mouse eggs in vitro. Nature 220, 592-593.

SURFACE CHANGES OF THE DEVELOPING TROPHOBLAST CELL

Thomas Ducibella

Department of Anatomy and
Laboratory of Human Reproduction
and Reproductive Biology
Harvard Medical School,
45 Shattuck Street,
Boston, Massachusetts 02115, U.S.A.

Recent studies have provided evidence that a series of cell surface changes during preimplantation development of the mouse embryo are required for normal morphogenesis, differentiation, and implantation. The first major period of membrane differentiation occurs at approximately the 8 cell stage. At this time, changes take place in membrane transport systems (Epstein, 1975; Biggers & Borland, 1976; Borland, this volume), surface glycoproteins (Pinsker & Mintz, 1973), surface antigens (Artzt et al., 1973; Muggleton-Harris & Johnson, 1976; Wiley & Calarco, 1975), intercellular junctions (Ducibella & Anderson, 1975a), and the morphology of the cell surface (Ducibella, Ukena, Karnovsky & Anderson, unpublished). Subsequent to these events, the cell membrane appears to further differentiate with the onset of an active transport system responsible for blastocoel fluid accumulation (Cross, 1973; Borland et al., 1976a); and, at the late blastocyst stage, changes are observed in the cell surface charge (Nilsson et al., 1973; Holmes & Dickson, 1973) and distribution of microvilli (Mayer et al., 1967; Smith & Wilson, 1974) which presumably prepare the embryo for attachment to the uterine lining during implantation.

During the preimplantation period, the outer cells of the mouse embryo differentiate into a well-developed epithelium, the trophoblast. In contrast, the inner cells which give rise to the inner cell mass, only differentiate overtly after implantation. Thus, the trophoblast is considered to be the first differentiated tissue of the mammalian embryo. It plays an important role in the formation of the blastocyst (Enders, 1971; Ducibella et al., 1975) and in establishing the "inside" conditions responsible for the development of the inner cell mass (for a review, see Herbert & Graham, 1974). In this chapter, these two events will be discussed in terms of the acquisition of new intercellular junctions, cell shape changes, reorganization of cell surface and cortical cytoplasmic structures, cell-cell recognition, and changes in surface antigens during development.

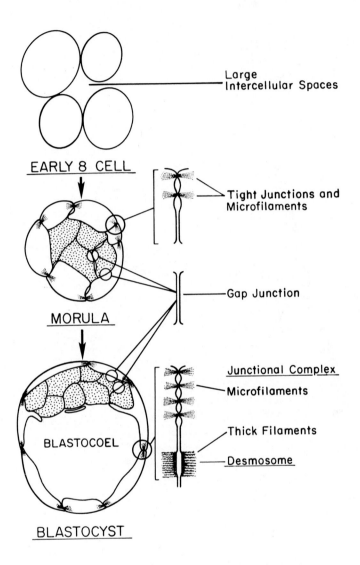

Large Intercellular Spaces

EARLY 8 CELL

Tight Junctions and Microfilaments

MORULA

Gap Junction

BLASTOCOEL

Junctional Complex

Microfilaments

Thick Filaments

Desmosome

BLASTOCYST

Figure 1: Schematic summary of the development of intercellular junctions in the preimplantation mouse embryo.

Figure 2: (opposite) Tight (a, b) and gap junctions (c, d) of the mouse embryo. Tight junctions in cross-sectional profile are observed as focal points of apparent membrane fusion in a stained section (a, arrowheads) which are outlined by lanthanum tracer in an unstained section (b, arrowheads). Gap junctions stained with lead and uranium solutions are observed to have a pentalaminar structure and reveal a narrow intercellular gap when infiltrated with lanthanum (approximately X 120,000).

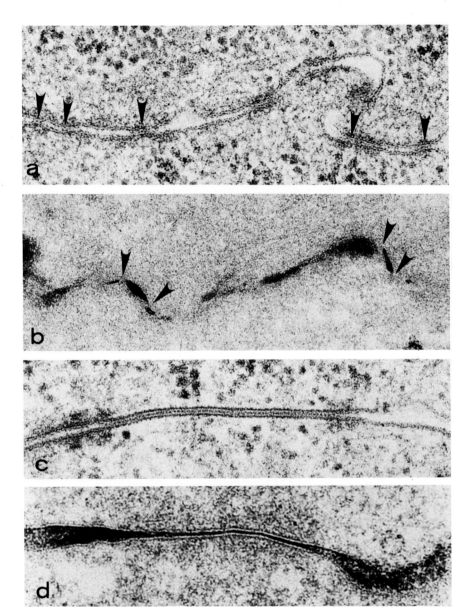

INTERCELLULAR JUNCTIONS

In the early mouse embryo, discrete types of intercellular junctions develop and provide an ultrastructural basis for the accumulation of blastocoel fluid, the cell position theory of determination, and interactions between the trophoblast and inner cell mass (Ducibella et al., 1975). The trophoblast junctional complex is composed of at least three structurally and functionally different kinds of junctions (Figure 1) in the rabbit and mouse (Hastings & Enders, 1975; Ducibella et al., 1975). Even after the advent of intercellular tracers and freeze fracture methods, many studies of mouse embryo junctions have failed to identify the components of the junctional complex and differentiate between tight and gap junctions. A brief discussion of these intercellular contacts is therefore necessary.

Tight junctions appear as focal fusions of the outer leaflets of the plasma membrane and locally exclude extracellular tracers, such as lanthanum and ruthenium red (Figure 2). When a tight junction develops circumferentially around the apical end of an epithelial cell, it is referred to as a zonular tight junction or a "zonula occludens". This junction acts as a permeability seal, limiting the passage of molecules across an epithelium via the inter- cellular spaces. Depending on the number of zonular tight junctions and their structural integrity, epithelia have been classified as relatively "tight" or "leaky" (Claude & Goodenough, 1973). In contrast to the tight junction, the gap junction (Figure 2) is characterized by an intercellular space of 4-5 nanometers when infiltrated with an extracellular tracer (Revel & Karnovsky, 1967; Goodenough & Revel, 1970). This junction is typically macular (not zonular and unable to act as a permeability barrier) and has been strongly implicated in the passage of small ions and molecules from one cell into another. Thus it may serve as a pathway for intercellular communication and metabolic coupling (Gilula et al., 1972). More detailed information on tight and gap junctions is available in reviews by Bennett (1973), McNutt and Weinstein (1973), and Gilula (1974).

In the mouse embryo, zonular tight junctions appear between the outer developing trophoblast cells of the morula. Since the inner cell mass cells of the blastocyst appear to have lost their capacity to become trophoblast (Gardner, 1975; Rossant, 1975a, b), they may undergo determination during the morula stage. The "inside-outside" or cell position theory of development proposes that the determination of the inner cells of the morula (presumptive inner cell mass) is due to their position inside the embryo where an internal microenvironment is created by the surrounding cells (Tarkowski & Wroblewska, 1967). Strong supporting evidence comes from the observations that zona-free embryos or blastomeres in various spatial arrangements form morphologically normal blastocysts (Tarkowski, 1961; Mintz, 1962; Hillman et al., 1972).

The experimental studies of Hillman et al. (1972) indicate that it may take as long as 8 hours of "inside" conditions to suppress the inner cells' ability to become trophoblast.   If determination does occur at the morula stage and requires an internal microenvironment maintained by tight junctions, the development of the zonula occludens would be expected to occur long before the blastocyst stage.   Two independent studies indicate that zonular tight junctions appear early in development and could maintain "inside" conditions for the required length of time.   First, it is at the morula stage (16 - 32 cells) that the tight junctions become impermeable to lanthanum tracer (Ducibella et al., 1975).   Second, in aggregation experiments, embryos which are paired less than 9 hours before cavity formation fail to aggregate (discussed later).   The failure of the cells to intermingle is probably due, in part, to the formation of zonular tight junctions (Burgoyne & Ducibella, unpublished).

The formation of a blastocoel requires that the trophoblast layer should be a tight epithelium, capable of moving fluid from the outside to the inside of the embryo without substantial leakage.   In transport studies, the rabbit has been the species of choice because of its rapid increase in volume from 2 nl on day 3-4 postcoitum to approximately 2.5 ml on day 10.

Zonular tight junctions not only serve to prevent the escape of fluid during blastocoel expansion but also are considered to be an integral part of the various models for epithelial fluid transport (Curran & McIntosh, 1962; Diamond & Bossert, 1967; Hill, 1975).   The classification of the trophoblast in the mouse and rabbit as a tight epithelium is warranted by a growing number of observations: (1) the presence of approximately 6 parallel tight junctional elements visualized in freeze fracture replicas of rabbit blastocysts (Hastings & Enders, 1975; Ducibella et al., 1975),   (2) a transtrophoblast resistance in the rabbit of 2600 ohms cm^2 (Cross, 1973,   (3) the ability of mouse blastocysts to concentrate certain ions, such as calcium, in blastocoel fluid in greater concentrations than the surrounding culture medium (Borland et al., 1976b), and (4) failure of mouse blastocysts to dissociate and collapse after long incubations in calcium-free medium (Ducibella, unpublished observations).

Because both the zonula occludens and release of cytoplasmic vesicles (Calarco & Brown, 1969) appear long before the onset of fluid accumulation, the release of these vesicles into the intercellular spaces is not a sufficient mechanism for blastocoel fluid accumulation.   It is likely, therefore, that an important component of the transport system is activated with the initiation of the blastocyst stage.   Although progress has been made in understanding the physiological mechanism of fluid transport (Borland et al., 1976a, b), the route of fluid transport through the trophoblast remains to be established.   An intracellular step in the development of the transport system or in the transport process itself is suggested by the formation of large vacuoles within dissociated cells (Sherman, 1975; Ducibella & Anderson, 1976).   Cytoplasmic vesicles recently have been proposed to be involved in the transport of ions across frog skin under stressed conditions (Voute et al., 1975).

The sequential appearance of the other two types of intercellular junctions, gap junctions and desmosomes, has been described previously (Ducibella & Anderson, 1975a; Ducibella et al., 1975).   Desmosomes are first observed with the onset of blastocoel formation and are localized between the trophoblast cells.   These junctions serve as strong intercellular spot welds.   Without them, the hydrostatic pressure thought to accompany blastocoel expansion would probably disrupt the zonula occludens resulting in fluid leakage.   Prior to implantation, gap junctions and small adhering junctions fixing the inner cell mass to the polar trophoblast lend support to the observation that the inner cell mass is stationary during blastocyst attachment to the uterus (Gardner, 1975).   An earlier theory advocating the migration of the inner cell mass (Kirby et al., 1967) would require the dissembly of these junctions.   Gap junctions are present during morphogenesis at the 8 cell stage (Ducibella & Anderson, 1975a).   Since the blastomeres are still large enough for electro-physiological coupling studies, it should be possible to determine if morpho-genesis is accompanied by the onset of intercellular coupling via gap junctions.

CELL SHAPE AND MORPHOGENESIS

The formation of the blastocyst is dependent upon the development of an outer layer of transporting epithelial cells associated by zonular tight junctions.   This trophoblast layer could not develop if the blastomeres were to retain the spherical shape which characterizes them in early cleavage stages, because round blastomeres can establish only focal or macular intercellular contacts.

In order for cells of the embryo to establish continuous close apposition at their lateral borders, they must alter their shape and spread on each other. In keeping with this requirement, a striking cell shape change, called compaction, has been observed at the 8 cell stage in the mouse embryo (Figures 3, 4).

Figure 3. (left opposite) The development of a single mouse embryo in vitro. (1: 2 cell; 2: 4 cell; 3: early 8 cell; 4: late 8 cell; 5: early morula; 6: late morula). Frame 4 demonstrates the compacted appearance of the late 8 cell embryo. (Courtesy of Dr. J.G. Mulnard, Archiv. Biol. 78, 107, 1967).

Figure 4. (right opposite) Scanning electron micrographs of uncompacted (a) and compacted (b) 8 cell mouse embryos. Note the change in cell shape and maximi-zation of cell-cell contact. The polar body (PB) does not participate in compaction (b). (From unpublished observations of Ducibella, T., Ukena, M.J Karnovsky, M.J., & Anderson, E).

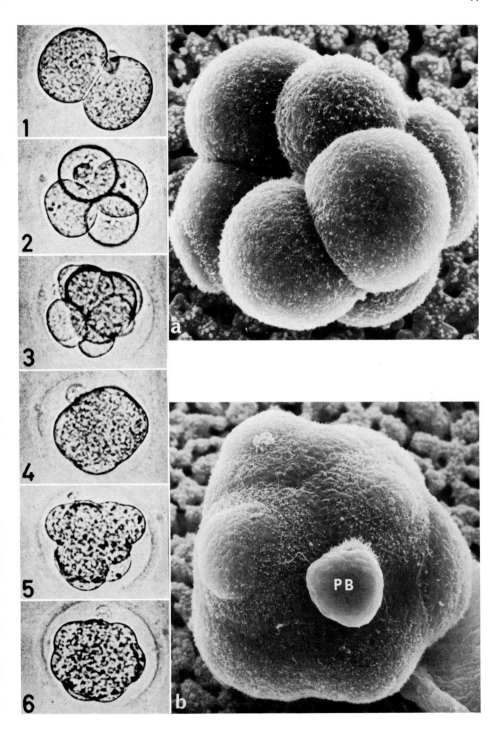

Photomicrographs of living primate embryos demonstrate that compaction occurs at
the 16 cell stage in the monkey (Lewis & Hartmann, 1933) and at a similar stage
in the human (Steptoe et al., 1971) and baboon (Panigel et al., 1975).   Close
cell packing also has been observed at the 8 cell stage in the rat (Schlafke &
Enders, 1967) and rabbit (Assheton, 1894; Lewis & Gregory, 1929).

In their observations on 8 cell mouse embryos, Lewis and Wright (1935)
noted the varying degrees of "compactness" or adhesion of the cells to each other.
Later, a cinematographic study by Mulnard (1967) of the development of a single
mouse embryo from the 2 cell stage clearly demonstrated that the blastomeres
undergo a dramatic shape change during the 8 cell stage (Figure 3).   Recently,
electron microscopic observations of mouse embryos flushed with fixative from
the reproductive tract demonstrate that compaction occurs in vivo and is
accompanied by the formation of focal tight junctions and gap junctions
(Ducibella & Anderson, 1975a,b).

Because compaction marks the onset of tight junction formation and provides
the necessary cell-cell apposition for the development of the zonula occludens
at the morula stage, it is considered to be the initial step in blastocyst
morphogenesis (Ducibella & Anderson, 1975a).   As the blastomeres maximize
contact during compaction, intercellular spaces narrow so dramatically that the
boundaries of individual cells become indistinguishable in the light microscope.
The close apposition of cell membranes of a compacted morula insures that the
inner cells do not have an intercellular "window" facing the zona pellucida.
Such a window would give inner cells access to the "outside" conditions which
are thought to be necessary for trophoblast development.

Thus compaction also is the morphogenetic mechanism which is responsible
for the segregation of the inner cells.   The necessity for complete enclosure
of the inner cells by the developing trophoblast is suggested by the observation
that in an artificial monolayer of morula-stage cells, both the peripherally
and centrally located cells appear to transport fluid (Stern, 1973).   However,
the inner cells of an early morula are probably not completely enclosed all of
the time.   During the mitoses subsequent to compaction, cells round up, and
after division resume their flattened appearance (Figure 3).   Before tight
junctions form between the new daughter cells, there is probably a short
interval in which the inside cells are exposed to "outside" conditions.   The
effect of these small windows in the outer cellular layer is probably minimized
by the asynchrony of cell division.

Studies recently have been completed which provide a new approach to the
study of the inside-outside theory of cell determination by preventing compaction.
Both calcium-free medium and cytochalasin reversibly inhibit compaction
(Figure 5) (Ducibella & Anderson, 1975a).   More recent studies have used a
subthreshold concentration of calcium, 0.02 mM (Ducibella & Anderson, 1976),

because cytochalasin affects cell division.   Eight- to sixteen-cell embryos
were pulsed in defined medium with this subthreshold dose of calcium.   They
continued cleavage, become uncompacted, and can be rescued in normal medium at
the end of the morula stage (Figure 6) at which time they compact and form
blastocysts (Ducibella, unpublished observations).   Whether or not the inner
cell mass of one of these embryos can give rise to normal postimplantation
embryonic structures awaits further study.   It should be possible to determine
whether cell position on the inside of the morula is the only prerequisite for
determination, or, whether the compacted state of the embryo and the tight
junctional permeability seal also are required.

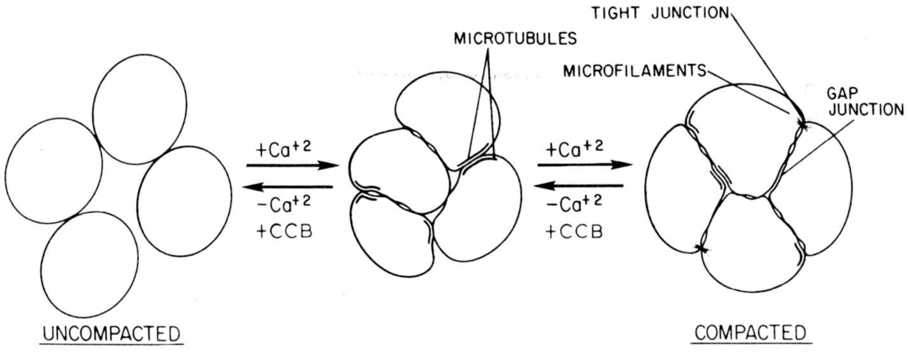

Figure 5: Summary of compaction and its reversible inhibition in vitro.
Diagrammatic sections of 8 cell mouse embryos (modified from Ducibella &
Anderson, 1975a). Calcium-free treatment of compacted embryos results in disruption of apical focal tight junctions; gap junctions and microtubules remain
(unpublished observations).

14

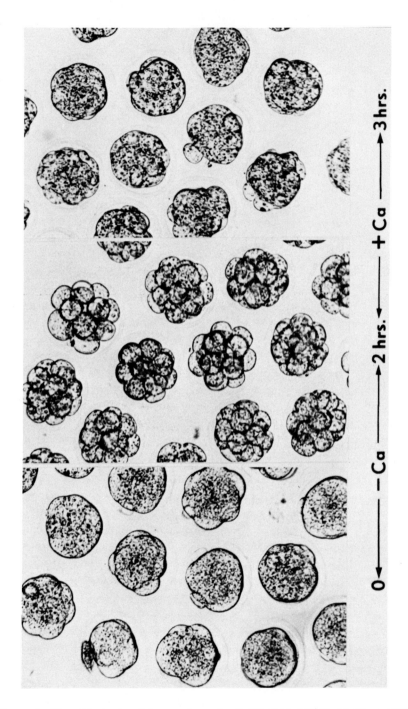

Figure 6: The effect of a 2 hour pulse of calcium-free medium on mouse early
morulae in vitro.

CELL SURFACE AND CORTICAL CYTOPLASMIC ORGANIZATION

The dramatic change in cell shape, the establishment of extensive areas of close membrane apposition, and the development of new intercellular junctions suggests an active role for the cell surface and cortex during compaction. The use of scanning electron microscopy and improved fixation for transmission electron microscopy have revealed that striking changes do occur during compaction in the structure of the blastomere surface and cortical cytoskeletal elements where new cell-cell contacts are established (Ducibella, Ukena, Karnovsky & Anderson, unpublished). Cytoskeletal elements, such as microtubules and microfilaments, are involved in establishing and maintaining cell shape and may play a role in the control of the organization of membrane-associated glycoproteins (Edelman, 1976) and transport systems (Berlin et al., 1974). Cell surface specializations, for example pseudopodia, filopodia (Gustafson & Wolpert, 1961; Albrecht-Buehler, 1976), and microvilli (Lesseps, 1963; Porter et al., 1973; Mooseker & Tilney, 1975) have motile properties and are frequently implicated in cell-cell and (or) cell-substrate interactions (also see Trinkaus, 1969).

At the onset of compaction, a reorganization of the blastomere surface is observed in which microvilli become restricted to an apical region and the basal zone of intercellular contact (Figures 7 and 8) whereas at the 4 and uncompacted 8 cell stages, a uniform distribution is present. The region of membrane between these two localizations of microvilli is relatively smooth. As the blastomeres spread on each other during compaction, many microvilli remain in the basal region of imminent cell-cell contact, as was also observed by Calarco and Epstein (1973), but few are present where the cells have completely spread on each other.

This sequence of events suggests that the basal microvilli play a role in approximating the plasma membranes of adjacent blastomeres and that a zone of microvilli moves apically in the intercellular furrow during compaction. When observed by transmission electron microscopy, these microvilli demonstrate linear arrays of microfilaments. Cross bridges are observed to connect these filaments with each other and to the membrane of the microvillus. This internal filamentous structure is considered to represent the basis for microvillar motility observed in differentiated epithelia (Mooseker & Tilney, 1975). In the compacting embryo, the motile properties of microvilli would allow them to

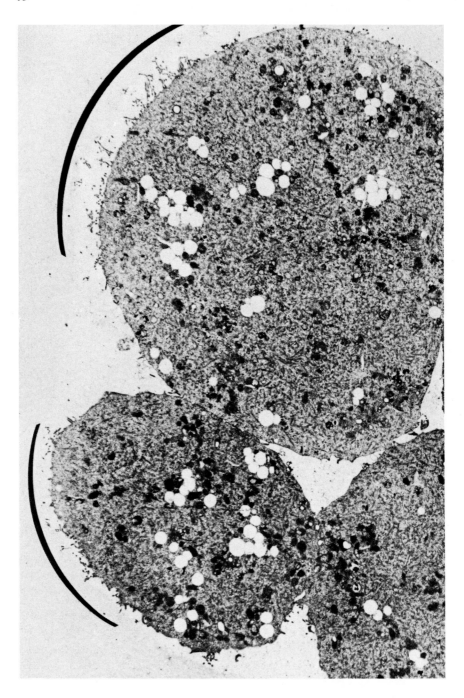

Figure 7: Apical localizations of microvilli (bracketed) on 2 blastomeres of an early 8 cell mouse embryo fixed in vivo (electron micrograph). (x 6,700).

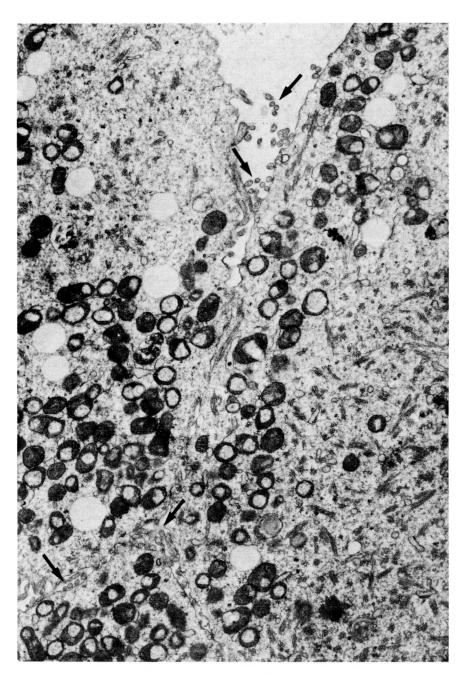

Figure 8: Basal microvilli (arrows) and localizations of
cortical mitochondria. Electron-micrograph of a tangential section from a compacting 8 cell embryo in the area where new intercellular contacts are being
established. (X 14,000).

extend from one cell across to the adjacent blastomere, leading to the development of intercellular adhesions.   Subsequent retraction of the microvilli would bring the membranes of adjacent cells into closer apposition.   If the intercellular cleft is a dynamic region of microvillus formation and/or disappearance, changes in microfilament organization would be expected to occur during compaction.

Beneath the closely apposed cell membranes of compacting embryos, microtubules and mitochondria become localized in the cortical cytoplasm (Figures 8 and 9).   Numerous microtubules are oriented parallel to the cell membrane, indicating that these structures may be involved in establishing or maintaining the compacted state of the embryo.   Similar parallel arrays of microtubules are not observed in the apical cytoplasm underneath the zona pellucida.   In tissue culture cells, microtubules are often observed underlying regions of the plasma membrane which has spread on an artifical substratum (Goldman et al., 1973) and they appear in this region when cells are stimulated to flatten on a substratum by analogues of cyclic adenosine monophosphate (Porter et al., 1974; Borman et al., 1975; Willingham & Pastan, 1975a).   Although compaction is not inhibited at the light microscope level when embryos are treated with low doses of the anti-microtubule agent colchicine (Ducibella & Anderson, 1975a), preliminary experiments demonstrate inhibition when the drug is applied prior to the 8 cell stage.

The localization of mitochondria in regions of new membrane apposition may function to provide locally high concentrations of adenosine triphosphate (ATP), by analogy with their localization near energy requiring transport systems of epithelial cells (Berridge & Oschman, 1972).   During preimplantation development, the turnover of labelled ATP is highest at the 8 cell stage (Ginsberg & Hillman, 1973) indicating an increased requirement for metabolic energy.   It is during this period that mitochondria begin to undergo transformation from the vacuolated type with few cristae to a more orthodox type with many cristae (Stern et al., 1971).   Mitochondria are more randomly distributed in the cytoplasm at the 4 cell stage and near the mitotic spindle during cell division subsequent to compaction. These redistributions of mitochondria reflect major organizational changes in the cytoplasmic matrix (as does the reorganization of microtubules) and possibly changes in metabolism as well.

By both scanning and transmission electron microscopy, it has been shown that, during compaction, cells of the embryo acquire polarity by virtue of a

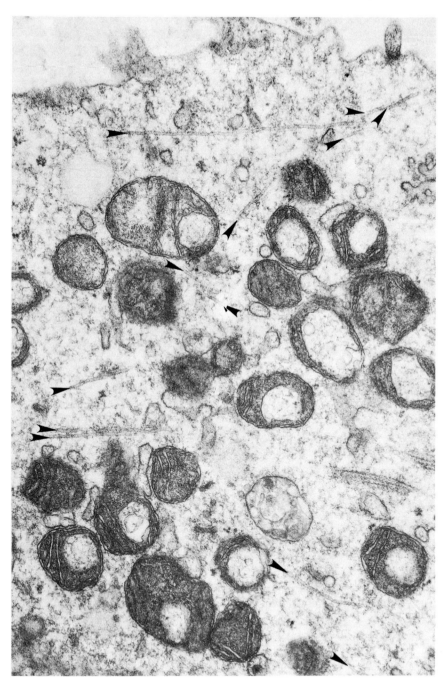

Figure 9: Mitochondria and microtubules (arrowed). Higher
magnification of a section similar to that shown in Figure 8. (X 43,200).

reorganization of their surface microvilli and associated cytoplasmic constituents, microtubules and mitochondria.   These cell surface and cortical
changes are the first manifestations of cell polarity in embryogenesis.   Normal
cell polarity established, in part, by tight junctions appears to be required for
normal transtrophoblast fluid transport at the blastocyst stage (Ducibella &
Anderson, 1976).

These observations on the acquisition of cell polarity also suggest that
morphogenesis involves the activation of a developmental program which
coordinately controls cortical cytoplasmic and cell surface organization.
During compaction, cells begin to acquire the morphological characteristics of
a mature epithelium: close cellular apposition, localizations of apical
microvilli, relatively non-microvillous lateral borders and tight junctions.
Therefore, our working hypothesis is that the early changes in cell structure
observed during compaction, represent the first signs of the development of the
outer epithelial trophoblast layer.   The observation that the cells loose some
features of polarity during post-compaction mitoses is consistent with the fact
that the blastomeres at the 8 cell stage are not yet determined and can give
rise to either inner cell mass or trophoblast (Tarkowski & Wroblewska, 1967;
Kelly, 1975).

CELL-CELL RECOGNITION

The maximization of cell-cell contact observed during compaction and in
aggregation studies (Tarkowski, 1961; Mintz, 1962; Hillman et al., 1972) indicate
that cell-cell recognition is an important process in blastocyst formation.
Observations on compaction, trophoblastic outgrowth in vitro, and delayed
implantation suggest that the outer cell surface of the developing trophoblast
cell undergoes a series of adhesive changes prior to implantation (Table 1).

In compacting embryos only a small number of intercellular junctions are
present and it is probably the non-junctional plasma membranes of adjacent cells
which are responsible for maintaining close cell-cell apposition.   The non-
junctional apposed membranes of these cells are usually within several hundred
angstroms, except where local exo- or endo-cytosis create small enlargements in
the intercellular spaces (Ducibella & Anderson, 1975a).   The non-junctional
cell surfaces in specimens prepared for electron microscopy may not be as closely
apposed as in living embryos due to possible tissue shrinkage or glycoprotein
cell surface coats which are not visualized by routine lead and uranyl acetate
staining.   The establishment of non-junctional intercellular adhesions requires
calcium, which is thought to act by complexing with anionic glycoprotein residues
from apposing cell surfaces or with a divalent intercellular adhesive glycoprotein (Steinberg, 1962; Kuhns et al., 1974).   Evidence that the apposed cell
surfaces of compacted mouse embryos are likely to be within 5-10 nanometers is

TABLE 1

CELL SPREADING

	Cell-Cell*	Cell-Substrate**
2-4 cells	-	-
8-16 cells	+	-
Late Blastocyst	-	+

* Within the embryo

** Zona-free embryos in culture

indicated by the failure of compacted embryos to dissociate in calcium-free medium after exposure to the tetravalent ligand concanavalin A (Ducibella, unpublished observations). This ligand also prevents compaction in vitro (Ducibella & Anderson, 1975b). Inhibition may occur by interfering with the establishment of intercellular adhesions or, as has been suggested for other cell types, by affecting the organization of cytoskeletal structures (Albertini & Clark, 1975; Edelman, 1976) or immobilizing the cell surface (Burger & Noonan, 1970).

In order that the embryo does not prematurely implant or aggregate with other embryos in utero, the surface of the embryo must loose its adhesiveness before hatching from the zona pellucida. During delayed implantation in lactating females, hatched blastocysts must forego attaching to the uterine lining until they receive a maternal signal. At this time, cell surfaces become adhesive. Immediately prior to attachment, changes in surface charge (Nilsson et al., 1973; Holmes & Dickson, 1973) and a decreased number of microvilli have been observed in the outer trophoblast surface (Mayer et al., 1967; Smith & Wilson, 1974). Since cyclic adenosine monophosphate can cause premature implantation in mouse embryos (Holmes & Bergstrom, 1975) and causes regression of microvilli in cultured mammalian cells (Willingham & Pastan, 1975b), the retraction of microvilli may be an important step in blastocyst attachment. Protease activity in the uterus may also play a role in altering the trophoblast cell surface (Mintz, 1971). Cell surface changes also can occur in the absence of the maternal environment. Mouse blastocysts which have developed in a defined medium can attach and spread on culture dishes forming large trophoblastic outgrowths. Suitable environmental conditions appear to be required, such as substrates of collagen (Jenkinson & Wilson, 1973) or cellular monolayers (Cole & Paul, 1965; Salomon & Sherman, 1975) or the culture medium can be modified to contain essential amino acids (Spindle & Pedersen, 1973). Thus, an adhesive change appears to occur both in vivo and in vitro which is responsible for a new phase of trophoblast differentiation.

   In order to determine when the outer surface of the embryo initiates
changes in cell-cell recognition, embryos between the 8 cell and blastocyst
stages were paired in defined medium and allowed to aggregate using the
techniques for making embryonic allophenics or chimeras (Tarkowski, 1961;
Mintz, 1962). When embryos of the same developmental age were used, a large
decrease in the percent of aggregating pairs is observed at the morula stage;
at the late morula - early blastocyst stages, 90-100% of the pairs failed to
aggregate (Burgoyne & Ducibella, unpublished).

   Further studies indicated that the change in the ability of embryos to
aggregate occurs during a period of one to two hours at the morula stage.
The number of hours prior to blastocoel formation was used to stage non-
aggregating pairs.   Greater than 90% of the embryos of pairs which failed to
aggregate, formed blastocoels within 9 hours of the onset of pairing.   This is
approximately the same time that tight junctions become zonular in the developing
trophoblast (Ducibella et al., 1975).   It could be argued that aggregation was
prevented solely by the immobilization of cells by intercellular junctions.
In order to explore this possibility, 8 cell embryos were paired with late
morulae.   If the outer surfaces of both embryos had the same adhesive characteristics during the course of aggregation, the younger embryo should spread around
the outside of the older one until the cells maximize contact and a spherical
aggregate is formed.   During these experiments, cells of the 8 cell embryo
flattened on each other (during compaction) but failed to spread around the
outer surface of the late morula.   These findings are interpreted as evidence
for a major adhesive change in the outer surface of the developing trophoblast
between the 8 cell and late morula stages.   This change may reflect the
beginning of the development of a non-adhesive outer surface which characterizes
epithelial cells (Vasiliev et al., 1975; DiPasquale & Bell, 1975; Elsdale &
Bard, 1975).   It is also likely that these new cell surface properties are
related to those which prevent the late blastocyst from prematurely attaching
to the uterine lining.

   The surface of the trophoblast cell is regionalized into adhesive and non-
adhesive areas separated by zonular tight junctions.   Therefore, certain
regions of the membrane become specialized to interact with other embryonic
cells, while other regions may lose this capability.   These regions are
probably biochemically different as well, since cell surface ATPase activity
appears to be located on the lateral borders of the outer cells of the morula
but not on the surfaces facing the zona pellucida (Kim & Biggers, unpublished
observations).

SURFACE ANTIGENS DURING EARLY EMBRYOGENESIS

Genetic studies have provided independent evidence for the importance of cell-cell recognition and adhesion, and for the sequential appearance during development of new cell surface components. In the mouse, several investigations of an increasing number of T-locus mutations indicate that the products of the dominant T-alleles are required for normal cell surface differentiation (Bennett, 1975). This contention is supported by evidence linking the T-locus and H-2 (Histocompatability) locus to a common ancestral origin as well as by analysis of the t^{12}/t^{12} mutant.

Most embryos of the genotype t^{12}/t^{12} die at the morula - blastocyst transition (Smith, 1956; Mintz, 1964) and, although early studies suggested an RNA defect, there is more compelling evidence that the primary cause is a defective cell surface component (Artzt et al., 1973; Artzt et al., 1974; Vitetta et al., 1975; Bennett, 1975; Jacob, 1975). These studies have utilized indirect antibody methods; however, it still remains to be established whether or not the antiserum which recognizes cells with the normal $+^{t12}$ products binds to t^{12}/t^{12} homozygous embryos. It is noteworthy that the normal antigen(s) are not detected on fertilized eggs but first become prominent at the 8 cell stage and continue to be present on the trophoblast of the blastocyst. Thus, new molecular components thought to be involved in cell-cell recognition appear during the same period in development as major morphological cell surface changes involved in morphogenesis. The observation that cleavage stage embryos of similar age are not a homogeneous population in cytotoxicity tests (Artzt et al., 1973) with the antiserum (above) may be due to the presence of embryos in different stages of compaction. Another mutant, t^{6}/t^{6}, dies in utero just after implantation and, in vitro, embryos develop small trophoblast outgrowths with inner cell masses which appear disorganized or even dissociated, suggesting decreased or abnormal adhesiveness (Pedersen, personal communication). The t^{9}/t^{9} mutation affects the mesodermal layer which lacks proper intercellular junction formation in 8-day embryos (Bennett, 1975). Bennett (1975) has proposed that T-locus dependent cell surface components play an important role in cell-cell recognition during new phases of morphogenesis.

Other surface antigens also become prominent at the 8 cell stage (Artzt et al., 1976) and appear to be of the non-H-2 type (Billington & Jenkinson, 1975; Muggleton-Harris & Johnson, 1976). Using an antiserum made to whole blastocysts, Wiley and Calarco (1975) have demonstrated the appearance of embryo-specific and cleavage-stage specific components appearing maximally at the 8-12 cell stages. Since the antiserum impairs normal development in vitro, it was assumed that its binding sites are required for normal development. Glycoprotein changes in the cell surface also have been detected between the 4-8 cell

24

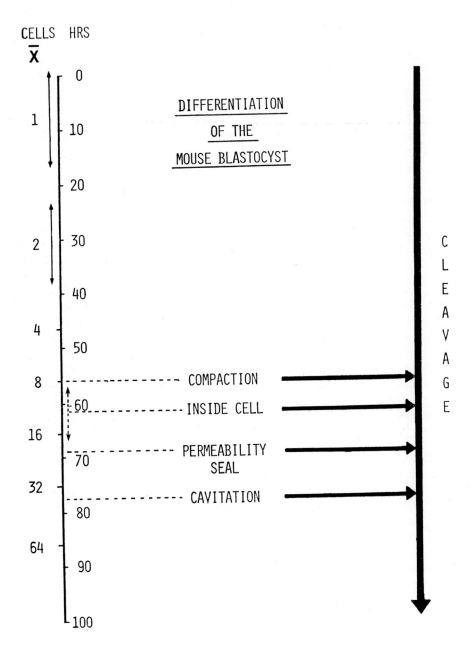

Figure 10: The events of embryogenesis (Courtesy of Dr. P. Burgoyne). The mean cell number and hours of development are taken from Barlow et al., 1972.

and blastocyst stages (Pinsker & Mintz, 1973). The appearance of new surface antigens at approximately the 8 cell stage may be due to new synthesis or unmasking. Although paternal antigens appear at the 6–8 cell stage (Muggleton-Harris & Johnson, 1975), the development of parthenogenetically activated embryos to the blastocyst stage (Graham, 1970; Tarkowski et al., 1970) indicates that maternally-derived cell surface components are sufficient for morphogenesis. When fertilized by normal sperm, the majority of ova from XO mothers undergo compaction later than controls and become abnormal (Burgoyne & Biggers, 1976). It is likely, therefore, that a maternally-derived program and/or metabolic substrates made during oogenesis are involved in normal cell surface differentiation during preimplantation development.

CONCLUSION

The development of the trophoblast begins long before the blastocyst stage and probably as early as the 8 cell stage in the mouse. Evidence has been presented that compaction, blastocyst expansion, and implantation are not merely accompanied by cell membrane changes, but that these changes are responsible for the maximization of cell-cell contact, fluid transport, and adhesion to the uterus.

From a combined morphological and antigenic point of view, the 8 cell stage embryo is one of the most thoroughly investigated stages in preimplantation mouse development. It is at this time that new antigens appear, one of which ($+^{t12}$) is thought to be involved in cell-cell recognition, and major alterations take place in the organization of the cell surface and cortical cytoskeletal elements. For the first 60 hours of development, cell shape and membrane changes are largely a result of the first three cleavage divisions. However, the ensuing 10 hours is an intense period of activity encompassing the first phase in the development of the trophoblast and creating an environment for the presumptive inner cell mass (Figure 10).

In contrast to cell reassociation studies involving enzymatic tissue dissociation and the disruption of normal intercellular relationships, the aggregation of cells during compaction occurs as a normal event during embryogenesis. Thus, the mouse embryo should prove to be an advantageous system for studying the time course of aggregation of embryonic cells and the influence of cell-cell contact on the organization of the cell surface and cytoskeletal structures.

ACKNOWLEDGEMENTS

I would like to thank Paul Burgoyne, Michael Borland, and Everett Anderson for their encouragement and valuable criticism. This work was supported by grant 5 F22 HD 03103-02 from the National Institutes of Health and by Center Grant HD 06645-05.

REFERENCES

ALBERTINI, D.F. & CLARK, J.I. (1975) Membrane-microtubule interactions: concanavalin A capping induced redistribution of cytoplasmic microtubules and colchicine binding proteins. Proc. Nat. Acad. Sci. USA 72, 4976-4980.

ALBRECHT-BUEHLER, G. (1976) Filopodia of spreading 3T3 cells: Do they have a substrate-exploring function? J. Cell Biol. 69, 275-286.

ARTZT, K., BENNETT, D., & JACOB, F. (1974) Primitive teratocarcinoma cells express a differentiation antigen specified by a gene at the T-locus in the mouse. Proc. Nat. Acad. Sci. USA 71, 811-814.

ARTZT, K., DUBOIS, P., BENNETT, D., CONDAMINE, H., BABINET, C. & JACOB, F. (1973) Surface antigens common to mouse cleavage embryos and primitive tetracarcinoma cells in culture. Proc. Nat. Acad. Sci. USA 70, 2988-2992.

ARTZT, K., HAMBURGER, L., JAKOB, H. & JACOB, F. (1976) Embryonic surface antigens: a "quasi-endodermal" teratoma antigen. Devel. Biol. 51, 152-157.

ASSHETON, R. (1894) A re-investigation into the early stages of the development of the rabbit. Quart. J. Micro. Sci. 37, 113-164.

BARLOW, P.B., OWEN, D.A.J. & GRAHAM, C. (1972) DNA synthesis in the pre-implantation mouse embryo. J. Embryol. exp. Morph. 27, 431-445.

BENNETT, D. (1975) T-locus mutants: suggestions for the control of early embryonic organization through cell surface components. In: M. Balls and A.E. Wild (Eds.) The Early Development of Mammals, Cambridge University Press, Cambridge, pp. 207-218.

BENNETT, M.V.L. (1973) Function of electrotonic junctions in embryonic and adult tissues. Fed. Proc. 32, 65-75.

BERLIN, R.D., OLIVER, J.M., UKENA, T.E. & YIN, H.H. (1974) Control of cell surface topography. Nature 247, 45-46.

BERRIDGE, M.J. & OSCHMAN, J.L. (1972) Transporting Epithelia. Academic Press, New York.

BIGGERS, J.D. & BORLAND, R.M. (1976) Physiological aspects of growth and development of the preimplantation mammalian embryo. Ann. Rev. of Physiol. 38, 95-119.

BILLINGTON, W.D. & JENKINSON, E.J. (1975) Antigen expression during early mouse development. In: M. Balls and A.E. Wild (Eds.), Cambridge University Press, Cambridge, pp. 219-232.

BORLAND, R.M., BIGGERS, J.D., & LECHENE, C.P. (1976a) Kinetic aspects of rabbit blastocoel fluid accumulation: an application of electron probe micro-analysis. Devel. Biol. 50, 201-211.

BORLAND, R.M., BIGGERS, J.D. & LECHENE, C.P. (1976b) Studies on the composition and formation of mouse blastocoel fluid using electron probe microanalysis. Devel. Biol. (in press).

BORMAN, L.S., DUMONT, J.N. & HSIE, A.W. (1975) Relationship between cyclic AMP, microtubule organization, and mammalian cell shape. Exp. Cell Res. 91, 422-428.

BURGER, M.M. & NOONAN, K.D. (1970) Restoration of normal growth by covering of agglutinin sites on tumor cell surface. Nature 228, 512-515.

BURGOYNE, P. & BIGGERS, J. (1976) The consequences of X-dosage deficiency in the germ line: impaired development in vitro of preimplantation embryos from XO mice. Devel. Biol. 51, 109-117.

CALARCO, P.G. & BROWN, E.A. (1969) An ultrastructural and cytological study of the preimplantation development in the mouse. J. Exp. Zool. 171, 253-284.

CALARCO, P.G. & EPSTEIN, C.J. (1973) Cell surface changes during pre-implantation development in the mouse. Devel. Biol. 32, 208-213.

CLAUDE, P. & GOODENOUGH, D.A. (1973) Fracture faces of zonulae occludentes from "tight" and "leaky" epithelia. J. Cell Biol. 58, 390-400.

COLE, R.J. & PAUL, J. (1965) Properties of cultured preimplantation mouse and rabbit embryos, and cell strains derived from them. In: G.E.W. Wolstenholme and M. O'Connor (Eds.) Little, Brown and Co., Boston, pp. 82-112.

CROSS, M. (1973) Active sodium and chloride transport across the rabbit blastocoel wall. Biol. Reprod. 8, 556-575.

CURRAN, P.F. & MCINTOSH, J.R. (1962) A model system for biological water transport. Nature 193, 347.

DIAMOND, J.M. & BOSSERT, W.H. (1967) Standing-gradient osmotic flow. A mechanism for coupling of water and solute transport in epithelia. J. gen. Physiol. 50, 2061-2083.

DIPASQUALE, A. & BELL, P.B. (1975) Comments on reported observations of cells spreading on the upper surfaces of other cells in culture. J. Cell Biol. 66, 216-218.

DUCIBELLA, T., ALBERTINI, D.F., ANDERSON, E., & BIGGERS, J.D. (1975) The preimplantation mammalian embryo: characterization of intercellular junctions and their appearance during development. Devel. Biol. 45, 231-250.

DUCIBELLA, T. & ANDERSON, E. (1975a) Cell shape and membrane changes in the eight-cell mouse embryo: pre-requisites for morphogenesis of the blastocyst. Devel. Biol. 47, 45-58.

DUCIBELLA, T. & ANDERSON, E. (1975b) Membrane and cell shape changes required for morphogenesis of the mouse embryo. J. Cell Biol. 67, 101a.

DUCIBELLA, T. & ANDERSON, E. (1976) The effect of calcium on tight junction formation and fluid transport in the developing mouse embryo. J. Cell Biol. 70, 95a.

EDELMAN, G.M. (1976) Surface modulation in cell recognition and cell growth. Science 192, 218-226.

ELSDALE, T. & BARD, J. (1975) Is stickiness of the upper surface of an attached epithelium in culture an indicator of functional insufficiency? J. Cell Biol. 66, 218-219.

ENDERS, A.C. (1971) The fine structure of the blastocyst. In: R.J. Blandau (Ed.), Biology of the Blastocyst, University of Chicago Press, Chicago, pp. 71-94.

EPSTEIN, C.J. (1975) Gene expression and macromolecular synthesis during preimplantation embryonic development. Biol. Reprod. 12, 82-105.

GARDNER, R.L. (1975) Analysis of determination and differentiation in the early mammalian embryo using intra- and inter-specific chimeras. In: C.L. Markert and J. Papaconstantinou (Eds.), The developmental biology of Reproduction, Academic Press, New York., pp. 207-236.

GILULA, N.B. (1974) Junctions between cells. In: R.P. Cox (Ed.), Cell Communication, Wiley & Sons, New York., pp. 1-29.

GILULA, N.B., REEVES, R.O. & STEINBACH, A. (1972) Metabolic coupling, ionic coupling, and cell contacts. Nature 235, 262-265.

GINSBERG, L., HILLMAN, N. (1973) ATP metabolism in cleavage-staged mouse embryos. J. Embryol. exp. Morph. 30, 267-282.

GOLDMAN, R.D., BERG, G., BUSHNELL, A., CHANG, C-M., DICKERMAN, L., HOPKINS, N., MILLER, M.L., POLLACK, R. & WANG, E. (1973) Fibrillar systems in cell motility. In: Locomotion of Tissue Cells. Elsevier, Amsterdam, pp. 83-107.

28

GOODENOUGH, D.A. & REVEL, J.P. (1970) A fine structural analysis of inter-cellular junctions in the mouse liver. J. Cell Biol. 45, 272-290.

GRAHAM, C.F. (1970) Parthenogenetic mouse blastocysts. Nature, 226, 165-167.

GUSTAFSON, T. & WOLPERT, L. (1961) Studies on the cellular basis of morpho-genesis in the sea urchin embryo. Directed movements of primary mesenchyme cells in normal and vegetalized larvae. Exptl. Cell Res. 24, 64-79.

HASTINGS, R. & ENDERS, A.C. (1975) Junctional complexes in the preimplantation rabbit embryo. Anat. Rec. 181, 17-34.

HERBERT, M.C. & GRAHAM, C.F. (1974) Cell determination and differentiation in the early mammalian embryo. Current Topics in Devel. Biol. 8, 151-179.

HILL, A.E. (1975) Solute-solvent coupling in epithelia: contribution of junctional pathway to fluid production. Proc. R. Soc. Lond. B. 191, 537-547.

HILLMAN, N., SHERMAN, M.I. & GRAHAM, C. (1972) The effect of spatial arrange-ment on cell determination during mouse development. J. Embryol. exp. Morph. 28, 263-278.

HOLMES, P.V. & DICKSON, A.D. (1973) Estrogen-induced surface coat and enzyme changes in the implanting mouse blastocyst. J. Embryol. Exp. Morph. 29, 639-645.

HOLMES, P.V. & BERGSTROM, S. (1975) Induction of blastocyst implantation in mice by cyclic AMP. J. Reprod. Fert. 43, 329-332.

JACOB, F. (1975) Mouse teratocarcinoma as a tool for the study of the mouse embryo. In: M. Balls and A.E. Wild (Eds.) The Early Development of Mammals, Cambridge University Press, Cambridge, pp. 233-242.

JENKINSON, E.J. & WILSON, I.B. (1973) In vitro studies on the control of trophoblast outgrowth in the mouse. J. Embryol. exp. Morph. 30, 21-30.

KELLY, S.J. (1975) Studies of the potency of the early cleavage blastomeres of the mouse. In: M. Balls and A.E. Wild (Eds.), The Early Development of Mammals, Cambridge University Press, Cambridge, pp. 97-105.

KIRBY, D.R.S., POTTS, D.M. & WILSON, J.B. (1967) On the orientation of the implanting blastocyst. J. Embryol. exp. Morph. 17, 527-532.

KUHNS, W.J., WEINBAUM, G., TURNER, R., & BURGER, M.M. (1974) Sponge aggregation: a model for studies on cell-cell interactions. Annals New York Acad. Sci. 234, 58-74.

LESSEPS, R.J. (1963) Cell surface projections: their role in the aggregation of embryonic chick cells as revealed by electron microscopy. J. Exptl. Zool. 153, 171-182.

LEWIS, W.H. & GREGORY, P.W. (1929) Cinematography of living developing rabbit eggs. Sci. 69, 226-229.

LEWIS, W.H. & HARTMAN, C.G. (1933) Early cleavage stages of the egg of the monkey (Macacus rhesus). Contrib. Embryol. Carnegie Inst. 24, 187-201.

LEWIS, W.H. & WRIGHT, E.S. (1935) On the early development of the mouse egg. Contrib. to Embryol. Carnegie Inst. 25, 115-143.

MAYER, G., NILSSON, O., & REINIUS, S. (1967) Cell membrane changes of uterine epithelium and trophoblasts during blastocyst attachment in rat. Z. Anat. Entwickl.-Gesch. 126, 43-48.

McNUTT, N.S. & WEINSTEIN, R.S. (1973) Membrane ultrastructure at mammalian intercellular junctions. Prog. Biophys. Molec. Biol. 26, 45-101.

MINTZ, B. (1962) Formation of genotypically mosaic mouse embryos. Am. Zool. 2, 432.

MINTZ, B. (1964) Formation of genetically mosaic mouse embryos, and early development of lethal (t^{12}/t^{12})-normal mosaics. J. exp. Zool. 157, 273-292.

MINTZ, B. (1971) Allophenic mice of multi-embryo origin. In: J.C. Daniel (Ed.) Methods in Mammalian Embryology. W.H. Freeman and Co., San Franciso, pp. 186-214.

MOOSEKER, M.S. & TILNEY, L.G. (1975) Organization of an actin filament-membrane complex. Filament polarity and membrane attachment in the microvilli of intestinal epithelial cells. J. Cell Biol. 67, 725-743.

MUGGLETON-HARRIS, A.L. & JOHNSON, M.H. (1976) The nature and distribution of serologically detectable alloantigens on the preimplantation mouse embryo. J. Embryol. exp. Morph. 35, 59-72.

MULNARD, J.G. (1967) Analyse microcinematographique du developpement de l'oeuf de souris du stade II au blastocyste. Archives de Biologie 78, 107-138.

NILSSON, O., LINDQVIST, I. & RONQUIST, G. (1973) Decreased surface charge of mouse blastocysts at implantation. Exptl. Cell Res. 83, 421-423.

PANIGEL, M., KRAEMER, D.C., KALTER, S.S., SMITH, F.C., & HEBERLING, R.L. (1975) Ultrastructure of cleavage stages and preimplantation embryos of the baboon. Anat. Embryol. 147, 45-62.

PINSKER, M.C. & MINTZ, B. (1973) Change in cell-surface glycoproteins of mouse embryos before implantation. Proc. Nat. Acad. Sci. USA 70, 1645-1648.

PORTER, K., PRESCOTT, D. & FRYE, J. (1973) Changes in surface morphology of Chinese hamster ovary cells during the cell cycle. J. Cell Biol. 57, 815-836.

PORTER, K., PUCK, T.T., HSIE, A.W. & KELLY, D. (1974) An electron microscope study of the effects of Bt_2cAMP on CHO cells. Cell 2, 145-162.

REVEL, J.P. & KARNOVSKY, M.J. (1967) Hexagonal array of subunits in inter-cellular junctions of the mouse heart and liver. J. Cell Biol. 33, C7-C12.

ROSSANT, J. (1975a) Investigation of the determinative state of the mouse inner cell mass. I. Aggregation of isolated inner cell masses with morulae. J. Embryol. exp. Morph. 33, 979-990.

ROSSANT, J. (1975b) Investigation of the determinative state of the mouse inner cell mass II. The fate of isolated inner cell masses transferred to the oviduct. J. Embryol. exp. Morph. 33, 991-1001.

SALOMON, D.S. & SHERMAN, M.I. (1975) Implantation and invasiveness of mouse blastocysts on uterine monolayers. Exptl. Cell Res. 90, 261-268.

SCHLAFKE, S. & ENDERS, A.C. (1967) Cytological changes during cleavage and blastocyst formation in the rat. J. Anat. 102, 13-32.

SHERMAN, M.I. (1975) The role of cell-cell interaction during early mouse embryogenesis. In: M. Balls and A.E. Wild (Eds.) The Early Development of Mammals. Cambridge University Press, Cambridge, pp. 145-165.

SMITH, A.F. & J.B. WILSON (1974) Cell interaction at the maternal-embryonic interface during implantation in the mouse. Cell Tissue Res. 152, 525-542.

SMITH, L.J. (1956) A morphological and histochemical investigation of a preimplantation lethal (t^{12}) in the house mouse. J. Exp. Zool. 132, 51-83.

SPINDLE, A.I. & PEDERSEN, R.A. (1973) Hatching, attachment, and outgrowth of mouse blastocysts in vitro: fixed nitrogen requirements. J. Exp. Zool. 186, 305-318.

STEINBERG, M.S. (1962) Calcium complexing by embryonic cell surfaces: relation to intercellular adhesiveness. In: M.I. Brennan and W.L. Simpson (Eds.) Biological Interactions in Normal and Neoplastic Growth. Little, Brown & Co., Boston, pp. 127-140.

STEPTOE, P.C., EDWARDS, R.G., & PURDY, J.M. (1971) Human blastocysts grown in culture. Nature 229, 132-133.

STERN, M.S. (1973) Development of cleaving mouse embryos under pressure. Differentiation 1, 407-412.

STERN, S., BIGGERS, J. & ANDERSON, E. (1971) Mitochondria and early development of the mouse. J. Exp. Zool. 176, 179-192.

TARKOWSKI, A.K. (1961) Mouse chimaeras developed from fused eggs. Nature 190, 857-860.

TARKOWSKI, A.K., WITKOWSKA, A.,& NOWICK, A. (1970) Experimental parthenogenesis in the mouse. Nature 226, 162-165.

TARKOWSKI, A.K. & WROBLEWSKA, J. (1967) Development of blastomeres of mouse eggs isolated at the 4- and 8-cell stage. J. Embryol. exp. Morph. 18, 155-180.

TRINKAUS, J.P. (1969) Cells into Organs: The Forces that Shape the Embryo. Prentice-Hall, New Jersey.

VASILIEV, J.M., GELFAND, J.M., DOMININA, L.V., ZACHAROVA, O.S. & LUBIMOV, A.V.L. (1975) Contact inhibition of phagocytosis in epithelial sheets: alterations of cell surface properties induced by cell-cell contacts. Proc. Nat. Acad. Sci. USA 72, 719-722.

VITETTA, E.S., ARTZT, K., BENNETT, D., BOYSE, E.A. & JACOB, F. (1975) Structural similarities between a product of the T/t-locus isolated from sperm and teratoma cells and H-2 antigens isolated from splenocytes. Proc. Nat. Acad. Sci. USA 72, 3215-3219.

VOUTE, C.L., MØLLGARD, K. & USING, H.H. (1975) Quantitative relationship between active sodium transport, expansion of endoplasmic reticulum and specialized vacuoles ("scalloped sacs") in the outermost living layer of the frog skin (Rana temporaria). J. Membrane Biol. 21, 273-289.

WILEY, L.M. & CALARCO, P.G. (1975) The effects of anti-embryo sera and their localization on the cell surface during mouse preimplantation development. Devel. Biol. 47, 407-418.

WILLINGHAM, M.C. & PASTAN, I. (1975a) Cyclic AMP and cell morphology in cultured fibroblasts. J. Cell Biol. 67, 146-159.

WILLINGHAM, M.C. & PASTAN, I. (1975b) Cyclic AMP modulates microvillus formation and agglutinability in transformed and normal mouse fibroblasts. Proc. Nat. Acad. Sci. USA 72, 1263-1267.

TRANSPORT PROCESSES IN THE MAMMALIAN BLASTOCYST

Raymond Michael Borland

Department of Physiology and
Laboratory of Human Reproduction
and Reproductive Biology
Harvard Medical School
45 Shattuck Street
Boston, Massachusetts 02115, U.S.A.

The metabolism of the preimplantation mammalian embryo, and the composition of oviducal and uterine fluids have been the object of extensive study (for reviews, see Ducibella & Borland, 1977; Biggers & Borland, 1976; Epstein, 1975; Johnson & Foley, 1974; Hamner, 1971). As discussed by Biggers and Borland (1976) three types of exchanges occur during early pregnancy within the female genital tract: (1) exchanges between the embryo and its microenvironment, (2) those between the microenvironment and the surrounding maternal tissues, and (3) those between the microenvironment and the neighbouring regions of the genital tract. As yet there have been relatively few studies of these processes. For example, Bellve and McDonald (1970) have investigated the pattern of fluid flow in the sheep oviduct. Studies on the active transport of C1 by the ampullary region of the rabbit oviduct (Brunton & Brinster, 1971), the electro-physiology of the rat endometrium (Levin & Edwards, 1968), and the elemental composition of oviducal and uterine fluids in the microenvironment of mouse and rabbit embryos (Roblero et al., 1976; Borland, Biggers & Lechene, 1975; Petzoldt, 1969, 1971) represent major advances in understanding the maternal contributions to the microenvironment. The quantitative aspects of exchange between the embryo and its in vivo microenvironments, however, are totally unknown. Nevertheless, major advances have been made in the elucidation of the transport capacities of the preimplantation embryo in vitro, particularly at the blastocyst stage.

There are two types of mammalian blastocysts: large, rapidly expanding blastocysts, such as those of the rabbit, ferret and pig, and minimally expanding blastocysts, such as those of the human, mouse, rat and hamster (Hafez, 1971). Because the rabbit blastocoele cavity undergoes an enormous increase in size from 2 nl on day 3 post-coitum (p.c.) to 66μ 1 on day 7 p.c., the day of implantation (Daniel, 1964), the rabbit blastocyst is largely composed of water. In contrast, the volume of the blastocoele cavity of the

mouse remains very small ~0.4 - 0.45 nl throughout the preimplantation period
(Dickson, 1966; Biggers, 1972).   Both types of blastocysts appear to form by
similar morphological mechanisms (see Ducibella, this volume).   At the 8 cell
stage, the blastomeres of the preimplantation embryo become closely apposed
during the process of compaction (Mulnard, 1967).   At this time, gap and
zonular apical tight junctions form between the presumptive trophoblast cells
(Ducibella et al., 1975) and are believed to create a permeability seal enabling
some cells to be isolated from the maternal environment and thereby permit the
differention of these cells into inner cell mass cells.   The outer cells of the
compacted embryo differentiate into trophoblast cells.   At the time of blastocyst formation, these cells become highly attenuated, organized into a simple
squamous epithelium, the trophectoderm (defined by Gardner & Papaioannou, 1975),
and acquire the capacity to translocate organic and inorganic solutes and water
from the environment into the blastocoele cavity.

HOMEOCELLULAR VERSUS TRANSCELLULAR TRANSPORT

Studies of the transport phenomena in preimplantation blastocysts have been
confined to mouse and rabbit embryos due to the fact that they can be readily
collected and cultured in vitro.   The differing capacities of the two types of
blastocysts to accumulate fluid have made each species suitable for different
types of transport studies.   Mouse blastocysts, due to their small size, have
been extensively used to study transport of organic solutes, such as nucleosides
and amino acids, into the embryonic cells, i.e., homeocellular transport, on the
assumption that accumulation of the solutes in blastocoelic fluid is small compared to their intracellular accumulation.   In contrast, the rabbit blastocyst,
due to its large size, has been especially suitable for transcellular transport
studies.   In many respects the rabbit blastocyst is a model system for such
studies.   During the preimplantation period it is free-living and maintains no
morphological relationships to the maternal organism.   Thus, it can be easily
flushed or mechanically removed from the female genital tract.   Furthermore,
the rabbit blastocyst is composed largely of a single cell layer surrounded by
a glycoprotein zona pellucida.   In contrast, most other epithelia that have
been used in transport studies, such as frog skin, toad bladder, kidney tubules
etc., contain multiple cellular and non-cellular layers and cannot be studied
in vitro without stripping them from the animal, thereby disrupting many
morphological relationships between the epithelium and the surrounding tissues.

In general, homeocellular transport is responsible for the creation and
maintenance of a high concentration gradient of a solute across the cell
membrane.   Transcellular transport, in contrast, may occur against low concentration gradients and is often involved in the movement of large amounts of
solute across the epithelium.   The distinction between the two types of

transport is not fundamental, however, since carrier-mediated transport mechanisms in either case are presumed to be located in the cell membrane and hence be primarily homeocellular.   Thus, transcellular transport can be viewed as a secondary phenomenon which appears if the cells become organized into an epithelium and are polarized with respect to the transport abilities or permeability properties of their cellular membrane.   The tight junctions that associate the trophoblast cells of the blastocyst (Ducibella et al., 1975; Hastings & Enders, 1975) are the morphological basis for physiological polarity that is established in the trophoblast cells.

CRITERIA FOR CARRIER-MEDIATED TRANSPORT

Postulation of carrier-mediated transport is based upon the assumption that there are a finite number of membrane-bound carriers with a finite rate of turnover.   This assumption has led to several criteria that are commonly used to establish that a carrier-mediated transport system exists (Stein, 1967).

(a)   The first and strongest argument for involvement of a carrier in solute translocation is the demonstration of movement of a substance in the absence of or against its chemical or electrochemical potential, as occurs in active transport (defined by Rosenberg, 1948).

(b)   Second, if the substrate concentration is raised, net translocation of the solute will approach a maximal attainable rate or be saturable.   This is in contrast to diffusion where translocation is linearly dependent upon the solute's gradient.

(c)   If a reversible solute-carrier complex is formed, followed by a relatively slow translocation step, an analogy between transport and Michelis-Menten enzyme kinetics can be made.   For example, a Lineweaver-Burk plot can be employed to study the unidirectional fluxes, and net fluxes under circum-stances where the forward flux vastly exceeds the back flux.

(d)   Just as self-saturation can occur, the addition of a second solute that uses the same transport carrier may reveal hetero-saturation, or com-petition for the same transport carrier.

(e)   Uphill transport of one solute may occur by the dissipation of a gradient for a second solute, i.e., countertransport, as long as the latter gradient persists.

(f)   Carrier-mediated transport processes are commonly affected by enzyme poisons, protein reagents, and hormones.

(g)   A high temperature dependence has often been used as a strong criterion of carrier-mediation but is now recognized to be weak.   While it is true that a carrier-mediated process may show a heat of activation in excess of the 4500 calories per mole characterizing free diffusion (Solomon, 1952), it is now known that the lipid membrane itself may represent a high energy barrier for pure diffusion of hydrophilic solutes.

NUCLEOSIDE TRANSPORT IN BLASTOCYSTS

The mean values of RNA per mouse embryo in freshly collected unfertilized
ova, in 1, 2, 8 to 16 cell embryos and in blastocysts increase significantly
after the 8 to 16 cell stage (Olds et al., 1973). Similarly, the total amount
of RNA per rabbit embryo changes little during early cleavage and then markedly
increases after the 16 cell stage (Manes, 1969). In preimplantation mouse
embryos an increasing amount of labeled uridine is incorporated into RNA between
the 8 cell and morula-blastocyst stages (Mintz, 1964; Monesi & Salfi, 1967;
Ellem & Swatkin, 1968; Woodland & Graham, 1969). These results were interpreted to mean that RNA synthesis is low during early cleavage and increases at
the morula and blastocyst stages. However, interpretation of the results
using labeled uridine was ambiguous in these early experiments, due to a lack
of knowledge of the unlabeled precursor pool in the embryo and the specific
activity of the immediate precursor, radioactive UTP, at the site of RNA
synthesis.

Since the specific activity of UTP would be determined by both the endogenous unlabeled pool of UTP and the rate of uptake of labeled uridine, efforts
were made to dissociate incorporation from the uptake process by determining
for each developmental stage the relationship between uptake of the precursor
into a soluble pool and its incorporation into macro-molecules (see Epstein,
1975). The technique was based upon the assumption that elevation of the
concentration of the labeled solute in the medium should eventually saturate
the endogenous pool with radioactive precursor and, thus, not limit the inherent
rate of RNA synthesis as detected by isotope incorporation. These studies led
to our current knowledge concerning nucleoside transport in the preimplantation
embryo. Daentl and Epstein (1971), for example, demonstrated that an increase
in the uptake of labeled uridine occurs concurrently with an increase in precursor incorporation between the 8 cell and morula-blastocyst stages. Uptake
of uridine is concentration-dependent and incorporation of labeled precursor is
maximal at substrate concentrations above 1μ M. Epstein and Daentl (1971)
noted a somewhat similar relationship between adenine and adenosine uptake and
incorporation in mouse embryos during 2 hr incubation periods. Elucidation of
the transport process involved in nucleoside uptake first came with the work of
Epstein et al. (1971). These investigators demonstrated that guanine uptake by
preimplantation mouse embryos increases during the first 3 days of development.
The major increase in guanine uptake occurs between the 2 and 8 cell stages, in
contrast to the pattern for uridine uptake (Daentl & Epstein, 1971), but net
uptake reached a maximum rate of 0.115 pmoles/hr/embryo at the early blastocyst
stage after 1-2 hr incubations. It was not shown whether uptake after such
long incubations is in a steady-state. Nevertheless, it was possible to

demonstrate that guanine uptake is followed by its rapid conversion to GMP by hypoxanthine-guanine phosphoryltransferase. The GMP is then further phosphorylated to GDP and GTP, the latter representing 65% of the soluble radioactivity in the embryo. At all stages of development very little intracellular radioactivity was recovered as guanine or guanosine using DEAE chromatography.

Subsequently, Daentl and Epstein (1973) characterized the mechanisms for uridine uptake by mouse blastocysts. Initial experiments revealed that the magnitude of uridine uptake is directly proportional to the number of blastocysts per sample and to the incubation time up to 20 minutes. Using 10 minute incubations which should reflect the initial rate of uridine uptake, these investigations observed saturation of the uridine transport system as the concentration of this nucleoside in the medium is raised. From a Lineweaver-Burk plot the apparent K_m and V_{max} were determined: 1.6μ M and 0.0063 pmoles/min/embryo, respectively. Unfortunately, these investigators were unable to determine K_m and V_{max} values at earlier developmental stages due to the large number of embryos required to obtain sizeable uptake. This made it impossible to distinguish whether the increased uptake of uridine after the 8 to 16 cell stage is secondary to activation and/or synthesis of transport systems or to variations in the volume of the embryos and size of the endogenous precursor pool. Nevertheless, uptake of uridine at both the morula and blastocyst stages is highly temperature sensitive and falls to $\sim$ 5% of control values at 4°C.

TABLE 1

URIDINE UPTAKE AND METABOLISM BY MOUSE BLASTOCYSTS*

Uptake (pmoles/embryo/30 min x 10^3)	Intracellular Metabolites (%)			
	UTP	UDP	UMP	Uridine Uracil
150 ± 15.5	53	24	8	16

* Mean ^{3}H-uridine uptake and percent of intracellular uridine metabolites after 30 minutes incubations at 37°C. Separations were achieved with DEAE-cellulose paper chromatography (From Daentl & Epstein, 1973).

As with guanine uptake, fractionation of intracellular radioactivity in the blastocyst revealed that most of the soluble radioactivity is present as the triphosphate and only a small fraction of the activity remains as uridine (Table 1). Uptake of uridine was inhibited by dinitrophenol, an uncoupler of oxidative phosphorylation, and by potassium cyanide, an inhibitor of cytochrome C, suggesting that phosphorylation of uridine to form UTP is an integral part of the uridine uptake mechanism. Other nucleosides, such as thymidine, adenosine, deoxyuridine, cytidine and deoxyadenosine, compete with labeled uridine for uptake and probably share the same carrier system.

The overall characteristics of uridine uptake in mouse blastocysts suggest that translocation is carrier-mediated. Similar carrier-mediated nucleoside transport systems have been found in other cells in culture in which nucleoside transport occurs via a single carrier-mediated system with broad specificity (Skehel et al., 1967; Scholtissek, 1968; Plagemann, 1971; Steck et al., 1969; Oliver & Paterson, 1971; Lieu et al., 1971).

As mentioned earlier, **carrier-mediated** transport systems are often affected by hormones. A limited amount of information is available suggesting that uridine uptake in mouse blastocysts is sensitive to steroids. Lau et al. (1973) demonstrated that physiological levels of estradiol (10^{-8} to 10^{-10} M) enhance labeled uridine uptake in 9 day delayed mouse blastocysts cultured in protein-free medium. The physiological basis for this apparent stimulation of uridine transport is not known but may involve changes in membrane permeability as suggested by Mohla and Prasad (1971). This question will not be resolved, however, until the kinetic parameters of labeled uridine transport are elucidated before and after exposure to the steroid.

AMINO ACID TRANSPORT

Recent interest in the amino acid transport of preimplantation embryos was stimulated by earlier studies in which the overall rate of protein synthesis was monitored in preimplantation embryos of different ages exposed for short time intervals to radioactive amino acids. The labeled proteins were detected by autoradiography or measured by scintillation counting of the radioactivity incorporated into a trichloroacetic acid insoluble fraction. The results show that synthesis of protein occurs at all stages of pre-implantation development of the mouse. The rate of protein synthesis is low until the 8 cell stage, after which it rapidly increases (for reviews, see Biggers & Borland, 1976; Epstein, 1975; Biggers & Stern, 1973). Observations that fertilized ova contain few ribosomes, whereas morulae and blastocysts contain many ribosomes and polyribosomes, support the interpretation that protein synthesis increases with development (Calarco & Brown, 1969).

As with nucleoside incorporation, an unambiguous interpretation of the pattern of labeled amino acid incorporation was impossible due to a lack of simultaneous data concerning the size of the endogenous amino acid pools and the transport capacity of the different stage embryos. Reports that total amino acid uptake per embryo increases between the 1-8 cell and morula stages (Tasca & Hillman, 1970; Brinster, 1971) suggested that permeability changes occur during the preimplantation period; however, the relationship between amino acid uptake and incorporation, and the characteristics of amino acid

accumulation remained largely obscure until the studies of Epstein and Smith
(1973) and Borland and Tasca (1974, 1975). These studies had certain
limitations. For example, Epstein and Smith (1973) used only 3-10 embryos
per sample and each uptake value represented the mean of merely 3 samples.
Furthermore, after labeling the embryos with radioactive amino acids, they
were washed 12 times in medium containing a non-radioactive form of the amino
acid at a high concentration of 1 mM. The washing procedure took 5 minutes
during which time there could have been considerable exchange diffusion of
intracellular labeled amino acid with the unlabeled amino acid in the wash
medium. Such a process could artificially lower the actual uptake values.
Both groups of investigators were also obliged to use relatively long
incubation periods (>15 minutes) due to the low uptake of radioactive amino
acids in limited numbers of embryos, and were unable to measure the endogenous
amino acid pools in the embryos under the various labeling conditions.
Measurement of the intracellular specific radioactivity of amino acids in
preimplantation embryos (Brinster et al., 1976) and the use of nonmetabolizable
radioactive amino acids (Heinz, 1972) hold promise of eliminating some of the
ambiguities of these uptake studies. Despite these possible limitations,
Epstein and Smith (1973) demonstrated that the rates of amino acid uptake and
incorporation in the mouse are low and relatively constant until the early
blastocyst stage at which time both rates markedly increase. Uptake continues
to increase in late blastocysts but incorporation remains constant. Thus
amino acids show an increase in transport capacity between the 8-16 cell and
early blastocyst stage similar to that shown for nucleosides. Amino acids,
however, show a further increase in uptake in late blastocysts that was not
evident for nucleoside transport.

Amino acid uptake in mouse embryos exhibits a number of characteristics
of carrier-mediated transport. Leucine uptake is saturable in morula-early
blastocyst stage embryos after 1 hr incubations when the concentration of
leucine in the medium is increased to 400μ M. Early blastocysts incubated for
2 hr in 212μ M leucine maintain a significant concentration gradient of
labeled leucine; soluble radioactivity within the embryo, 67% of which remains
unmetabolized as leucine, is 16-fold more concentrated than in the medium.
The uptake of both leucine and lysine is temperature sensitive with a Q_{10} of
3-4. Finally, both leucine and lysine uptake is inhibited by competition with
structurally similar amino acids (Epstein & Smith, 1973).

Borland and Tasca (1974) confirmed some of these findings and further
elucidated the mechanisms by which amino acids are accumulated by 4 cell, late
morulae and early mouse blastocysts. Thirty 4 cell embryos or twenty early
blastocysts were incubated at 37°C in a protein-free medium containing either
labeled leucine or methionine for 15 to 60 minutes. The embryos were then

washed in ice-cold amino acid-free medium to limit loss of soluble radioactivity from the embryo.   Uptake during the labeling period was linear with
leucine and methionine concentrations from 10^{-7} to 10^{-4} M, so that 15 minute
uptake values were used to estimate the initial rate of amino acid uptake.
Since uptake did not reach a steady state, the net isotope uptake in 15 minutes
was assumed to approximate the unidirectional influx.   Lineweaver-Burk double
reciprocal plots of these 15 minute uptake values were then used to calculate
the apparent K_m and V_{max} values (Table 2). ^{3}H-leucine uptake at both the
4 cell and early blastocyst stages exhibits similar K_m values, but a 2.5-fold
increase in V_{max} occurs at the blastocyst stage.   Methionine uptake in early
blastocysts exhibits complex kinetics suggestive of two transport systems
(Christensen, 1966), one with a K_m similar to the K_m at the 4-cell stage.   The
V_{max} for methionine transport in one of the blastocyst systems is approximately
2-fold greater than the V_{max} at the 4-cell stage.   The increased V_{max} values
for both leucine and methionine at the early blastocyst stage suggest that
there is a change in the number or mobility of carrier sites, as has been
suggested in other cell preparations (Isselbacher, 1972; Martin et al., 1971;
Foster & Pardee, 1969).

TABLE 2

UPTAKE OF ^{3}H-LEUCINE AND ^{3}H-METHIONINE BY PREIMPLANTATION MOUSE EMBRYOS*

Amino Acid	Four-cell stage		Early blastocysts	
	K_m	V_{max}	K_m	V_{max}
	(M)	(pg/embryo/15 min)	(M)	(pg/embryo/15 min)
^{3}H-leucine	33	20	50	50
^{3}H-methionine	63	56	890	22
			62	104

*   Embryos were incubated at 37°C in protein-free medium containing the
     labeled amino acid.   Lineweaver-Burk double reciprocal plots of the 15 min
     uptake of each labeled amino acid were performed to calculate the kinetic
     parameters, K_m and V_{max} (From Borland & Tasca, 1974).

     Both leucine and methionine uptake in mouse blastocysts exhibit a number of
other characteristics of carrier-mediated active transport.   As shown in
Table 3 early blastocysts accumulate both amino acids against sizeable chemical
gradients.   Furthermore, other large neutral amino acids, such as valine,
phenylalanine and ethionine compete for leucine and methionine uptake.   Most
importantly, methionine and leucine uptake at the morula and blastocyst stages
were shown to be Na^+-dependent, whereas both amino acids are transported by
Na^+-independent processes at the 4 cell stage (Borland & Tasca, 1974).   The

TABLE 3

^{3}H-LEUCINE AND ^{3}H-METHIONINE ISOTOPE DISTRIBUTION RATIOS IN EARLY MOUSE
BLASTOCYSTS*

^{3}H-Leucine Concentration (μM)	Isotope Distribution Ratio	^{3}H-Methionine Concentration (μM)	Isotope Distribution Ratio
0.1	66	0.125	90
1	68	1.25	161
10	47	12.5	72
50	25	62.5	36
100	16	125.0	22

* The isotope distribution ratio equals dpm/embryo/hr divided by dpm contained
in a volume of medium equivalent to the embryo's volume (From Borland &
Tasca, 1974).

kinetics of methionine uptake by early blastocysts in Na$^+$-depleted media (23,
83 and 141 mM Na$^+$) during 15 minute incubation periods indicate that Na$^+$
decreases the K_m for methionine transport. The uptake of methionine is
relatively resistant to ouabain, a specific inhibitor of Na$^+$-K$^+$ ATPase, and
unaffected by K$^+$-free medium. In contrast, both these treatments inhibit
uptake in late morula-stage embryos. These results suggested a dependency
upon the transmembrane Na$^+$-gradient in late morulae (Borland & Tasca, 1975).
The relative insensitivity of early blastocysts to the experimental treatments
may be due to localization of Na$^+$-K$^+$ ATPase on the juxtacoelic surface of the
mouse trophoblast cells and high levels of K$^+$ stored in blastocoele fluid, as
discussed later in this chapter.

The data suggest that the role of Na$^+$ in methionine transport by mouse
blastocysts may be similar to its role in glycine transport in Ehrlich ascites
tumor cells (Lin & Johnstone, 1971; Johnstone, 1972). The linear dependency
of methionine and leucine uptake upon the extracellular Na$^+$ concentration in
early blastocysts and the role of Na$^+$ in decreasing the K_m of methionine
transport are consistent with Lin and Johnstone's (1971) proposal that extracellular Na$^+$ acts directly on a membrane-bound carrier to increase the
affinity of the carrier for the amino acid.

The developmental importance of the activation or synthesis of Na$^+$-
dependent amino acid transport systems in morulae and early mouse blastocysts
is not known. Two cell embryos cultured in defined medium in the absence of
any amino nitrogen source can develop into early blastocysts. However, when
the period of culture in such medium is extended for 5 days, fewer of the

embryos undergo hatching and a large proportion of the embryos collapse (Cholews & Whitten, 1970). Furthermore, mouse blastocysts developing in vitro in amino acid-free medium are also smaller than blastocysts of the same age in vivo (Bowman & McLaren, 1970). These data and the fact that activation or synthesis of Na^+-dependent amino acid transport systems occurs just prior to blastocyst formation raise the possibility that such transport mechanisms may facilitate blastocoele fluid accumulation. For example, if co-transport of Na^+ and amino acid occurs (defined in Stein, 1967), with energy from the Na^+ gradient driving the uphill movement of amino acids into the cell, inward movement of Na^+ could activate electrogenic sodium pumps (Glitsch, 1972) and supply Na^+ to Na^+-K^+ ATPase located on the juxtacoelic surface of mouse trophoblast cells (Moon & Biggers, unpublished). Co-transport of Na^+ and amino acids would result in depolarization of the trophoblast cell in amino acid-containing media, as demonstrated in other cell preparations (Laris et al., 1976; Rose & Schultz, 1971; Maruyama & Hoshi, 1972). Electrophysiological evidence for co-transport in blastocysts is still lacking, but the presence and role of Na^+ pumps in blastocyst expansion is now understood and will be discussed later in this chapter. Alternatively, activation or synthesis of amino acid transport systems may be required for post-implantation development. As shown by both Gwatkin (1966) and Spindle and Pedersen (1973), amino acids are required for hatching, attachment and outgrowth of mouse blastocysts in vitro.

In the previous discussion, amino acid transport has been assumed to be primarily homeocellular and involved in concentrating amino acids within the embryonic cells for protein synthesis. It is reasonable to expect that the active transport of these amino acids will produce amino acid gradients across the trophoblast cell membrane that are much greater than the isotope distribution ratios calculated for the whole embryo. Such gradients would be favourable for the passive movement of these amino acids from the cell into the blastocoele fluid. Such transcellular movement of amino acids in mammalian blastocysts does occur as demonstrated by studies of the composition of rabbit blastocoele fluid (Lesinski et al., 1967). These investigators demonstrated that seventeen amino acids are present in rabbit blastocoele fluid on the 6th and 7th days p.c. and that the total α-amino nitrogen content in blastocoele fluid is 5-fold or more higher than in maternal serum. Between the 7th and 8th days p.c. a considerable drop in total amino acid content of blastocoele fluid occurs. This drop is correlated with increasing concentrations of protein in blastocoele fluid after implantation. For example, up to the 7th day p.c. blastocoele fluid is relatively protein-free (Hamana & Hafez, 1970; McCarty & Kehwick, 1949; Zimmerman et al., 1963; Lesinski et al., 1967). As pointed out by the latter investigators, near the time of implantation a progressive increase in the concentration of protein in rabbit blastocoele fluid occurs, reaching half maternal serum levels at 7 days 18 hr.   After implantation on days 8 and 9 p.c. blastocoele protein levels are only slightly lower than maternal serum levels. Such a rapid increase in blastocoelic protein concentration occurring simultaneously with a rapid decrease in α-amino nitrogen content could suggest that blastocoele fluid is a reservoir of amino acids that are utilized for protein synthesis.   Alternatively, this pattern could merely indicate an overall permeability change in rabbit trophectoderm as it moves from its uterine fluid microenvironment and comes into direct contact with maternal serum.   The mechanisms involved in homeocellular movement of amino acids in rabbit blastocysts may not be similar to those in the mouse blastocyst and must be studied before any conclusion can be made concerning the actual mechanism of transcellular amino acid transport in this species.

TRANSLOCATION OF MACROMOLECULES

Large molecules synthesized in the maternal body are transferred nearly intact from the blood into ovarian eggs in a wide number of species, ranging from insects and fish to birds and mammals (for a review, see Glass, 1970). Schechtman (1955) has called the transferred molecules heterosynthetic because they are not made in the egg but are presynthesized by the mother and then transferred.   Transport of heterosynthetic molecules is also believed to occur into mammalian blastocysts and has stimulated extensive research on the nature and possible importance of uterine specific proteins.

Uptake of heterosynthetic molecules by blastocysts first involves passive diffusion of the molecules across the zona pellucida.   In rat and mouse blastocysts the zona pellucida does not appear to screen molecules on the basis of molecular weight (Hastings et al., 1972).   For example, ferritin, with a molecular weight larger than that of any uterine or oviducal protein, easily gains access to the blastomeres of all preimplantation stages.   Uptake into the embryo has been demonstrated primarily by analyzing rabbit blastocoele fluid. Based upon such analyses, Zimmerman et al. (1963) concluded that albumin and antibodies do not enter the rabbit blastocyst until 7 days 12 hr p.c.   In contrast, Beier (1968a,b) demonstrated that 6 day p.c. blastocysts contain $8-14\mu$ g of protein whose electrophoretic pattern is quite different from that of maternal serum proteins but nearly identical to that of proteins in uterine fluid.   Blastocoele fluid was shown to contain uteroglobin, albumin, β-globulin, and α-globulin.   Hafez and Sugawara (1967) and Hamana and Hafez (1970) also reported that albumin is present in rabbit blastocoele fluid, as early as day 5 p.c., and that on day 7 p.c. at the time of implantation of all protein bands characteristic of maternal serum are present in the blastocoele. Kulangara and Crutchfield (1973) showed that bovine serum albumin readily enters

5-6 days p.c. blastocysts exposed to this substance in vitro or in utero but not after the protein was given intravenously to the mother.

There is evidence that protein accumulation by the preimplantation rabbit blastocyst occurs via pinocytosis and is highly selective (Hastings & Enders, 1974). These investigators flushed rabbit embryos from the uterine horns with 0.9% NaC1 21-167 hr p.c. and incubated embryos for 3-60 minutes in saline, containing either horseradish peroxidase (a glycoprotein of 40,000 molecular weight), ferritin (molecular weight of 480,000) or myoglobin (a protein of 17,000 molecular weight). Blastocysts showed a greater uptake of peroxidase than morulae as evident by a greater number of product-containing vesicles and vacuoles and by greater rapidity of uptake, i.e. increased number of labeled vesicles/area at shorter time intervals. Smaller spherical images within the larger vacuoles suggested that fusion of smaller vesicles with multivesicular bodies was occurring. Peroxidase did not transverse the intertrophoblast junctions, and uptake was largely eliminated by incubating the blastocysts at 4°C. Myoglobin was accumulated by mechanisms similar to those used by peroxidase. In contrast, ferritin in the saline medium showed very little uptake. Quite surprisingly, however, injection of peroxidase and ferritin into the blastocoele cavity resulted in endocytosis of both products, indicating that the trophoblast cell is polarized with respect to its transport capacity for ferritin.

It has been proposed that accumulation of progesterone in the rabbit blastocyst may be related to the appearance of the uterine protein, uteroglobin, in rabbit blastocoele fluid and that uteroglobin may act as a carrier protein to transport this steroid into the embryo (Beier & Maurer, 1975). In opposition to this hypothesis and the presence of heterosynthetic molecular transport in blastocysts in vivo, Kulangara (1975) has reported that maternal proteins are not present in 5-7 day p.c. blastocoele fluid. Rabbit serum albumin (RSA) and gamma globulin (RGG) were measured by radial immuno-diffusion in fluid obtained from single blastocysts at 5, $5\frac{1}{2}$, 6 and $6\frac{1}{2}$ days p.c. after systemic injection of these protein tracers into the doe. Samples of blastocoele fluid showed little or no RSA and RGG, unless blastocoele fluid was artificially contaminated. Three sources of contamination were identified: (1) blastocyst contact with aqueous solutions used to flush or rinse them, (2) carrying in proteins by large pipette tips, and (3) withdrawal of trophoblast from the zona pellucida. The high pH and high viscosity of uterine fluid (containing 68-96 mg/ml total protein) are believed to limit protein passage by pinocytosis in vivo.

Recent measurements of the sulphur concentration in rabbit blastocoele fluid from freshly collected rabbit blastocysts tend to support the idea that the protein concentration in blastocoele fluid is low and does not change during preimplantation development (Borland et al., 1976a). In these studies·rabbit blastocysts were flushed from the uterine horns with mineral oil and sample of blastocoele fluid collected by micropuncture using pipettes with tip diameters <15μ m. Samples of fluid were analyzed for their sulphur concentration using electron probe analysis. Since the electron probe measures total sulphur content, both organic and inorganic, the measured sulphur concentrations reflect the protein content of the fluid. The sulphur concentrations at 110, 135 and 159 hr p.c. are 3.44, 2.77 and 2.96 mM/1, respectively, and were not significantly different. In contrast, the sulphur concentration of rabbit serum is approximately 20 mM/1 (Burgoyne, Borland, Biggers & Lechene, unpublished). In light of these observations and those of Kulangara (1975), past analyses of the protein content of blastocoele fluid using collection procedures that could produce contamination, must be viewed with cautious reservation at this time.

ION TRANSPORT AND FLUID ACCUMULATION IN RABBIT BLASTOCYSTS

Advances in our understanding of ion transport and fluid accumulation in the mammalian blastocyst have been largely confined to work with rabbit embryos. The rabbit blastocyst forms on the 3rd day p.c. and remains freeliving in the uterus until the 7th day p.c. At that time the.trophoblast cells begin to penetrate the uterine epithelium over-lying blood vessels and uterine fluid ceases to separate blastocysts from the uterine epithelium and blood. On day 8 p.c. trophoblast cells are in direct contact with maternal capillaries and with blood itself at some points (Böving, 1962).

In an attempt to elucidate the physical forces causing blastocyst expansion, Tuft and Böving (1970) measured the freezing point depressions of plasma, uterine fluid and blastocoele fluid on days 4-8 p.c. As shown in Table 4, prior to implantation blastocoele fluid is isotonic or slightly hypertonic to uterine fluid. In contrast, after implantation, on days 7 and 8 p.c., blastocoele fluid is hypotonic to maternal plasma. Based upon the assumption that maternal plasma is the microenvironment of 7 and 8 day blastocysts, Tuft and Böving (1970) concluded that blastocoele fluid is hypotonic to the blastocyst's environment. This, they believed, was evidence for "active water transport" since active solute transport with water flowing passively would produce a fluid that is isotonic or hypertonic to plasma. Gamow and Daniel (1970) substantiated these measurements of osmolarities of blastocoele and uterine fluids and lent support to the hypothesis of active water transport by claiming that blastocysts in culture have less than half the tonicity of freshly collected blastocoele fluid; however, no data were shown to document this latter finding.

TABLE 4

OSMOLARITIES OF MATERNAL PLASMA, UTERINE FLUID AND BLASTOCOELE FLUID
IN THE RABBIT[a]

Fluid sampled	Day post-coitum	Osmolarity (mOsm/l)
Maternal Plasma	4-8	307
Uterine Fluid	4-6	288
Blastocoele Fluid	4	290
"	5	296
"	6	298
"	7	298
"	8	296

[a] The osmolarities were calculated from freezing point depression measurements (From Tuft & Böving, 1970).

Evidence for carrier mediated transport of ions in rabbit blastocysts was initially very indirect and not correlated with blastocoele fluid accumulation. Lutwak-Mann et al. (1960) obtained the first evidence suggesting control of ion movement across the rabbit **trophectoderm** by administering labeled ions ($^{32}PO_4^{-3}$, $^{35}SO_4^{-2}$, $^{24}Na^+$, $^{42}K^+$ and $^{131}I^-$) parenterally to pregnant rabbits on days 0 to 12 of gestation. Measurements of radioactivity were made 45 minutes after ion injection and the results were expressed as the proportion of the total radioactive ion injected found present per mg of tissue or fluid. In the 6 day p.c. blastocyst proportional uptake values for all ions were low and showed a marked divergence. In contrast the endometrial secretions contained all ions with proportional uptake values greatly exceeding those in the blasto-cyst and blastocoele fluid. These results suggested that the rabbit blastocyst is relatively impermeable to certain ions and demonstrates selective permeability.

Smith (1970) measured the Na^+, Cl^-, K^+ and HCO_3^- content and accumulation rate of blastocoele fluid in 6 day p.c. blastocysts. Fluid and NaCl accumulations were markedly reduced by temperature reduction and by the presence of ouabain at a concentration of 5×10^{-4} M in the culture medium. Neither results strongly support carrier mediated ion transport nor do they disprove active water transport.

Support for the notion that some ions are actively transported across the rabbit trophectoderm came with the demonstration that a transtrophectoderm potential difference (TPD) exists in $5\frac{1}{2}$ and 6 day p.c. blastocysts (Cross & Brinster, 1969; Gamow & Daniel, 1970). The observed TPD values differed in the two studies but **both** reports indicated that the blastocoele cavity is

negative with respect to the outside (Table 5) and that the TPD is energy
dependent and vanishes if the blastocyst is exposed to anoxia or such metabolic
inhibitors as dinitrophenol, sodium cyanide and iodoacetic acid.   The TPD is
also significantly affected by ionic substitutions.   For example, the TPD
decreases to a positive potential when $SO_4^=$ is substituted for Cl^-; and
replacement of Cl^- by Br^-, I^- and $NO_3^=$ decrease the TPD to 49%, 48% and 36%,
of its control value, respectively.   Replacement of Na^+ with Li^+ results in
an increase in the TPD.   Removal of K^+ or increasing the K^+ concentration by
25 mM does not affect the TPD (Cross & Brinster, 1970).

TABLE 5

TRANSTROPHECTODERM POTENTIAL OF FREE-LYING RABBIT BLASTOCYSTS*

Age of Blastocysts (days p.c.)	Transtrophectoderm Potential Difference (mV)
5	-7.6 ± 0.55
6	-11.9 ± 0.65
7	-2.5 ± 0.55

* Values are means $\pm$ SEM. The blastocoele cavity is negative with respect
to culture medium (From Cross & Brinster, 1969).

The strongest argument for carrier-mediated transport, the demonstration
that a solute moves in the absence of or against its electrochemical potential,
was applied to the blastocyst by Cross (1973a).   In electrophysiology this
criterion is met by the use of the Ussing and Zerahn (1951) short-circuit
technique as shown in Figure 1A.   The epithelial membrane is perfused with
identical solutions on both sides, and the spontaneous potential generated by
the membrane is reduced to zero by a source of electromotive force which is
used to pass a variable current through the membrane.   Under these conditions
there are no electrical or chemical potentials capable of causing ion movement.
Any current flowing in the external circuit and measured by the microammeter
is the short-circuit current (SCC), and represents the algebraic sum of all
active ion transport.   Under short-circuit conditions the net flow of any ion,
as determined from the unidirectional fluxes of isotopic tracers, is due to
active transport.

To apply this technique to rabbit trophectoderm, Cross (1973a) perfused
and short-circuited 6 day blastocysts as shown in Figure 1B.   In such
preparations the resistance of the bathing solution, measured with the
blastocyst in position, was always less than 40 ohms cm^2, and the resistance
of the trophectoderm measured 2647 ohms cm^2.   Thus, the rabbit trophectoderm

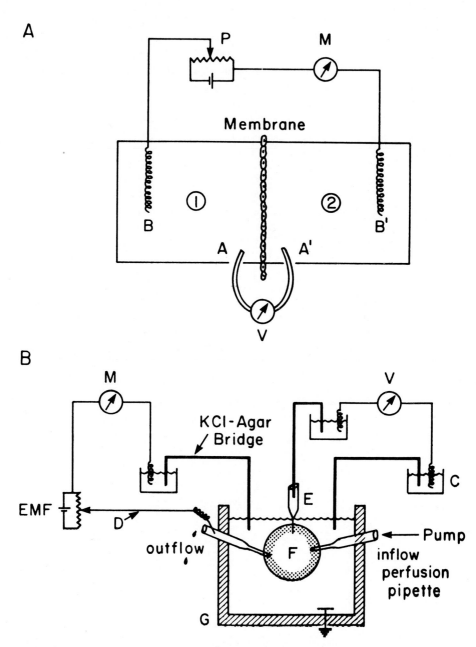

Figure 1: Experimental setup for the study of active transport across the epithelial membranes (A) and the rabbit blastocyst (B). A and A' are potential detecting electrodes. B and B' are current passing electrodes. V, milli-voltmeter; M, microammeter; P, potential divider; C, Ag-AgCl-KCl half-cell, D, Ag-AgCl wire; E, microelectrode; F, blastocyst; G, chamber used to secure the blastocyst. (Courtesy of Cross, 1973a).

is a tighter epithelium than frog skin, which has a resistance of 2000 ohms
cm^2 (Ussing & Zerahn, 1951).

TABLE 6

Na AND C1 FLUXES ACROSS SIX DAY p.c.
RABBIT TROPHECTODERM*

Ion	J_{mb}	J_{bm}	J_{net}	SCC ($\mu M/cm^2/hr$)	PD (mV)
C1	1.73 $\pm$ 0.10	1.14 $\pm$ 0.11	0.59 $\pm$ 0.14	0.18	-12.7
Na	1.39 $\pm$ 0.09	0.81 $\pm$ 0.13	0.58 $\pm$ 0.15	0.20	-13.9

* Values are means $\pm$ SEM. Unidirectional fluxes, outside medium to
blastocoele (J_{mb}) and blastocoele to medium (J_{bm}), of either Na or C1
were determined on separate blastocysts. J_{net} equals net ion flux.
Fluxes are expressed as $\mu M/cm^2/hr$. SCC = short circuit current, the
algebraic sum of all active ion transport. PD = potential difference
across trophectoderm, inside with respect to outside. (Crom Cross, 1973a).

The unidirectional fluxes (J) of ^{22}Na and $^{36}C1$ ions from the exterior
medium into the blastocoele cavity (J_{mb}) and from the blastocoele cavity to the
exterior medium (J_{bm}) were determined under steady state conditions after the
TPD and SCC had remained stable for at least 30 minutes. As shown in Table 6,
influx of Na^+ and $C1^-$ into the blastocoele is larger than efflux and results
in net accumulation of Na^+ and $C1^-$ in the blastocoele fluid. Unfortunately,
J_{mb} and J_{bm} were measured in separate blastocysts, and the actual magnitudes
of J_{net} may be in error due to rapid changes in ion accumulation rates with
development (Figure 3). Assuming these errors are small, the net Na and C1
fluxes are approximately equal, are both 3 fold larger than the short-circuit
current and together account for only 5% of the current. Replacement of outside medium with $C1^-$-free medium resulted in a reversal of orientation of the
TPD and consequently of the SCC. In contrast, when the outside bathing
solution was changed to Na^+-free medium, a higher negative TPD was recorded and
the positive SCC increased in magnitude. With either ion-replacement, however,
the absolute change in the SCC was smaller than predicted if Na and C1 are
transported by independent transport processes (Table 7). This quantitative
discrepancy could be due to partial neutral and/or electrical coupling between
the net fluxes of the two ions. Electrical coupling could occur within the
epithelial layer itself and not be reflected by the TPD between the two bathing
media. To rigorously demonstrate independent Na and C1 active transport
across the rabbit trophectoderm, unidirectional Na and C1 fluxes should be
measured after replacing the media on both sides of the epithelia with $C1^-$-free
or Na^+-free media, respectively. If $^{Na}J_{net}$ and $^{C1}J_{net}$ approach zero after ion
replacement, neutral coupled transport must be assumed. As an example of

TABLE 7

EFFECT OF ION SUBSTITUTIONS IN THE OUTSIDE BATHING MEDIUM ON TPD
AND SCC IN RABBIT BLASTOCYSTS*

Control (normal Ringer solution)		Outside bathing medium	Experimental	
TPD** (mV)	SCC*** ($\mu A/cm^2$)		TPD (mV)	SCC ($\mu A/cm^2$)
-12.0	7.47	C1-free	+ 5.1	-7.67
-14.5	6.25	Na-free	-27.6	11.01

* (From Cross, 1973a).

** The blastocoele is negative with respect to outside medium.

*** Positive SCC indicates a net flow of positive change from blastocoele to outside bathing medium.

electrical coupling, active Na transport could occur into an intraepithelial compartment, such as the lateral intercellular spaces, that is bounded by highly conductive plasma membranes or extracellular pathways and, therefore, possess a short electrical space constant. Under these circumstances, a local intraepithelial potential difference, that is not reflected by the TPD, and is caused by active Na transport, may serve as the driving force for passive C1 absorption (Frizzell et al., 1975).

Regardless of the exact relationships between Na and C1 transport, the results strongly suggest that active transport processes are involved in the transtrophectoderm movement of Na and C1. Similar data suggestive of active Na and C1 transport have been reported in the rabbit ileum (Field et al., 1971), bullfrog intestine (Quay & Armstrong, 1969), flounder intestine (Huang & Chen, 1971), and cow and goat rumen (Stevens, 1964).

Cross (1973a) noted that bicarbonate-free medium greatly reduces the TPD and SCC in 6 day blastocysts, suggesting that bicarbonate movement may be the source of the SCC. To test this hypothesis, Cross (1974) perfused rabbit blastocysts and measured the pH and total CO_2 content of the bathing medium and of the blastocoele fluid at the beginning and end of 5 hr incubation periods. By assuming that both blastocoele fluid and the culture medium are in equilibrium with respect to the H_2CO_3-CO_2 reaction, the free CO_2 and HCO_3^- content of each fluid was estimated using the Henderson-Hasselbach equation.

At the start of the perfusion there were equal HCO_3^- and free CO_2 concentrations on each side of the trophectoderm. At the end of the incubation period the average HCO_3^- and free CO_2 concentrations in the blastocoele fluid increased 30% and 26%, respectively (Table 8). The pH of blastocoele fluid remained the same as the pH of the culture medium as expected from the

TABLE 8

BICARBONATE AND FREE CO_2 CONCENTRATIONS IN 6-DAY p.c. RABBIT BLASTOCYSTS*

Sample	Blastocoele Fluid			External Medium			Final Transtrophectoderm Concentration Difference	
	pH	Free CO_2 (mM)	HCO_3 (mM)	pH	Free CO_2 (mM)	HCO_3 (mM)	Free CO_2 (mM)	HCO_3 (mM)
Initial	7.40	1.17	23.1					
Final	7.41	1.49	30.0	7.42	1.18	23.2	0.31	6.9
Change	+0.01	+0.32	+6.9					

* Twenty rabbit blastocysts were perfused with medium and incubated for approximately 5 hrs. pH was determined with an antimony microelectrode. Total CO_2 was determined with a Natelson microgasometer. The HCO_3^- concentration was derived from the Henderson-Hasselbach equation. (From Cross, 1974).

demonstration by Cross (1973b) that H^+ acts as if it is passively distributed across the trophectoderm.

The accumulation of HCO_3^- could be due to either or all of three mechanisms (Cross, 1974) (Figure 2):

1. Active transport of bicarbonate ions per se from the medium into the blastocoele fluid.

2. Hydration of CO_2 after CO_2 diffusion into the blastocoele fluid, with active H^+ transport from the cell into the medium.

3. Intracellular hydration of CO_2 with subsequent active bicarbonate ion transport into the blastocoele.

Experimental techniques for distinguishing H^+ transport from HCO_3^- transport are currently available and could be applied to the rabbit blastocyst (for a review, see Brodsky & Schilb, 1974). Such techniques as using P_{CO_2} electrodes for differentiating between dissolved CO_2 and bicarbonate (Severinghaus et al., 1969) and the mass spectrometer (Gurtner et al., 1973) would facilitate these studies and enable detection of possible displacements from equilibrium of the $H_2CO_3 - CO_2$ reaction in the medium and blastocoele fluid (Brodsky & Schilb, 1974), and distinguish H^+ from bicarbonate ion transport. Regardless of the mechanism(s) causing bicarbonate accumulation, such accumulation always occurs against an electrochemical potential, since the TPD is negative with respect to the exterior medium, and the process must be via active transport of either HCO_3^- or H^+.

Based on the numerical values of the Na and Cl net fluxes and the SCC, Cross (1973a) postulated that the amount of active solute transport could account for

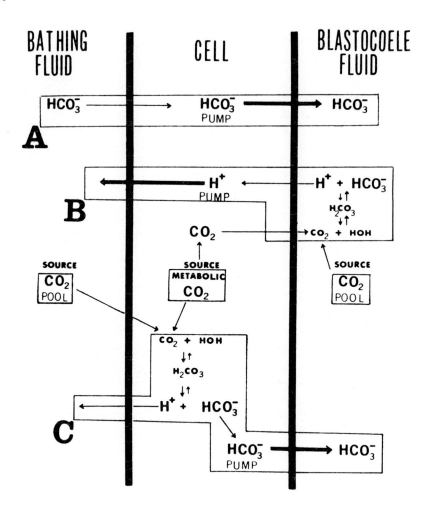

Figure 2: Possible transport mechanisms responsible for bicarbonate accumulation in the 6-day post-coitum rabbit blastocysts. (Courtesy of Cross, 1974).

the fluid accumulation that occurs in the rabbit blastocoele prior to
implantation.   As previously mentioned, the use of separate blastocysts to
determine the unidirectional fluxes decreases the reliability of these net flux
measurements.   It is very unlikely that influxes and effluxes per unit area
are constant in all 6-day blastocysts, since the net ion accumulation rates
increase rapidly as the blastocysts develope.   This latter point was clearly
demonstrated by Borland et al. (1976a) who used X-ray spectrometry by electron
probe excitation to analyze Na, C1, K, Ca, Mg, S and P concentrations in
picoliter samples of blastocoele fluid obtained from single rabbit blastocysts
110 to 159 hr p.c.   The rates of accumulation (net fluxes) of the solutes into
the rabbit blastocoele in vivo were estimated from the measured concentrations
and previously published rates of fluid accumulation (Daniel, 1964) according
to the equation:

$$dA/dt = C.dV/dt + V.dC/dt$$

where dA/dt is the rate of accumulation of a solute with time, C.dV/dt is the
rate of accumulation attributable to the rate of change in the volume of fluid
transported (assuming the concentration of the solute in the transported fluid
remains constant in the infinitesimal interval dt), and V.dC/dt is the rate of
accumulation attributable to the rate of change of the composition of fluid
transported (assuming the volume of the blastocoele remains constant in the
interval dt).   Determinations of dC/dt are not available but satisfactory
estimates of the rates of accumulation can be calculated for the three predominant electrolytes that were measured (Na, C1 and K) using only the first
term of the equation.   For justification of this see Figure 1 in Borland et al.
(1976a).   By calculating the rate of accumulation of these ions per unit area
it was possible to show that accumulation increases with development (Figure 3).
Furthermore , since translocation of these three elements occurs through a
squamous epithelium held together by tight junctions (Ducibella et al., 1975;
Hastings & Enders, 1975), all three ions must pass through the trophoblast cell
and/or their intercellular junctions.   Because the area of the individual
trophoblast cells of the rabbit blastocyst remains approximately constant
(0.31μ m^2) at all ages (Daniel, 1964), the rate of passage of these elements
must increase with development.   It was hypothesized that these increases may
be due to: (1) activation or synthesis of additional transport systems,
(2) permeability changes in the trophoblast cells and/or (3) permeability
changes in the intercellular pathways between the trophoblast cells (Borland et
al., 1976a).

   The transport mechanisms controlling fluid accumulation in rabbit blastocysts were directly determined by Borland et al. (1976a) by incubating $5\frac{1}{2}$ day
blastocysts in culture media supplemented with various concentrations of
sucrose to which blastocysts are highly impermeable.   Blastocysts continued

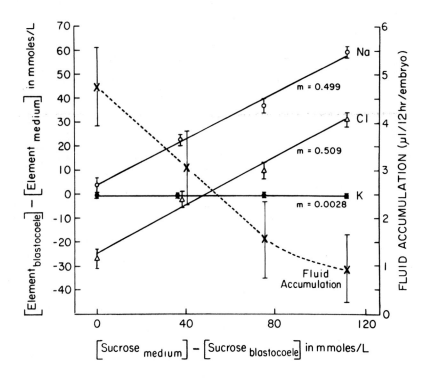

Figure 3: The rate of accumulation of Na, C1 and K in moles cm^{-2} h^{-1} in 4-6 days post-coitum rabbit blastocysts. (From Borland, Biggers & Lechene, 1976a).

to expand, but at a diminishing rate, in the presence of up to 80 mM sucrose. Electron probe microanalysis of the blastocoele fluid under these conditions showed that the blastocoele concentrations of both Na and C1 (but not K, Ca, Mg, S and P) increase 1 mM in response to every 2 mM sucrose gradient across the trophectoderm (Figure 4). These results indicate that the accumulation of fluid in the blastocoele is secondary to the transport of Na and C1, as formerly hypothesized by Cross (1973a). Furthermore, the data do not support the views of Tuft and Böving (1970) and Gamow and Daniel (1970) that the rabbit blastocyst transports water or a hypotonic fluid since such mechanisms would require that less than 1 mM NaC1 be transported for every 2 mM sucrose gradient.

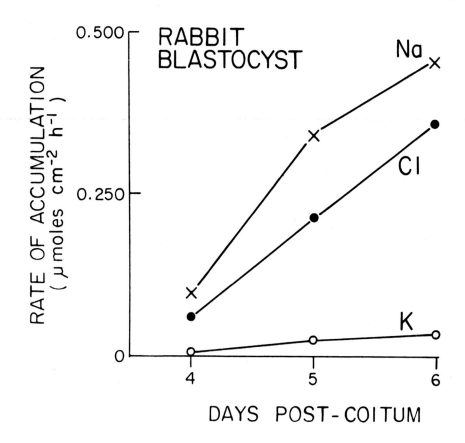

Figure 4: Effects of sucrose concentration gradients (X-axis) on the concentration gradients of Na, Cl, and K (Y-axis) across rabbit trophectoderm. The increases in rabbit blastocyst volumes in the presence of increasing concentrations of sucrose are shown by the dotted line. Na, Cl and K concentrations in blastocoele fluid and culture medium were determined by electron probe microanalysis. (Courtesy of Borland, Biggers & Lechene, 1976a).

ROLE OF NA^+-K^+ ATPASE IN BLASTOCYST TRANSPORT PROCESSES

Recently, Kim and Biggers (unpublished) have used the techniques of Ernst (1972) for the ultrastructural localization of K^+-dependent, ouabain-sensitive nitrophenyl phosphatase (Na^+-K^+ ATPase) activity in mouse and rabbit blastocysts. Sections of paraformaldehyde-fixed tissue were incubated in medium containing: 5 mM p-nitrophenyl phosphate, 10 mM $MgCl_2$, 10 mM KCl,

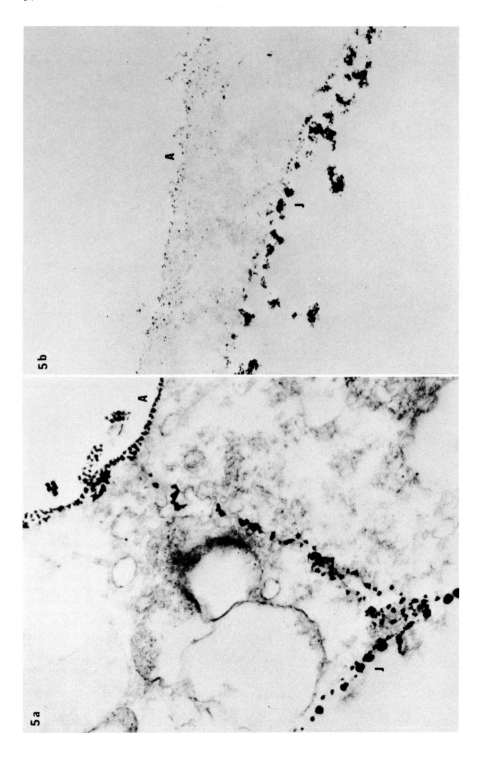

100 mM Tris-HC1 buffer (pH 8.5 or 9.0) and 20 mM $SrCl_2$ to precipitate hydrolyzed phosphate. After incubation at room temperature the sections were treated with $Pb(NO_3)_2$ to convert SrP_i to PbP_i precipitates for visualisation in the electron microscope. The reaction product of this enzyme activity is located on both the abcoelic and juxtacoelic surfaces of rabbit trophoblast cells and on the cell membranes lining the intercellular spaces between the cells (Figure 5a). However, the distribution of reaction product on each surface of the cell is different. In contrast to this pattern in the rabbit blastocyst, mouse blastocysts exhibit Na^+-K^+ ATPase activity on the juxtacoelic surface, but only slightly on the abcoelic surface, of the trophoblast cells (Figure 5b). The reaction product in both species of blastocysts was inhibited by K^+-free medium and ouabain proving the specificity of the reaction for Na^+-K^+ ATPase.

To elucidate the role of this enzyme in blastocyst physiology, Borland et al. (1976c) exposed 6 day p.c. rabbit blastocysts to ouabain. Figure 6 is a schematic presentation of the rabbit trophectoderm and the predicted response after exposing trophectoderm cells to ouabain on their juxtacoelic or abcoelic surfaces. Rabbit blastocysts have been shown to be highly impermeable to K, so K is shown leaking from the trophoblast cells into the blastocoele fluid (Borland et al., 1976c; Cross & Brinster, 1970). Both abcoelic and juxtacoelic Na^+-K^+ ATPase (Na pumps) are presumed to regulate the Na and K levels of the intracellular fluid. In addition to their role in cellular homeostasis, sodium pumps on the juxtacoelic surface of the cell and on the membranes facing the intercellular spaces are shown to regulate blastocoele K levels by pumping K from the blastocoele into the intracellular space with the co-ordinant pumping of Na from the intracellular space into the blastocoele. If this pump is electrogenic, fluid transport should accompany Na^+-K^+ ATPase activity. It should be emphasized that ATPase activity on the membrane facing the inter-cellular spaces is probably the critical activity for fluid movement, as proposed in various models for transepithelial fluid movement (Diamond & Bossert, 1967; Curran & McIntosh, 1962; Hill, 1975). Such fluid-filled intercellular spaces have been implicated in water movement across a variety of epithelia (Machen & Diamond, 1969; Grantham et al., 1969; Kaye et al., 1966; Diamond & Tormey, 1966). Presently no morphological or physiological evidence exists for the role of intercellular channels in fluid movement across the trophecto-derm. However, such channels do exist (Figure 7) and could represent the site for local osmosis or electro-osmosis, as defined by Diamond and Bossert (1967) and Hill (1975), respectively.

Figure 5a,b: Localization of Na^+-K^+ ATPase activity in rabbit (5a) and mouse (5b) blastocysts. The abcoelic surfaces of the rabbit and mouse trophoblast cells are designated by the letter A; juxtacoelic surfaces of the cells facing the blastocoele cavity are designated by the letter J. (Courtesy of Kim & Biggers, unpublished).

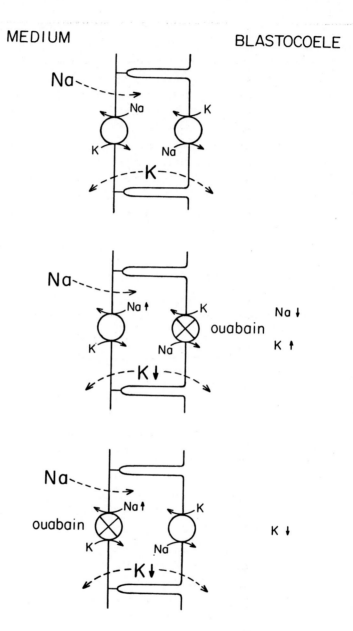

Figure 6: A schematic presentation of the 6 day p.c. rabbit trophectoderm and
the predicted response after exposing the trophoblast cells to ouabain on their
juxtacoelic or abcoelic surfaces. Increases and decreases in ionic concentrations are shown by arrows facing upwards and downwards, respectively.
(Courtesy of Borland, Biggers & Lechene, unpublished).

To test the model shown in Figure 6, ouabain was either injected into the blastocoele cavity or placed in the culture medium surrounding blastocysts that were injected with medium alone. After 4 hr incubations, ouabain injected blastocysts lose significant amounts of fluid, whereas control blastocysts and blastocysts exposed to abcoelic ouabain continue to expand (Borland, Biggers & Lechene, unpublished). Thus the effect of ouabain on fluid accumulation by the rabbit blastocyst is similar to the effect of this drug on amphibian blastulas in which ouabain in the medium does not affect blastula formation, whereas ouabain injected into the blastocoele causes blastula collapse (Slack & Warner, 1973). The effect of these treatments on the Na, C1 and K concentrations in blastocoele fluid after 12 hr incubations is shown in Table 9. As predicted in Figure 6, juxtacoelic ouabain causes the blastocoele K concentration to increase but Na concentration to decrease. Abcoelic ouabain produces a slight decrease in blastocoele K but this concentration was not significantly different from blastocoele fluid K levels in controls.

TABLE 9

EFFECT OF OUABAIN ON THE ELEMENTAL COMPOSITION OF RABBIT BLASTOCOELE FLUID*

Treatment	Concentrations (mM/1)		
	Na	C1	K
Control	160 ± 4	132 ± 4	7.24 ± 1.17
Juxtacoelic Ouabain ($19.2 \pm 6.12\mu$ M)	144 ± 4***	121 ± 3**	21.1 ± 1.06***
Abcoelic Ouabain (20μ M)	156 ± 4**	127 ± 3	6.03 ± 1.01**

* Six-day post-coitum rabbit blastocysts were incubated in vitro for 12 hrs. Samples of blastocoele fluid were analyzed by electron probe microanalysis. Values are means $\pm$ SEM.

** Significantly different from control values (P 0.05).

*** Significant difference between values in the two ouabain experiments (P 0.05). (Courtesy of Borland, Biggers & Lechene, 1976c).

The data conclusively establish that a ouabain sensitive transport mechanism, presumably Na^+-K^+ ATPase, is located on the juxtacoelic surface of rabbit trophoblast cells and is involved in transtrophectoderm fluid movement and regulation of blastocoele K levels. The role of abcoelic Na^+-K^+ ATPase in rabbit blastocyst physiology has yet to be determined. Regardless of its location, both abcoelic and juxtacoelic Na^+-K^+ ATPase may be involved in maintaining low intracellular Na concentrations which would be a prerequisite for any solute transport energized by the Na^+ gradient across the cell membrane.

Na pumps in the mouse blastocyst (Figure 5b) are presumed to have similar functions to juxtacoelic pumps in the rabbit. DiZio and Tasca (1974) have

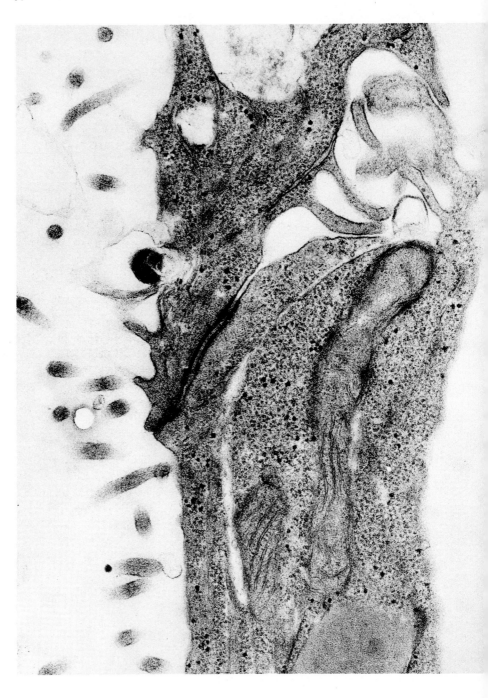

reported that re-expansion of early mouse blastocysts collapsed with cytochalasin B is Na^+-K^+-dependent, and inhibited by ouabain. These investigators have speculated that cytochalasin B may weaken junctional complexes and permit ouabain access to Na^+-K^+ ATPase on the juxtacoelic surface of the mouse trophoblast cell. This has not been substantiated with freeze fracture or by transmission electron microscopy. Thus, the effects of Na^+- and K^+-free media and of ouabain may be on the cytochalasin B recovery process and not necessarily related to normal blastocyst expansion. Further studies are needed to clarify the role of Na^+-K^+ ATPase activity in mouse blastocyst solute transport and fluid accumulation.

ION TRANSPORT IN MOUSE BLASTOCYSTS

Cross et al. (1973) measured the average TPD of partially expanded mouse blastocysts. The mouse blastocyst maintains a TPD of -5.0 ± 0.5 mV, blastocoele negative to culture medium, suggestive of transcellular active ion transport. Borland et al. (1976b) have obtained blastocoele fluid from single mouse blastocysts and analyzed picoliter aliquots of fluid for Na, Cl, K, Ca, Mg, S and P using electron probe microanalysis. As shown in Table 10, freshly collected mouse blastocysts maintain higher concentrations of Na, K, Ca, S and P than mouse uterine proestrous fluid. Preliminary data by Hazra, Biggers and Lechene (unpublished) indicate that estrogen and progesterone affect the quantity of mouse uterine fluid but not its elemental composition. Thus, it is reasonable to assume that the composition of mouse proestrous uterine fluid reflects the composition of the scant amount of uterine fluid in the blastocysts' microenvironment.

In order to confirm that mouse blastocysts maintain significant elemental gradients across the trophectoderm, samples of blastocoele fluid were collected from blastocysts raised in vitro from the two-cell stage and analysed by electron probe microanalysis (Table 11). Like freshly collected mouse blastocysts, blastocysts in vitro maintain significant concentration gradients across the trophectoderm: Na, K, Ca, Mg and S concentrations are greater in blastocoele fluid than in culture medium.

If the absorption of Na, Cl and K into the blastocoele were entirely passive, the maximum concentration gradient for these elements across the trophectoderm would be related to the TPD according to the Nernst equation. The absolute values of the mean equilibrium potentials for Na, Cl and K can be compared with

Figure 7: A transmission electron micrograph of rabbit trophectoderm from blastocysts in vitro from 135 to 159 hr post-coitum. Apical tight junctions and a desmosome are seen binding two trophoblast cells together immediately below microvilli facing the exterior medium. A long, tortuous intercellular channel is seen between the cells and leads to the blastocoele cavity. (Courtesy of Ducibella, unpublished).

TABLE 10

MEAN CONCENTRATIONS (mM/L) OF SODIUM, CHLORINE, POTASSIUM, MAGNESIUM, SULPHUR AND PHOSPHORUS IN MOUSE BLASTOCOELE FLUID AND PROESTROUS UTERINE FLUID*

| Element | Blastocoele Fluid** | | Proestrous Uterine Fluid*** |
	84-87 hr p.c.	94-97 hr p.c.	
Na	167	173	152
Cl	150	150	160
K	32.6	23.0	14.1
Ca	6.61	6.36	0.86
Mg	4.98	3.09	6.24
S	8.70	7.46	5.54
P	3.80	3.36	2.24

*   Samples of blastocoele fluid and proestrous uterine fluid were analyzed by electron probe microanalysis.

**   From Borland, Biggers & Lechene (1976b).

***   From Borland, Hazra, Biggers & Lechene, unpublished.

the absolute value of the observed TPD (ψ_{obs}) measured by Cross et al. (1973). The comparisons show that $\psi_K > \psi_{obs}$ and $\psi_{Cl} < \psi_{obs}$, thus suggesting that Cl and K are actively accumulated in the blastocoele. ψ_{Na} is slightly smaller than ψ_{obs} indicating that movement of Na is energetically favourable, and may be accumulated either passively or actively.   These claims that Cl and K are actively transported rest upon the assumption that the observed TPD of the mouse is real and not systematically biased due to technical artifacts.

Significantly more Ca but less K and Mg is present in blastocoele fluid of blastocysts grown in vitro than in freshly collected blastocysts.   Despite these changes, blastocysts raised in vitro from the 2 cell stage are fully functional, since they can be transferred to pseudopregnant hosts and develop into normal mice (Biggers et al., 1965).   However, the doubling time of cells is less in cultured blastocysts than in blastocysts developing in vivo with the result that blastocysts of the same age are smaller in vitro (Bowman & McLaren, 1970).   Perhaps the lower K and Mg concentrations in the blastocoele fluid in vitro are not optimal for normal cell division rates in the blastocyst.   The high Ca concentration in mouse blastocoele fluid may also be important in regulating cell division rates.   Barlow et al. (1972) have demonstrated that inner cell mass cells multiply at a significantly higher rate than trophoblast cells.   An increase in extracellular Ca levels controls the initiation of DNA synthesis and proliferation of a number of cultured cells, (Whitfield et al., 1973; Boynton et al., 1974) and may play a similar role in the mouse blastocyst.

TABLE 11

MEAN CONCENTRATIONS (mM/L) OF SODIUM, CHLORINE, POTASSIUM, CALCIUM, MAGNESIUM, SULPHUR AND PHOSPHORUS IN MOUSE BLASTOCOELE FLUID IN VITRO*

Element	Blastocoele Fluid	Culture Medium
Na	172	150
Cl	147	147
K	13.0	6.02
Ca	8.94	2.86
Mg	2.61	1.60
S	6.04	3.37
P	1.96	2.06

* Samples of blastocoele fluid and culture medium were analyzed by electron probe microanalysis (From Borland, Biggers & Lechene, 1976b).

Further studies are needed to understand the mechanisms by which mouse blastocysts concentrate elements in the blastocoele cavity and the role of these elements in blastocyst growth and differentiation.

ACKNOWLEDGEMENTS

The author is indebted to Mrs. Mary Forte for preparation of the typescript. The preparation of this review has been made possible by grants from the Rockefeller Foundation RF-65040, the NICHD grant HD-06916-01A1, and the NIH Centers Grant 1-P01 HD-06645-03.

REFERENCES

BARLOW, P., OWEN, D.A.J. & GRAHAM, C. (1972) DNA synthesis in the preimplantation mouse embryo. J. Embryol. exp. Morph. 27, 431-445.

BEIER, H.M. (1968a) Biochemisch-entwicklungs-physiologische Untersuchungen am Protein-milieu fur die Blastozystenwicklung des Kaninchens (Oryctolagus cuniculus) Zool. FB. Anat. Bd. 85, 72-77.

BEIER, H.M. (1968b) Protein patterns of endometrial secretion in the rabbit. Proc. Ind. Int. Seminar Reprod. Physiol. Sexual Endocrinology Ovo-Implantation, Human Gonadotropins and Prolactin, Brussels, Karger.

BEIER, H.M. & MAURER, R.R. (1975) Uteroglobin and other proteins in rabbit blastocyst fluid after development in vivo and in vitro. Cell Tissue Res. 159, 1-10.

BELLVE, A.R. & McDONALD, M.F. (1970) Directional flow of Fallopian tube secretion in the ewe at onset of the breeding season. J. Reprod. Fert. 22, 147-149.

BIGGERS, J.D. (1972) Mammalian blastocyst and amnion formation. In: A.C. Barnes & A.E. Seeds (Eds.), The Water Metabolism of the Fetus, Thomas, Springfield, Illinois, pp. 1-31.

BIGGERS, J.D. & BORLAND, R.M. (1976) Physiological aspects of growth and development of the preimplantation mammalian embryo. Ann. Rev. Physiol. 38, 95-121.

BIGGERS, J.D., MOORE, B.D. & WHITTINGHAM, D.G. (1965) Development of mouse embryos in vivo after cultivation from two-cell ova to blastocysts in vitro. Nature 206, 734-735.

BIGGERS, J.D. & STERN, S. (1973) Metabolism of the preimplantation mammalian embryo. Adv. Reprod. Physiol. 6, 1-60.

BORLAND, R.M., BIGGERS, J.D. & LECHENE, C.P. (1975) Electrolyte gradients across the mouse trophoblast and mechanism of rabbit blastocyst fluid accumulation. J. Cell. Biol. 67, 38a (Abstr.).

BORLAND, R.M., BIGGERS, J.D. & LECHENE, C.P. (1976a) Electron probe microanalysis of picoliter samples of rabbit blastocoele fluids: Kinetic aspects of blastocyst solute accumulation. Devel. Biol. 50, 201-211.

BORLAND, R.M., BIGGERS, J.D. & LECHENE, C.P. (1976b) Studies on the composition and formation of mouse blastocoele fluid using electron probe microanalysis. Devel. Biol., in press.

BORLAND, R.M., BIGGERS, J.D. & LECHENE, C.P. (1976c) Regulation of rabbit blastocoele potassium by Na^+ - K^+ ATPase on the juxtacoelic surface of the trophoblast cells. Soc. Study of Reprod. (Abstr.), in press.

BORLAND, R.M. & TASCA, R.J. (1974) Activation of a Na^+-dependent amino acid transport system in preimplantation mouse embryos. Devel. Biol. 36, 169-183.

BORLAND, R.M. & TASCA, R.J. (1975) Na^+-dependent amino acid transport in preimplantation mouse embryos. II. Metabolic inhibitors and nature of the cation requirement. Devel. Biol. 46, 192-201.

BÖVING, B.G. (1962) Anatomical analysis of rabbit trophoblast invasion. Contrib. Embryol. 37, 33-55.

BOWMAN, P. & McLAREN, A. (1970) Cleavage rate of mouse embryos in vivo and in vitro. J. Embryol. exp. Morph. 24, 203-207.

BOYNTON, A.L., WHITFIELD, J.F., ISAACS, R.J. & MORTON, H.J. (1974) Control of 3T3 cell proliferation by calcium. In Vitro 10, 12-17.

BRINSTER, R.L. (1971) Uptake and incorporation of amino acids by the preimplantation mouse embryo. J. Reprod. Fert. 27, 329-338.

BRINSTER, R.L., WIEBOLD, J.L. & BRUNNER, S. (1976) Protein metabolism in preimplanted mouse ova. Devel. Biol. 51, 215-224.

BRODSKY, W.A. & SCHILB, T.P. (1974) The means of distinguishing between hydrogen secretion and bicarbonate reabsorption: Theory and application to the reptilian and mammalian kidney. In: F. Bronner & A. Kleinzeller (Eds.), Current Topics in Membranes and Transport, Vol. 4, Academic Press, New York, pp. 161-228.

BRUNTON, W.J. & BRINSTER, R.L. (1971) Active chloride transport in the isolated rabbit oviduct. Am. J. Physiol. 221, 658-661.

CALARCO, P.G. & BROWN, E.H. (1969) An ultrastructural and cytological study of preimplantation development of the mouse. J. Exp. Zool. 171, 253-284.

CHOLEWA, J.A. & WHITTEN, W.K. (1970) Development of two-cell mouse embryos in the absence of a fixed-nitrogen source. J. Reprod. Fert. 22, 553-555.

CHRISTENSEN, H.N. (1966) Methods for distinguishing amino acid transport of a given cell or tissue. Fed. Proc., Fed. Amer. Soc. Exp. Biol. 25, 850-853.

CROSS, M.H. (1973a) Active sodium and chloride transport across the rabbit blastocoele wall. Biol. Reprod. 8, 566-575.

CROSS, M.H. (1973b) Rabbit blastocoele pH. J. Exp. Zool. 186, 17-23.

CROSS, M.H. (1974) Rabbit blastocoele bicarbonate: accumulation rate. Biol. Reprod. 11, 654-662.

CROSS, M.H. & BRINSTER, R.L. (1969) Trans membrane potential of the rabbit blastocyst trophoblast. Exp. Cell Res. 58, 125-127.

CROSS, M.H. & BRINSTER, R.L. (1970) Influence of ions, inhibitors and anoxia on transtrophoblast potential of rabbit blastocyst. Exp. Cell Res. 62, 303-309.

CURRAN, P.F. & McINTOSH, J.R. (1962) A model system for biological water transport. Nature 193, 347-348.

DAENTL, D.L. & EPSTEIN, C.J. (1971) Developmental interrelationship of uridine uptake, nucleotide formation and incorporation into RNA by early mammalian embryos. Devel. Biol. 24, 428-442.

DAENTL, D.L. & EPSTEIN, C.J. (1973) Uridine transport by mouse blastocysts. Devel. Biol. 31, 316-322.

DANIEL, J.C. (1964) Early growth of rabbit trophoblast. Amer. Naturalist 98, 85-98.

DIAMOND, J.M. & BOSSERT, W.H. (1967) Standing-gradient osmotic flow. A mechanism for coupling of water and solute transport in epithelia. J. Gen. Physiol. 50, 2061-2083.

DIAMOND, J.M. & TORMEY, J.M. (1966) Role of long extracellular channels in fluid transport across epithelia. Nature 210, 817-820.

DICKSON, A.D. (1966) The form of the mouse blastocyst. J. Anat. 100, 335-348.

DIZIO, S.M. & TASCA, R.J. (1974) Ion-dependent, ouabain-sensitive re-expansion of mouse blastocysts collapsed with cytochalasin B. Proc. Am. Soc. Cell Biol. 63, 85a (Abstr.).

DUCIBELLA, T., ALBERTINI, D.F., ANDERSON, E. & BIGGERS, J.D. (1975) The preimplantation mammalian embryo: characterization of intercellular junctions and their appearance during development. Devel. Biol. 45, 231-250,

DUCIBELLA, T. & BORLAND, R.M. (1977) The preimplantation mammalian embryo. In: D. Hamilton (Ed.), The Essentials of Reproductive Biology: A Basis for Clinical Practice, M.I.T. Press, Boston.

ELLEM, K.A.O. & GWATKIN, R.B.L. (1968) Patterns of nucleic acid synthesis in the early mouse embryo. Devel. Biol. 18, 311-330.

EPSTEIN, C.J. (1975) Gene expression and macromolecular synthesis during preimplantation embryonic development. Biol. Reprod. 12, 82-105.

EPSTEIN, C.J. & DAENTL, D.L. (1971) Precursor pools and RNA synthesis in preimplantation mouse embryos. Devel. Biol. 26, 517-524.

EPSTEIN, C.J., DAENTL, D.L., SMITH, S.A. & KWOK, L.W. (1971) Guanine metabolism in preimplantation mouse embryos. Biol. Reprod. 5, 308-313.

EPSTEIN, C.J. & SMITH, S.A. (1973) Amino acid uptake and protein synthesis in preimplantation mouse embryos. Devel. Biol. 33, 171-185.

ERNST, S.A. (1972) Transport adenosine triphosphatase cytochemistry. II. Cytochemical localization of ouabain-sensitive, potassium-dependent phosphatase activity in the secretory epithelium of the avian salt gland. J. Histochem. Cytochem. 20, 23-38.

FIELD, M., FROMM, D. & McCOLL, I. (1971) Ion transport in rabbit ileal mucosa. I. Na and Cl fluxes and short-circuit current. Am. J. Physiol. 220, 1388-1396.

FOSTER, D.O. & PARDEE, A.B. (1969) Transport of amino acids by confluent and nonconfluent 3T3 and polyoma virus-transformed 3T3 cells growing on glass cover slips. J. Biol. Chem. 244, 2675-2681.

FRIZZELL, R.A., DUGAS, M.C. & SCHULTZ, S.G. (1975) Sodium chloride transport by rabbit gallbladder. Direct evidence for a coupled NaCl influx process. J. Gen. Physiol. 65, 769-795.

GAMOW, E. & DANIEL, J.C. (1970) Fluid transport in the rabbit blastocyst. Wilhelm Roux' Archiv. 164, 261-278.

GARDNER, R.L. & PAPAIOANNOU, V.E. (1975) Differentiation in the trophectoderm and inner cell mass. In: N. Balls & A.E. Wild (Eds.), The Early Development of Mammals, Cambridge University Press, London, pp. 107-132.

GLASS, L.E. (1970) Translocation of macromolecules. In: O.A. Schjeide & J. DeVellis (Eds.), Cell Differentiation, Van Nostrand and Rheinhold Company, New York, pp. 42-118.

GLITSCH, H.G. (1972) Activation of the electrogenic sodium pump in guinea-pig auricles by internal sodium ions. J. Physiol. 220, 565-582.

GRANTHAM, J.J., GANOTE, C.E., BURG, M.B. & ORLOFF, J. (1969) Paths of trans-tubular water flow in isolated renal collecting tubules. J. Cell Biol. 41, 562-576.

GURTNER, G.H., BURNS, B., SCIUTO, A.M. & DAVIES, D.G. (1973) Relationship between blood and extravascular P_{CO_2}, H^+ and HCO_3^- in lung, cerebrospinal fluid, and brain. In: G. Nahas & K.E. Schaefer (Eds.), Carbon Dioxide and Metabolic Regulations, Springer-Verlag, New York, pp. 35-45.

GWATKIN, R.B.L. (1966) Amino acid requirement for attachment and outgrowth of mouse blastocysts in vitro. J. Cell. Physiol. 68, 335-343.

HAFEZ, E.S.E. (1971) Some maternal factors affecting physicochemical properties of blastocysts. In: R.J. Blandau (Ed.), The Biology of the Blastocyst, University of Chicago Press, Chicago, Illinois, pp. 139-191.

HAFEZ, E.S.E. & SUGAWARA, S. (1967) Protein distribution in the blastocoelic fluid of lactating rabbits. Fertil. Steril. 18, 565-569.

HAMANA, K. & HAFEZ, E.S.E. (1970) Disc electrophoretic patterns of uteroglobin and serum proteins in rabbit blastocoelic fluid. J. Reprod. Fert. 21, 555-558.

HAMNER, C.E. (1971) Composition of oviductal and uterine fluid. In: G. Raspe (Ed.), Advances in the Biosciences, Vol. 6, Pergamon Press, Vieweg, New York, pp. 143-164.

HASTINGS, R.A. & ENDERS, A.C. (1974) Uptake of exogenous protein by the preimplantation rabbit. Anat. Rec. 179, 311-330.

HASTINGS, R.A. & ENDERS, A.C. (1975) Junctional complexes in the preimplantation rabbit embryo. Anat. Rec. 181, 17-34.

HASTINGS, R.A., ENDERS, A.C. & SCHLAFKE, S. (1972) Permeability of the zona pellucida to protein tracers. Biol. Reprod. 7, 288-296.

HEINZ, E. (1972) Transport of amino acids by animal cells. In: L.E. Hokin (Ed.), Metabolic Transport, Academic Press, New York, pp. 455-501.

HILL, A.E. (1975) Solute-solvent coupling in epithelia: contribution of the junctional pathway to fluid production. Proc. R. Soc. Lond. B. 191, 537-547.

HUANG, K.C. & CHEN, T.S.T. (1971) Ion transport across intestinal mucosa of winter flounder (Pseudopleuronectes americanus). Amer. J. Physiol. 220, 1734-1738.

ISSELBACHER, K.J. (1972) Increased uptake of amino acids and 2-deoxy-D-glucose by virus transformed cells in culture. Proc. Nat. Acad. Sci. U.S.A. 69, 585-589.

JOHNSON, A.D. & FOLEY, C.W. (1974) The Oviduct and its Function, Academic Press, New York, 369pp.

JOHNSTONE, R.M. (1972) Glycine accumulation in the absence of Na^+ and K^+ gradients in Ehrlich ascites cells. Biochim. Biophys. Acta 282, 366-373.

KAYE, G.I., WHEELER, H.O., WHITLOCK, R.T. & LANE, N. (1966) Fluid transport in the rabbit gall bladder, a combined physiological and electron microscopic study. J. Cell Biol. 30, 237-265.

KULANGARA, A.C. (1975) Absence of maternal proteins in 5-7-day blastocyst fluid indicates limited protein passage before implantation. J. Exp. Zool. 193, 101-108.

KULANGARA, A.C. & CRUTCHFIELD, F.L. (1973) Passage of bovine serum albumin from the mother to rabbit blastocysts. II. Passage from uterine lumen to blastocyst fluid. J. Embryol. exp. Morph. 30, 471-482.

LARIS, P.C., PERSHADSINGH, H.A. & JOHNSTONE, R.M. (1976) Monitoring membrane potentials in Ehrlich ascites tumor cells by means of a fluorescent dye. Biochim. Biophys. Acta 436, 475-488.

LAU, N.I.F., DAVIS, B.K. & CHANG, M.C. (1973) Stimulation of in vitro 3-H-uridine uptake and RNA synthesis in mouse blastocysts by 17 -estradiol. Proc. Soc. Exp. Biol. Med. 144, 333-336.

LESINSKI, J., JAJSZCZAK, S., BENTYN, K. & JANACZARSKI, I. (1967) Concentration of free amino acids in rabbit blastocyst fluid. Am. J. Ob. Gyn. 99, 280-283.

LEVIN, R.J. & EDWARDS, F. (1968) The transuterine endometrial potential difference, its variation during the oestrous cycle and its relation to uterine secretion. Life Sciences 7, 1019-1036.

LIEU, T.S., HUDSON, R.A., BROWN, R.K. & WHITE, B.C. (1971) Transport of pyrimidine nucleosides across human erythrocyte membranes. Biochim. Biophys. Acta 241, 884-893.

LIN, K.T. & JOHNSTONE, R.M. (1971) Active transport of glycine by mouse pancreas: evidence against the Na^+-gradient hypothesis. Biochim. Biophys. Acta 249, 144-158.

LUTWAK-MANN, C., BOURSNELL, J.C. & BENNETT, J.P. (1960) Blastocyst-uterine relationships: uptake of radioactive ions by the early rabbit and its environment. J. Reprod. Fert. 1, 169-185.

MACHEN, T.E. & DIAMOND, J.M. (1969) An estimate of the salt concentration in the lateral intercellular spaces of rabbit gall-bladder during maximal fluid transport. J. Membrane Biol. 1, 194-213.

MANES, C. (1969) Nucleic acid synthesis in preimplantation rabbit embryos. I. Quantitative aspects, relationship to early morphogenesis and protein synthesis. J. Exp. Zool. 172, 303-310.

MARTIN, G.S., VENUTA, S., WEBER, M. & RUBIN, H. (1971) Temperature-dependent alterations in sugar transport in cells infected by a temperature-sensitive mutant of Rous Sarcoma virus. Proc. Nat. Acad. Sci. U.S.A. 68, 2739-2741.

MARUYAMA, T. & HOSHI, T. (1972) The effect of D-glucose on the electrical potential profile across the proximal tubule of newt kidney. Biochim. Biophys. Acta 282, 214-225.

McCARTY, E.F. & KEHWICK, R.A. (1949) The passage into the embryonic yolk-sac cavity of maternal plasma proteins in rabbits. Addendum. J. Physiol. 108, 184-185.

MINTZ, B. (1964) Synthetic processes and early development in the mammalian egg. J. Exp. Zool. 157, 85-100.

MOHLA, S. & PRASAD, M.R.N. (1971) Early action of oestrogen on the incorporation of H^3 uridine in the blastocyst and uterus of rat during delayed implantation. J. Endocr. 49, 87-92.

MONESI, V. & SALFI, V. (1967) Macromolecular synthesis during early development in the mouse embryo. Exp. Cell Res. 46, 632-635.

MULNARD, J.G. (1967) Analyse microcinematographique du developpement de l'oeuf de souris du Stade II au blastocyste. Archives de Biologie 78, 107-138.

OLDS, P.J., STERN, S. & BIGGERS, J.D. (1973) Chemical estimates of the RNA and DNA contents of early mouse embryos. J. Exp. Zool. 186, 39-47.

OLIVIER, J.M. & PATERSON, A.R.P. (1971) Nucleoside transport. I. A mediated process in human erythrocytes. Can. J. Biochem. 49, 262-270.

PETZOLDT, U. (1969) Bestimmung anorganische Ionen in Uterus und Blastozyste des Kaninchens während der fruhen Trachtigkeit. Zool. Anz., Suppl. 33, 128-133.

PETZOLDT, U. (1971) Untersuchung uber das anorganische Milieu in Uterus und Blastozyste des Kaninchens. Zool. Jb. Physiol. 75, 547-593.

PLAGEMANN, P.G.W. (1971) Nucleoside transport by Novikoff rat hepatoma cells growing in suspension culture. Specificity and mechanism of transport reactions and relationship to nucleoside incorporation into nucleic acids. Biochim. Biophys. Acta 233, 688-701.

QUAY, J.F. & ARMSTRONG, W. McD. (1969) Sodium and chloride transport by isolated bullfrog small intestine. Amer. J. Physiol. 217, 694-702.

ROBLERO, L., BIGGERS, J.D. & LECHENE, C.P. (1976) Electron probe micro-analysis of the elemental microenvironment of oviducal cleavage stages of the mouse. J. Reprod. Fert. 46, 434-454.

ROSE, R.C. & SCHULTZ, S.G. (1971) Studies on the electrical potential profile across rabbit ileum. Effects of sugar and amino acids on transmural and trans-mucosal electrical potential differences. J. Gen. Physiol. 57, 639-663.

ROSENBERG, T. (1948) On accumulation and active transport in biological systems. I. Thermodynamic considerations. Acta. Chem. Scandinav. 2, 14-33.

SCHECHTMAN, A.M. (1955) Ontogeny of the blood and related antigens and their significance for the theory of differentiation. In: E.G. Butler (Ed.), Biological Specificity and Growth, Princeton University Press, Princeton, New Jersey, pp. 3-31.

SCHOLTISSEK, C. (1968) Studies on the uptake of nucleic acid precursors into cells in tissue culture. Biochim. Biophys. Acta 158, 435-447.

SEVERINGHAUS, J.W., HAMILTON, F.N. & COTEV, S. (1969) Carbonic acid production and the role of carbonic anhydrase in decarboxylation in brain. Biochem. J. 114, 703-705.

SKEHEL, J.J., HAY, A.J., BURKE, D.C. & CARTWRIGHT, L.N. (1967) Effect of actinomycin D and 2-mercapto-1-(-4-pyridethyl) benzimidazole on the incor-poration of H^3 uridine by chick embryo cells. Biochim. Biophys. Acta 142, 430-439.

SLACK, C. & WARNER, A.E. (1973) Intracellular and intercellular potentials in the early amphibian embryo. J. Physiol. 232, 313-330.

SMITH, M.W. (1970) Active transport in the rabbit blastocyst. Experentia 26, 736-738.

SOLOMON, A.K. (1952) The permeability of the human erythrocyte to sodium and potassium. J. Gen. Physiol. 36, 57-110.

SPINDLE, A.I. & PEDERSEN, R.A. (1973) Hatching, attachment, and outgrowth of mouse blastocysts in vitro: Fixed nitrogen requirements. J. Exp. Zool. 186, 305-318.

STECK, T.L., NAKATA, Y. & BADER, J.P. (1969) The uptake of nucleosides by cells in culture. I. Inhibition by heterologous nucleosides. Biochim. Biophys. Acta 190, 237-249.

STEIN, W.D. (1967) The Movement of Molecules across Cell Membranes, Academic Press, New York, 369 pp.

STEVENS, C.E. (1964) Transport of sodium and chloride by the isolated rumen epithelium. Amer. J. Physiol. 206, 1099-1105.

TASCA, R.J. & HILLMAN, N. (1970) Effects of actinomycin D and cyclo-heximide on RNA and protein synthesis in cleavage stage mouse embryos. Nature 225, 1022-1025.

TUFT, P.H. & BÖVING, B.G. (1970) The forces involved in water uptake by the rabbit blastocyst. J. Exp. Zool. 174, 165-172.

USSING, H.H. & ZERAHN, K. (1951) Active transport of sodium as the source of electric current in the short-circuited isolated frog skin. Acta Physiol. Scandinav. 23, 110-127.

WHITFIELD, J.F., RIXON, R.H., MacMANUS, J.P. & BALK, S.D. (1973) Calcium, cyclic adenosine 3',5' monophosphate and the control of cell proliferation: a review. In Vitro 8, 257-278.

WOODLAND, H.R. & GRAHAM, C. (1969) RNA synthesis during early development of the mouse. Nature 221, 327-332.

ZIMMERMAN, W., GOTTSCHEWSKI, G.H.M., FLAMM, H. & KUNZ, C. (1963) Experimentelle Untersuchungen uber die Aufnahme von Eiweiss, Viren und Bakterien während der Embryogenese des Kaninchens. Devel. Biol. 6, 233-249.

PROTEIN SYNTHESIS AND GENE EXPRESSION
IN PREIMPLANTATION RABBIT EMBRYOS

Gilbert A. Schultz and Edward B. Tucker

Division of Medical Biochemistry
Faculty of Medicine
University of Calgary
Calgary, Alberta, Canada T2N 1N4

The understanding of control processes underlying early mammalian development will, in general, come from the gradual accumulation of genetic and biochemical information obtained from mammalian experimental systems. Embryogenesis and differentiation are processes whereby cells believed to be of identical genetic constitution develop into phenotypically distinct entities. Since conversion of genetic information into cellular phenotype requires the synthesis of mRNA molecules and their subsequent translation into polypeptides, an understanding of transcriptional and translational events is basic to the elucidation of the mechanisms controlling the developmental process.

As indicated previously (Manes, 1975), most of the published data regarding genetic and biochemical activities in preimplantation embryos have been derived from studies on the mouse and rabbit. Thus, the information may not be general to all mammals. Additionally, many studies have involved hormonal superovulation to increase embryo yields and have monitored synthetic activity by incorporation of radioactive materials during in vitro culture. Because of this, some of the biochemical events observed may be unique to manipulations of embryos in culture and not characteristic of development within the oviduct or uterus. However, without these technical procedures it is not possible to investigate biochemical parameters in any detail and since embryos will continue to develop normally when returned to the reproductive tract after periods of in vitro maintenance, some justification can be made that information derived from in vitro studies is characteristic of embryos in vivo. A distinctive feature of early mammalian development is the formation of the blastocyst whose structure is essentially similar in all mammals. While there is marked variation from one species to another in size of eggs, rate of cleavage, length of the preimplantation period and length of the gestational period, detailed study of a few laboratory species may provide some patterns of genetic activity which are general to many species. Differences which do occur may subsequently be explainable on the basis of the variables mentioned above. Recently, several comprehensive reviews of genetic and biochemical aspects of mouse and rabbit preimplantation development have been published (Wales, 1975; Epstein, 1975; Manes, 1975; Schultz & Church, 1975). It will be the purpose of this paper to analyze recent developments related to the kinds and amount of protein synthesized during preimplantation rabbit development and to discuss possible transcriptional and translational control mechanisms. The paper will deal principally with recent studies in our own laboratory on rabbit embryos and, when applicable, with studies from other laboratories on mouse and rabbit embryos.

PATTERN OF PREIMPLANTATION RABBIT DEVELOPMENT

General

As in other sexually reproducing organisms, reproduction and early development in the rabbit begins with the production of gametes. The mammalian egg is microlecithal in that it contains very little yolk material and shows no apparent polarity. The first meiotic division, resulting in halving of the nuclear DNA, occurs in the Graafian follicle. The second meiotic division begins just before ovulation but is not completed until the egg is penetrated by a sperm during fertilization. The ovulated but unfertilized rabbit egg is metabolically active and can incorporate precursors of both nucleic acid and proteins into acid insoluble products (Manes, 1975; Schultz, 1975). Thus, at the gamete level, the ovulated egg appears to contain all the necessary components of the protein synthetic apparatus which allow it to make an important contribution to successful early development following fertilization. The process of spermatogenesis, on the other hand, yields a terminally differentiated sperm cell whose nucleus is highly condensed and whose genetic expression is nearly completely repressed. In the trout, this is due to a process in which the histones characteristic of the somatic cell are displaced from their tight combination with DNA and replaced by a new series of small arginine-rich, sperm-specific proteins called protamines (see Dixon, 1972, for review). While less detail is known in mammalian systems, synthesis of an arginine-rich protein (protamine) in middle-stage mouse spermatids has been demonstrated (Lam et al., 1970) with concomitant restriction of transcription at the spermatid stage (Kierszenbaum & Tres, 1975). Following fertilization, there is a rapid removal of protamine accompanying the activation of the male pronucleus. Recently, Ecklund and Levine (1975) studied mouse sperm which had been labelled with ^{3}H -arginine autoradiographically through the fertilization process. They found that the label was completely lost from the sperm nucleus during the completion of the second meiotic division by the maternal chromosomes. This step of protamine removal is probably necessary for the patterned addition of histone and non-histone proteins to the male DNA under the control of the egg cytoplasm such that its genetic activity fits into the developmental program of the embryo.

In the rabbit, fertilization takes place 10-12 hours post-coitum. The first cleavage begins at about 12 hours later with subsequent cleavages occurring at about 8 hour intervals. The blastomeres become progressively smaller and the nucleocytoplasmic ratio is increased during the cleavage period. There is little overall growth of the embryo until the third day of development post-coitum. At day 3, the morula begins to cavitate to form the blastocyst and moves from the oviduct to the uterus. The blastocyst, with distinguishable trophoblast and inner cell mass cells, then increases in size due to fluid uptake and an increase in cell number to about 80,000 (Manes & Daniel, 1969) before implanting at 6.5 to 7 days post-coitum. This timetable of development is summarized in Table 1 and, unless indicated otherwise, is the reference for staging embryos throughout this paper.

TABLE 1

DEVELOPMENT OF THE EARLY RABBIT EMBRYO

Day*	0	1	2	3	4	5	6	7
RNA content** (μg/embryo)	.020	.028	.034	.069	.123	.414	2.800	-
Cell Number***	1	2	16	128	1024	9000	80,000	-

* Day 0, coitus; day 0.5, HnRNA and 4sRNA synthesis detected; day 3, rRNA synthesis detected and embryo leaves oviduct for uterus; day 7, implantation.

** Data from Manes (1969).

*** Data from Manes & Daniel (1969).

At the light microscope level, the first time more than one cell type can be recognized is the early blastocyst stage in which two distinct cellular compartments exist: the inner cell mass and the trophoblast. These two cell types are committed to mutually exclusive paths of development in the sense that they cannot be interconverted by experimental manipulation (Gardner, 1975). This determination of cells for trophoblast versus inner cell mass may, however, occur earlier. Available data support the conclusion that blastomeres of both the cleaving mouse (Tarkowski & Wroblewska, 1967) and rabbit (Moore et al., 1968) embryo are totipotent up to the 8 cell stage. In the mouse, experiments have demonstrated that after the 8 cell stage, the fate of a blastomere depends on whether it occupies an inside or outside position in the morula (see Gardner, 1975 for review). The outer cells will end up as trophoblast while the inner cells will end up as inner cell mass. Careful ultrastructural examination of the mouse morula shows that the outermost blastomeres possess focal tight junctions which become zonular and exclude lanthanum, thereby separating the inner cells from the maternal environment (Ducibella et al., 1975). This compartmentalization may create a microenvironment inside the embryo which in turn may be the first observable differentiation of cells.

Transcription

Transcriptional patterns in preimplantation rabbit embryos have been recently reviewed in considerable detail by Manes (1975) and Schultz and Church (1975). However, since synthesis of the major RNA classes is relevant to supply of template molecules for translation (mRNA) and the translation machinery (tRNA and rRNA) a brief review is justified. In short, 4S (tRNA) and heterogeneous RNA synthesis are detectable from the fertilized egg stage and onwards (Table 1). Ribosomal RNA synthesis, however, is first detectable at the 3-day (64 to 128 cell) stage, the time at which the morula begins to transform into the blastocyst and the embryo moves from the oviduct to the uterus. Presumptive messenger RNA synthesis, identified as RNA-containing poly(A)sequences at the 3'-OH terminus of the molecule, can be detected during cleavage (Schultz et al., 1973) and newly synthesized poly(A)-containing RNAs are shown to be associated with polysomes as early as the 16 cell stage (Schultz, 1973a). Thus, early development in the rabbit, as in the mouse, is characterized by transcription from the zygote genome nearly immediately following fertilization (Manes, 1975; Schultz & Church, 1975). The contribution of maternal mRNA molecules stored in the egg cytoplasm during early mammalian development is not yet clearly defined. Molecular hybridization of ^{3}H -polyuridylic acid to RNA obtained from unfertilized rabbit eggs demonstrates the poly(A)-containing RNA species (putative mRNAs) as present (Schultz, 1975). The content of poly(A) within the RNA of ovulated but unfertilized rabbit eggs is not detectably different. However, a small shift in the distribution of poly(A)-containing RNA from the subribosomal to the ribosomal fraction of the cell was found to accompany fertilization (Schultz, 1975). Recent re-evaluation of the requirement for participation of the embryonic genome in early rabbit development by use of the toxin α-amanitin to inhibit synthesis of heterogeneous RNA has revealed that 1-2 cell embryos can continue to cleave and develop for 2 to 3 cell generations in the absence of RNA synthesis (Manes, 1973). It is possible that the poly(A)-containing RNA demonstrated in unfertilized eggs (Schultz, 1975) is mobilized and utilized as a template for protein synthesis until the 8 to 16 cell stage, the time at which development arrests in the presence of α-amanitin. No experimental proof is available to support this contention at this time, nor is there evidence to support the suggestion that stored maternal mRNAs are selectively activated and translated at various times during preimplantation

development. However, such phenomena do occur in comparative systems and they
should not be ruled out.   Thus, the relative roles of maternal and zygotic
development have not yet been clearly defined.   The experiments with chemical
enucleators such as α-amanitin (Manes, 1973) indicate that zygote genome transcription is required for development to proceed past the 16 cell stage.
Genetic variants of β-glucuronidase (Wudl & Chapman, 1976) and glucose phosphate
isomerase (Brinster, 1973a) in mouse strains have been used as tools to time
paternal gene expression as an experimental means of estimating when the embryonic genome becomes functional in the mouse.   In both cases, the paternal gene
product is detected during cleavage at or shortly after the 8 cell stage,
supporting the observation that zygote genome transcripts are being utilized by
this stage.

Qualitative Changes in Translation Products During Early Development

While studies on poly(A)-containing RNA have provided a general picture of
messenger RNA metabolism in early rabbit development, a more precise analysis
of changes in gene expression might presently come from studies on polypeptides,
the gene end-products.   Because polypeptides are coded for by messenger RNAs
and because high resolution techniques are available which allow analysis of
large numbers of polypeptides within a complex biological mixture, these gene
end-products presently provide more sensitive probes to estimate qualitative
differences in mRNA populations from one developmental stage to another than a
study of mRNA molecules themselves.

It is clear that protein synthesis is actively taking place and is required
for oogenesis.   In the mouse, if primary oocytes are incubated in a medium
containing puromycin, the germinal vesicle breaks down and chromosomes condense,
but further meiotic progression is stopped (Wassarman & Letourneau, 1976).
Histone F_1 was identified as one of the proteins synthesized at this time.
Many other enzymes and structural proteins must also be synthesized during
oogenesis.   At least 20 enzymes have been reviewed by Epstein (1975) in terms
of their activity between fertilization and the blastocyst stage of the mouse.
Presumably, nearly all of these proteins present in the 1 cell fertilized egg
originated during oogenesis.   While metabolic activity of the ovulated but
unfertilized egg is low (as measured by O_2 consumption) compared to post-
fertilization stages, the egg does have the machinery to support early energy
metabolism (Wales, 1975), DNA replication and cell division, in addition to
having a complete protein synthetic apparatus (Epstein, 1975; Manes, 1975).
Mouse oocytes also appear to be able to repair DNA damaged by X-rays (Pedersen &
Cleaver, 1975).   Thus mature ova contain a complex mixture of proteins and are
also active in protein synthesis (Manes, 1975) although much of the synthesis

detected at this stage and in very early post-fertilization stages may represent replacement of proteins degraded by endogenous mechanisms.

The most-detailed study of qualitative, stage specific aspects of protein synthesis in preimplantation rabbit embryos published to date is that of Van Blerkom and Manes (1974). Using the technique of autoradiography of thin SDS-polyacrylamide gel slabs to resolve proteins which had incorporated ^{35}S - methionine, they were capable of resolving as many as a hundred discrete bands from stages immediately following fertilization through blastocyst formation and maturation. The newly synthesized proteins varied in size from less than 10,000 to greater than 200,000 daltons. Interestingly, the major qualitative changes in the pattern of protein synthesis occurred between the 2 cell stage and morula (i.e. prior to blastocyst formation) rather than at and after blasto-cyst formation, the time at which embryonic cells are differentiating into inner cell mass and trophoblast. A similar result occurs in the mouse in which the major changes in kinds of proteins synthesized occurs between the 1 and 8 cell stage (Epstein & Smith, 1974; Van Blerkom & Brockway, 1975). While the protein synthetic pattern is complex, and differences do occur, the large majority of proteins produced during preimplantation rabbit development are shown to be identical from one stage to another (Van Blerkom & Manes, 1974) and may represent enzymes and proteins required for viability of any growing and dividing cell population. Using the high-resolution two-dimensional electrophoretic technique developed by O'Farrell (1975), Van Blerkom et al. (1976) have studied the proteins synthesized by trophoblast and inner cell mass cells of the mouse blastocyst. While each tissue did reveal a number of tissue-specific poly-peptides, again the majority of the several hundred newly-synthesized poly-peptides resolvable by this procedure appear to be common to the 2 cell types although quantitative differences in particular polypeptides are evident.

We have applied the O'Farrell technique to the resolution of ^{14}C-amino acid-labelled polypeptides during early stages of rabbit development (Figures 1 and 2). As expected, the pattern of polypeptides resolved by this procedure is complex at all stages of development studied. Some 72 and 53 polypeptides were resolved in the 1 cell unfertilized egg and 16 cell cleavage embryo, respectively (Figures 1A and 1B) while about 156 and 537 polypeptides were resolved in the early (4-day) and late (6-day) blastocyst, respectively (Figures 2A and 2B) under the con-ditions we employed. Because protein synthetic rates are greater in later stages and because cell numbers per embryo are markedly increased, the lower number of polypeptides resolved in the 1 cell and 16 cell samples likely reflects the smaller amount of radioactive sample applied. That is, many polypeptides at earlier stages are probably synthesized in amounts below the levels of detection by the procedures we have used.

The patterns of polypeptides in the four embryonic stages presented in Figures 1 and 2 are remarkably distinct. This is not so much due to the presence of a large number of stage-specific polypeptides as to differential synthesis of polypeptides common to all stages. For example, the major spots in the 1 cell unfertilized egg designated by arrows in Figure 1A are also present in 16 cell and later stages but in diminished intensity relative to other polypeptides. Arrows in Figure 1B designate two major polypeptides in the 16 cell embryo which are also predominant in the 4-day blastocyst (Figure 2A). These two polypeptides are less predominant in the 1 cell egg (Figure 1A) and 6-day blastocyst (Figure 2B). In the 6-day blastocyst, the predominant protein designated by the arrow appears to be actin (Van Blerkom & Manes, 1974). While these are but a few examples, careful analysis of the plates reveals many other variations in the quantitative proportions of particular polypeptides at different stages of preimplantation development.

Apart from the apparent quantitative differences between individual spots in the various patterns, the other striking feature is the apparent increase in the number of polypeptides resolved as development proceeds. However, as indicated above, we feel that we are not detecting many of the polypeptides produced at the 1 cell, 16 cell, and even the 4-day blastocyst stages. Hence, we can conclude little about stage-specific spots. The majority of polypeptides in the 1 cell and 16 cell embryo patterns seem to be shared in common (Figure 1) and presumably represent proteins required for general cell function. Similarly, most of the polypeptides in the 4-day blastocyst are also present in the 6-day blastocyst (Figure 2). However, many stage-specific polypeptides representing differential expression of genes may in fact exist and are missed because they fall outside the isoelectric focussing pH range we have used or are so weakly radioactive we have not been able to detect them. In the one-dimensional studies of Van Blerkom and Manes (1974), the major differences in protein synthetic patterns were shown to occur during cleavage. We need to do more experimentation at more intermediate stages and under more sensitive conditions before we can contribute further detail to this finding.

A final point of consideration is that many of the proteins synthesized by the 1 cell ovulated, but unfertilized egg, are still being synthesized (although in many cases at a reduced level) at the 16 cell and blastocyst stages. The 1 cell proteins must be considered as polypeptides derived from maternal messenger RNA. Hence, these identical polypeptides in post-fertilization states are still either coded for by long-lived maternal mRNA or else at least some of the transcription products from the zygote genome are the same species as those in the egg. This aspect could presumably be clarified by analysis of polypeptide patterns of fertilized 1 cell eggs cultured to the 8 to 16 cell stage in the presence and absence of a transcriptional inhibitor like α-amanitin. Such

experiments remain to be done and may cast some light on the relative contributions of maternal and zygote genome transcription products to polypeptide
synthesis in early rabbit development.

Ontogeny of Lactate Dehydrogenase Isoenzymes

Our analysis of newly-synthesized polypeptides by two-dimensional
electrophoresis in the previous section has provided evidence for some stage-
specific polypeptide patterns which are molecular markers of differentiation in
the rabbit embryo.   Analysis of gene expression and cellular differentiation
has also been facilitated through the study of changes in multiple molecular
forms of enzymes (isoenzymes) in the developmental sequence.   The enzyme
lactate dehydrogenase (LDH) has been a particularly useful index protein in a
number of developmental studies.

The early rabbit embryo has low levels of LDH activity until the fourth
day of development (early blastocyst stage) at which time activity increases
(Brinster, 1973b; Schultz & Browder, 1975).   Electrophoretic analysis has
demonstrated that LDH-5 (A_4 subunit composition) is the predominant isoenzyme
in the unfertilized egg and early cleavage stages (Figure 3).   As activity
increases at the 4-day blastocyst stage, there is a concomitant appearance of
more B-subunits of LDH resulting in a shift in the LDH pattern toward the more
anodal LDH-1 type (Figure 3).   By the 6-day blastocyst, the contributions of

Figure 1: Two-dimensional gel electrophoresis separation of rabbit
preimplantation embryo proteins. Embryos were labelled for 4 h at 37°C in Ham's
F-10 medium containing 10% fetal calf serum and 200μ Ci/ml of ^{14}C-amino acid
mixture (40 Ci/mmole, New England Nuclear Corp.). Samples were prepared and
electrophoresed according to O'Farrell (1975). In the first dimension, proteins
were separated by isoelectric focussing on 110 mm cylindrical gels. In the
second dimension, proteins were separated according to molecular weight on sodium
dodecyl sulfate, exponential-gradient (9-15% acrylamide) polyacrylamide slab gels.
The pH range was determined by cutting a one-dimensional gel after isoelectric
focussing into 5 mm pieces, eluting the pieces in degassed distilled water, and
measuring the pH of each eluate. Approximate molecular weights are given on the
right and were determined from gels calibrated with proteins of known molecular
weight. The gels were impregnated with PPO according to Bonner and Laskey (1974)
prior to fluorography with X-ray film. Dot diagrams (A_1 and B_1) on the right are
presented to diagramatically illustrate the distribution of spots independent of
intensity. A, sample of 73 unfertilized eggs (40,000 cpm) applied to the isoelectric focussing gel. Arrows designate polypeptides which constitute a major
portion of the total incorporated material at this stage of development but a
markedly decreased proportion in later stages. B, sample of forty-five 16 cell
embryos (36,000 cpm) applied to the isoelectric focussing gel. Arrows designate
two very intense spots which are only faintly present by comparison in the 1 cell
egg.

Figure 2: Two-dimensional gel electrophoresis separation of proteins of rabbit
blastocysts. Procedures are as described in Figure 1, except that PPO impreg-
nation of the gel prior to exposure of X-ray film was performed on A only. A,
sample of seventeen 4-day blastocysts (500,000 cpm); B, sample of three 6-day
blastocysts (232,000 cpm). Arrows designate spots which demonstrate that the
predominant incorporation occurs at different spot positions in the early and
late blastocyst, respectively.

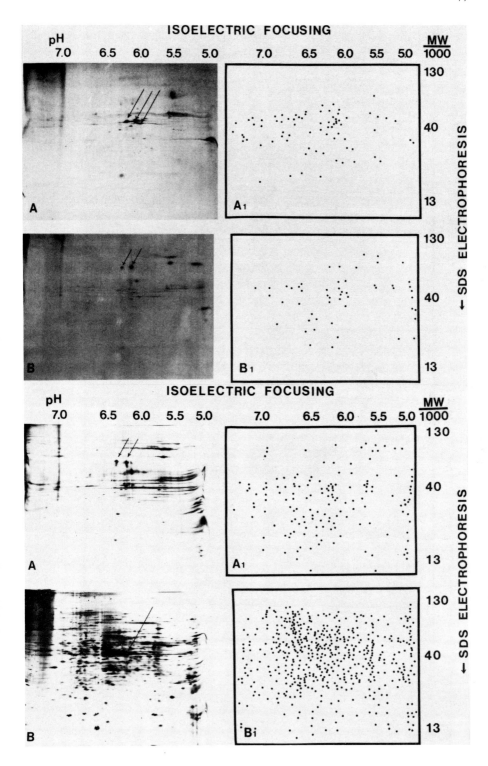

ISOELECTRIC FOCUSING

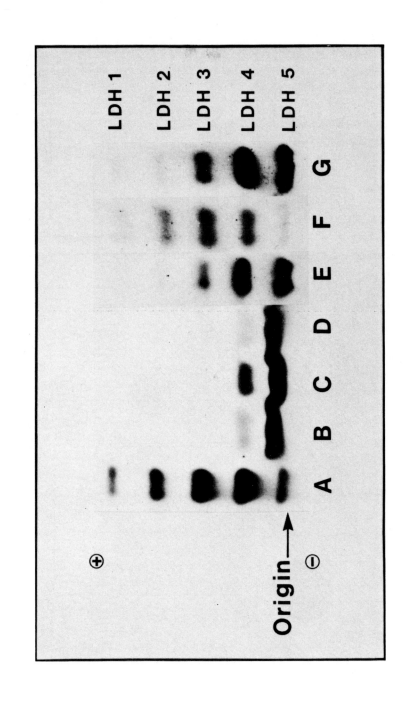

A and B polypeptides to LDH isoenzymes appear to be about equal as demonstrated by the roughly binomial pattern of staining intensity of the five LDH isoenzymes at this stage (Figure 3). Similar findings have been reported by Brinster (1973b). With respect to this enzyme, rabbit embryos differ considerably from the mouse. Quantitatively, mouse LDH activity is high in unfertilized eggs and then decreases during the preimplantation period (Brinster, 1965). Preimplantation mouse embryo LDH is composed almost exclusively of B polypeptides while implantation marks an increase in synthesis of A subunits and a resultant shift in the isozyme pattern in the opposite direction to that observed in rabbits (Auerbach & Brinster, 1967; Rapola & Koskimies, 1967). Engel et al. (1975) have suggested that LDH-1 (B_4) is also the predominant isozyme in unovulated rabbit oocytes and that there is a shift to LDH-5 (A_4) activity as a consequence of hormonal stimuli required for ovulation. We have dissected oocytes from rabbit ovaries and separated attached follicular cells from the egg (Peacock & Schultz, unpublished results). The follicular cells contain predominantly LDH-1 (A_4) while the oocyte itself has LDH-5 (B_4) just as do pronase-cleaned, ovulated, unfertilized eggs (Figure 3). While hormonal effects cannot be ruled out, the LDH-1 pattern observed by Engel et al. (1975) would seem to represent that of contaminating follicular cells.

The LDH activity increase and appearance of B subunits at day 4 of development suggests that translation of the message for this enzyme is initiated just prior to this stage. While again, it appears that such a message would be derived from transcription of the zygote genome, we cannot rule out the possibility of the recruitment of a stored maternal mRNA or some other form of post-transcriptional or translational control. It is difficult to assign any causal role to the appearance of LDH polypeptides at the time of blastocyst formation with the differentiation process. Similarly, we do not know if stage-specific polypeptides synthesized in the cleavage and preblastocyst stages play a determinative role or provide an index of commitment to differentiation of one or the other cell type in the blastocyst. However, continued elucidation of

Figure 3: Electrophoresis patterns of LDH isoenzymes from rabbit liver, eggs and embryos on cellulose acetate. The photograph is a composite of various developmental stages, all of which were run adjacent to liver extracts as a marker pattern. A) Rabbit liver LDH pattern. B) Extract of 20 unfertilized rabbit eggs cleaned with hyaluronidase. C) Extract of 20 unfertilized rabbit eggs cleaned with pronase. D) Extract of 20 pronase-cleaned 2-day (16 cell) embryos. E) Extract of six 4-day blastocysts. F) Sample containing the equivalent LDH activity of one-twentieth of one 6-day blastocyst. G) Extract of 20 uncleaned unfertilized rabbit eggs. (Reproduced by permission of Plenum Publishing Corp. from Schultz & Browder, 1975).

changes in various proteins during the developmental process will help to provide
a molecular map of the ontogenetic events occurring during preimplantation
stages.

QUANTITATIVE ASPECTS OF PROTEIN SYNTHESIS
Changes in Rates of Protein Synthesis in Early Rabbit Development

A factor which may be as important to early development and differentiation
as changes in the kinds of proteins produced is the overall rate of synthetic
activity.   Absolute measurement of protein synthetic rates at different stages
of development are difficult because incorporation of precursors is related to
uptake and, thus, knowledge of precursor transport is basic to any assessment of
quantitative aspects of protein synthesis (Epstein, 1975).   Some idea of
changing rates of synthesis are important to assess the relative roles of transcriptional and translational regulation mechanisms in control of the developmental process.

Detailed studies by Epstein and Smith (1973) in mice have shown that there
is a significant increase in the rate of protein synthesis on a per embryo basis
between the 8 cell and blastocyst stages which supported the suggestions of
earlier papers (Monesi & Salfi, 1967; Tasca & Hillman, 1970).   Studies on ^{3}H -
amino acid uptake and incorporation (acid soluble and insoluble radioactivity)
in the early rabbit embryo by Karp et al. (1974) have revealed that there is a
three-fold increase in ^{3}H -amino acid uptake and an eleven-fold increase in
incorporation between the 2-day (16 cell) embryo and the 3-day (128 cell) embryo.
That is, there appears to be a marked increase in rate of protein synthesis at
the time of transition of the morula to the blastocyst.   Subsequently, the
available data suggests that there is a further gradual continued increase in
the rate of protein synthesis on a per embryo basis as the blastocyst matures
and expands between 4 and 6 days of development (Manes & Daniel, 1969; Karp et
al., 1974).   Radioautographic analysis of rabbit embryos pulse-labelled with
^{3}H -amino acids for 10 minutes supports the biochemical data mentioned above
(Karp et al., 1974).   Grain densities were low in the 2 and 16 cell embryos
but markedly increased in the late morula.   In order to gain some insight into
what factors contribute to this increase in rate of synthesis in the early rabbit
embryo, we have conducted some studies on the transcriptional availability of
messenger RNA, ribosome availability, and the levels and functional ability of
the initiator tRNA.   These studies are discussed in the following three sub-
sections.

Availability of messenger RNA

One of the major factors which can determine both the kinds and amounts of
proteins produced in the embryo is the amount of template RNA available for

translation. While there is no information available on specific mRNA species, some information on total RNA synthesis during preimplantation rabbit development is published. When ^{3}H -uridine incorporation into acid insoluble material is measured and expressed on a per embryo basis, there is a steady increase from the fertilized egg through to the 6-day blastocyst (Manes, 1969). On a per cell basis, however, ^{3}H -uridine incorporation over a 2 h labelling period was found to be greatest in the day 3 (128 cell) and day 4 (1024 cell) embryos, that is, the late morula to early blastocyst stages (Manes, 1969). Karp et al. (1973) have confirmed this marked increase in ^{3}H -uridine incorporation at the onset of cavitation and blastocyst formation using radioautographic analysis on pulse-labelled embryos. The increase in ^{3}H -uridine incorporation per cell at the 3-day (128 cell) stage is in part due to increased uptake of ^{3}H -uridine from the medium by these embryos at this stage. However, uptake per cell again declines by the 1028 cell (day 4) blastocyst, but incorporation of ^{3}H -uridine remains at a high level (Karp et al., 1973; Manes, 1969). Thus, on a per cell basis, total RNA synthesis appears to increase as development proceeds from the morula to the blastocyst, peaking at the 4-day (1024 cell) stage before declining again as the blastocyst expands to the 6-day (80,000 cell) stage. This increased rate of synthesis appears to be due largely to the activation of ribosomal RNA synthesis which is delayed until the 3-day stage in rabbit embryos. This is corroborated both by polyacrylamide gel analysis of the kinds of RNA made at this stage (Manes, 1971) and predominant localization of label in nucleoli in autoradiographs of early blastocysts pulse-labelled with ^{3}H -uridine (Karp et al., 1973).

While the information above provides a picture of quantitative rates of total RNA synthesis, it presents little insight into the amounts of messenger RNA in the rabbit embryo at progressive stages of development. Clearly, some understanding of the synthesis and stability of messenger RNA (mRNA) is pre-requisite to the assessment of the relative importance of transcriptional and translational processes in the regulation of early mammalian development. Measurement of the synthesis of the total mRNA population of a cell is difficult because of the heterogeneity in molecular weights of molecules in this RNA class. However, the presence of poly(A) sequences covalently linked to the 3'-OH terminus of the major portion of mRNA molecules from eukaryotic organisms provides a useful probe for purification and study of metabolism of this class of RNA. A wealth of literature on poly(A)-containing RNAs has appeared in the last five years. The fact that the major part of polysomal heterogeneous RNA contains poly(A) tracts, that the size of poly(A) tracts of cytoplasmic mRNA is very similar in a wide range of eukaryotic organisms, that no other population of polyribonucleotides other than HnRNA and mRNA contain long poly(A) tracts at their 3'-OH ends, and that several purified mRNAs with demonstrated template activity for their appropriate polypeptides in heterologous or cell-free systems contain poly(A) tracts, is sufficient evidence to argue that a study of poly(A)-containing RNA is representative of putative mRNA (see Greenberg, 1975 for review).   Clearly, as much as 30% of the mRNA population of cells may lack poly(A) (Milcarek et al., 1974; Nemer et al., 1975), and this RNA class is not represented in methods which select for poly(A)-containing RNAs.   Nonetheless, at least the poly(A) plus class of mRNA can be analyzed.

Molecular hybridization between ^{3}H -polyuridylic acid and unlabelled RNA prepared from unfertilized eggs and 10-hour post-fertilization stage embryos has been used to demonstrate that poly(A)-containing RNA (putative mRNA) is present (Schultz, 1975).   The quantity of poly(A) was found to be about 0.25% of the total RNA in both stages, there being no marked increase as a consequence of fertilization.   Additionally, it has been shown that poly(A)-containing RNAs are synthesized during early cleavage and later developmental stages (Schultz et al., 1973).   The newly synthesized poly(A)-containing mRNA does become associated with polysomes at least as early as the 16 cell cleavage stage embryo and presumably is utilized for translation into polypeptides (Schultz, 1973a). In blastocysts, the newly synthesized poly(A)-containing RNA was shown to have considerable stability, the majority of the population having a half-life of about 18 hours (Schultz, 1974).

However, absolute quantitative analysis of amounts of poly(A)-containing RNA synthesized at various stages of development must be based on measurements of precursor pools and the kinetics of incorporation of precursors into product. Recently, elegant methods have been developed to measure rates of synthesis and turnover of RNA in sea urchin embryos through the analysis of the entry of ^{3}H -adenosine into ATP and RNA (Emerson & Humphreys, 1970; Brandhorst & Humphreys, 1971).   We have applied this technique to a study of the synthesis and turnover of polysomal poly(A)-containing RNA in preimplantation rabbit embryos by analyzing the kinetics of entry of ^{3}H - uridine into UTP and into RNA.   The procedure involved labelling groups of 2-, 4- or 6-day rabbit embryos for various periods of time in vitro in Ham's F-10 medium containing 200μ Ci/ml of ^{3}H -5, 6-uridine (40 Ci/mmole).   Embryos were then washed three times in ice-cold medium lacking isotope and were divided into two equal groups.   The one group of each labelling time and each stage of development was used for determination of the specific activity of UTP within the nucleotide pool, and the other for analysis of uridine incorporation into polysomal poly(A)-containing mRNA.

The entry of ^{3}H-uridine into UTP in 2-day, 4-day and 6-day preimplantation rabbit embryos is presented in Figure 4d.   In general, the specific activity of the UTP pool in these embryos was found to reach a maximum by 1 to 2 hr of labelling and then gradually declined toward 48 hr of labelling.   The molar

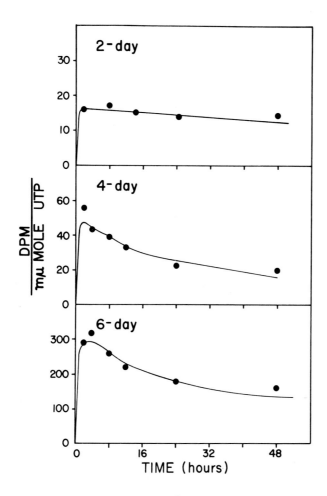

Figure 4: Kinetics of incorporation of ^{3}H-5-uridine into UTP in preimplantation rabbit embryos. To determine the specific activity of UTP, total nucleotides were prepared by the procedure of Emerson and Humphreys (1971) in which cold-acid-soluble nucleotides were adsorbed to acid washed norite and subsequently recovered by incubation in 50% ethanol/0.1 N NH_4OH. UTP was resolved by two-dimensional ion-exchange thin-layer chromatography on polyethylene-imine cellulose according to Neuhard et al. (1965). The UTP was eluted from the chromatogram and quantitated spectrophotometrically (extinction coefficient is 10.020×10^3 for UTP at neutral pH; Hurlburt, 1957) prior to determination of radioactivity by liquid scintillation counting corrected for quenching by the external standard method.

accumulation of poly(A)-containing mRNA was determined by dividing the radio-activity in UMP in the respective RNA molecules by the average specific activity of the UTP pool over each labelling period (Figure 5). The rate of accumulation of poly(A)-containing mRNA gradually decreases with time indicating that it has a limited stability (Figure 5). The accumulation curves approach a plateau or

84

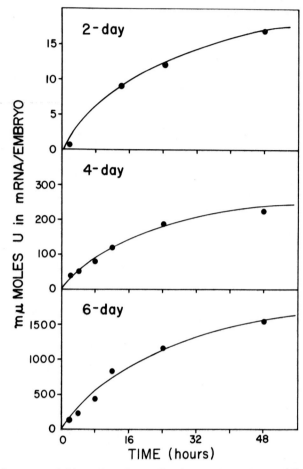

Figure 5: Molar accumulation of newly synthesized polysomal poly(A)-containing
mRNA in various stages of development of preimplantation rabbit embryos. The
curves show the accumulation of uridine into RNA calculated from incorporation
data of uridine into polysomal poly(A)-containing mRNA and the average specific
activity of the UTP pool in Figure 4. To obtain uridine incorporation data,
an equivalent number of embryos which had been simultaneously labelled with the
group used for UTP specific activity (Figure 4) were mixed with rabbit liver
tissue and RNA extracted from ribosomal pellets as described previously (Schultz
et al., 1973). Poly(A)-containing RNA was purified by affinity chromatography on
oligo-dT cellulose according to the procedure of Aviv and Leder (1972). Radioactivity in the mRNA was determined in a solubilizing scintillation cocktail and
was again corrected for quenching by the external standard method. Base analysis
of the RNA by the procedure of Kleppe et al. (1970) and Brown and Littna (1966)
revealed that about 96% of the radioactivity in the poly(A) RNA was in UMP after
1 hr of labelling. This decreased to 87-90% after 24 hr and to 78-81% after
48 hr of labelling due to nucleotide interconversions (primarily UTP to CTP).
The data have been corrected for the amount of radioactivity in ribonucleoside
monophosphates other than UMP. More complete details of the use of the procedures described above in cultured rabbit lung cells have been published
previously (Schultz, 1973b).

steady-state level after 48 hr of labelling.   From the shape of the curves in
Figure 5, it is estimated that steady-state values for accumulation of poly(A)-
containing mRNA for 2-day, 4-day and 6-day embryos are 19, 250 and 1715 n moles
UMP in mRNA per embryo, respectively.

    If an equilibrium situation exists in which the rate of synthesis of the
mRNA molecule is equal to its rate of decay, the accumulation curve of radio-
actively-labelled RNA to steady-state is the inverted curve for decay for
unlabelled molecules beginning at the time of addition of isotope (Brandhorst &
Humphreys, 1971).   After the addition of isotope, every mRNA molecule
synthesized would be labelled and whenever an unlabelled molecule decayed, it
would be replaced by a labelled molecule.   Since decay follows first-order
kinetics (Fritz et al., 1969), the curve for decay of unlabelled molecules can
be subjected to standard decay analysis and if all the mRNA has a uniform
stability, the decay curve should approximate a straight line directly
proportional to the decay constant.   The data of Figure 5 has been subjected
to decay analysis on the above basis (see Figure 6).   From Figure 6, the

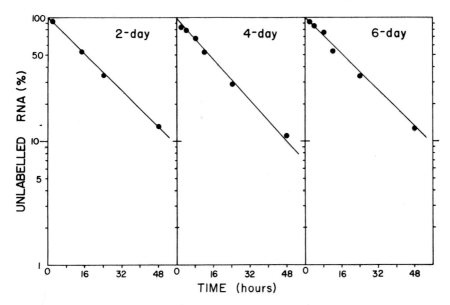

Figure 6: Standard decay analysis of the accumulation of radioactive poly(A)-
containing mRNA in various stages of development of preimplantation rabbit
embryos. The data is replotted from Figure 5 on the basis that the accumulation
curve of radioactive mRNA is the inverted curve for decay of unlabelled molecules
beginning at the time of addition of isotope as described in the text. The
logarithm of the percentage of total unstable mRNA remaining unlabelled is
plotted against time. The slopes of the lines are proportional to the decay
constants of the RNA.

average half-life or turnover rate of poly(A)-containing mRNA is estimated to be about 16 hr for each of 2-day, 4-day and 6-day embryos.

Having established the steady-state level of RNA (Figure 5) and the half-life (Figure 6), the instantaneous rate of synthesis can be calculated as the product of the decay constant times the steady-state level (Brandhorst & Humphreys, 1971). For the 2-day (16 cell) embryo this value is 4.5×10^{-15} moles of U/embryo/min or on a cellular basis, 2.8×10^{-16} moles of U/cell/min. For the 4-day embryo the corresponding values are 6.26×10^{-14} moles of U/embryo/min and 6.1×10^{-17} moles of U/cell/min. Finally, for the 6-day blastocyst, the values are 4.32×10^{-13} moles of U/embryo/min or 5.4×10^{-18} moles of U/cell/min. In summary, on an embryo basis, there is about a 100-fold increase in the rate of poly(A)-containing mRNA synthesis from the 2-day cleavage embryo to the mature 6-day blastocyst. On a cellular basis, however, the rate of synthesis of this class of molecules declines about 50-fold over the same developmental period. A similar decline per cell is observed during early sea urchin development (Emerson & Humphreys, 1970). In contrast to a decrease in RNA synthesis, the rate of protein synthesis per cell remains relatively the same through blastocyst formation and expansion (Manes & Daniel, 1969). Therefore, the differentially greater rate of protein synthesis would not seem to be due to a greater availability of mRNA molecules for translation in the cells since the rate of synthesis of at least the poly(A)-plus class of mRNA is decreasing at these stages. Presumably, then, the rate of protein synthesis may be regulated by other means such as ribosome numbers, initiation factors, or other translation factors.

Availability of Ribosomes

In the radioautographic studies of Karp et al. (1974), 16 cell cleavage stage embryos pulse-labelled with ^{3}H -amino acids for 10 minutes show low levels of incorporation with no preferential accumulation of newly synthesized proteins within the nucleus. The 128 cell morula, however, shows a marked increase in incorporation with a rapid accumulation of isotope within the nuclei of all cells during a 10 minute pulse label (Karp et al., 1974). Since this period of development is characterized by a sharp increase in rRNA synthesis, these authors suggest that the nucleolar labelling likely reflects the incorporation of a large fraction of the isotope into ribosomal proteins. To assess whether or not this is true, we have conducted experiments to assess when newly synthesized ribosomal proteins are detectable biochemically in the ribosomal fraction of the cells of preimplantation rabbit embryos.

Embryos were collected from rabbits at 2.5 days (32 to 64 cell stage), 3 days (128 cell stage) and 4 days post coitum and were labelled in vitro in the presence of ^{14}C -amino acids. The embryos were washed in medium lacking radioactive precursor, homogenized along with carrier rabbit liver tissue, and ribosomal fractions prepared as described previously (Schultz, 1973a). To distinguish whether any newly synthesized proteins were in fact associated with ribosomes and not with other ribonucleoproteins which can cosediment with ribosomes, the ribosomal fractions were fixed in formaldehyde and banded in CsC1 buoyant density gradients according to procedures described previously (Chen et al., 1971; Schultz, 1973a). Analysis of the gradients reveals that there is essentially no radioactivity banding with the absorbance profile (1.500 - 1.600 g/cc) characteristic of ribosomes in the 2.5 day old embryo (Figure 7). However, by the third and fourth day of development there clearly is newly synthesized protein associated with the ribosomes (Figure 7). This labelled protein appears to represent ribosomal proteins rather than nascent polypeptides on polysomes since the 2.5-day stage which is also active in protein synthesis has negligible radioactivity co-banding at densities characteristic of ribosomes. These findings are consistent with the suggestion of Karp et al. (1974) that ribosomal protein synthesis is stimulated at the 3-day (128 cell) stage, the time at which nucleolar synthesis of rRNA is activated (Manes, 1971; Karp et al., 1973).

Electron microscopy of sections of the unfertilized egg and early cleaving rabbit embryo demonstrates low cytoplasmic ribosome density with scattered small clusters of 3 to 6 ribosomes which are presumed to be polysomes (Van Blerkom et al., 1973; Karp et al., 1974). Manes (1975) has suggested that since the early embryo is synthesizing some proteins much larger than would be predicted from the polysome size, available messenger templates are not fully loaded with ribosomes. Following the activation of the embryonic nucleolus at the onset of blastocyst formation, cytoplasmic ribosomes are much more numerous and polysomes visible by electron microscopy are much larger (Van Blerkom et al., 1973; Karp et al., 1974). Hence, it is possible that the amount of protein produced in embryos during cleavage may be limited by the number of ribosomes available especially since the instantaneous rate of synthesis of poly(A)-containing mRNA on a per cell basis is greater during cleavage than in blastocyst stages. The increase in rate of protein synthesis at day-3 could be due to a rise in synthesis related to ribosome production. Subsequent accumulation of ribosomes may provide the machinery for further increases in rate of protein synthesis as the blastocyst expands and matures although other factors (see below) may also be important and make a contribution to the overall quantitative amount of protein produced.

Levels and modification of methionyl-transfer RNA

In the previous sections, the role of transcription in the production of template RNAs for translation and availability of ribosomes in relation to

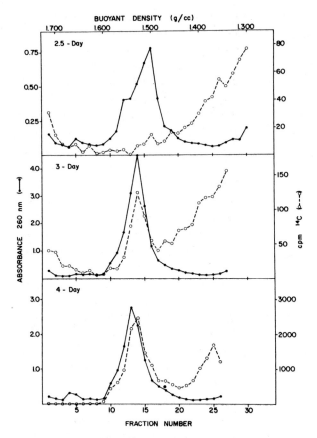

Figure 7: CsC1 buoyand density gradients of formaldehyde fixed polysomes from 2.5-, 3- and 4-day preimplantation rabbit embryos. 111 2.5-day embryos, 91 3-day embryos and 79 4-day embryos were labelled in vitro with 100μ Ci/ml ^{14}C -amino acids for 4 hours. The embryos were mixed with carrier rabbit liver tissue and ribosomal fractions prepared. The ribosomes were suspended in 0.03 M sodium phosphate buffer, pH 7.2, 0.003 M MgC1$_2$. The suspension was made 6% with respect to formaldehyde and left to fix at 4°C for 24 hours. The fixed ribosomes were applied to CsC1 gradients and centrifuged to equilibrium. The gradients were fractionated and assayed for absorbance at 260 nm (●—●) and TCA-precipitable radioactivity (o---o). Buoyant density was calculated from refractive index measurements made on a small aliquot of every fifth fraction.

protein synthesis has been discussed.   It has also been suggested that ribosome numbers may be the rate-limiting component for protein synthesis during cleavage. However, rates of protein synthesis in developing embryos may also be limited by other translational factors, particularly those affecting the process of initiation.   In eukaryotes, this process relies on the binding of the initiator tRNA, tRNA$_f^{Met}$ to small ribosome subunits before the binding of mRNA containing

an appropriate initiator codon (AUG) can occur.   Thus, levels of initiator tRNA
in embryonic cells or its degree of methylation may be important in regulating
amounts of protein synthesized by embryonic cells.   The methylation of tRNA is
known to serve an important role in the interaction of tRNA species with their
respective amino-acyl rRNA synthetases (Shugart et al., 1968; Marmor et al.,
1971), the appropriate coding response (Capra & Peterkofsky, 1968) and the
ability of tRNA to attach to ribosomes (Gefter & Russell, 1969).   A study on
the levels and methylation of tRNA^{Met} extracted from preimplantation rabbit
embryos has been published (Clandinin & Schultz, 1975) and might have some
implications relevant to the molecular basis for the increase in rate of protein
synthesis in late blastocyst stages in mammalian embryos.

The quantity of methionyl tRNA was determined by purification of total
tRNA following extraction with phenol on Sephadex G-50, aminoacylation of the
tRNA preparation with L-^{35}S -methionine using purified amino acyl-tRNA
synthetase from rabbit liver, and subsequent purification of methionyl-tRNA by
reverse phase chromatography on RPC-3 columns.   The molar quantity of methionyl-
tRNA was calculated from the molar quantity of ^{35}S methionine acylated to RNA
(Clandinin & Schultz, 1975).   The quantity of methionyl-tRNA increases about
50-fold as development proceeds from 2 days post-coitum to 6 days.   However,
the amount of methionyl-tRNA per cell declines some 100-fold from 1990 x 10^{-18}
moles per cell in the 2-day (16 cell) embryo to 19.1 x 10^{-18} moles per cell in
the 6-day blastocyst.   This parallels a 60-fold decrease on a cellular basis
for total RNA.   However, methylation of purified methionyl-tRNA by a rabbit
liver methylase extract using S-adenosyl-L-methionine as the methyl donor
illustrated that 2-day preimplantation embryo tRNA is highly hypomethylated
relative to tRNA from later stages of development or relative to rabbit liver
and E. coli tRNA (Table 2).

As the methylation of tRNA may be critical to the ability of tRNA to form
initiation complexes with ribosomes, binding experiments with rabbit liver
ribosomes and synthetic Ap:Up:G were conducted.   It was observed that 6-day
embryo methionyl-tRNA is initially 17-fold more effective in ribosome binding
than is methionyl-tRNA from 2-day and 4-day stages (Figure 8).   The lower
binding rate displayed by methionyl-tRNA of 4-day embryos suggests that it may
not yet be modified in the correct site for optimum rate of ribosome binding to
occur, even though this tRNA is more highly methylated than methionyl-tRNA of
2-day embryos.   The possibility that the transition in the degree of methylation
of methionyl-tRNA during rabbit embryo development also affects the affinity of
methionyl-tRNA synthetase for tRNA^{Met}, thus regulating the amount of charging
tRNA^{Met} species was not investigated but cannot be ruled out.   It may also be
of significance that although the quantity of methionyl-tRNA per genome declines
during development, its biological activity increases.   In summary, the increased

ability of tRNA^Met to bind to synthetic Ap:Up:G message and ribosomes in 6-day
rabbit blastocyst correlates with the general pattern of increase in protein
synthesis during blastocyst maturation (Manes & Daniel, 1969; Karp et al., 1974).
This suggests that this altered rate of synthesis may be regulated in part, at
the translational level via the initiator tRNA.   While increase in ribosome
numbers may also be a major factor in controlling the rate of protein synthesis
in the rabbit blastocyst, transitional changes in tRNA during embryo development
of a number of other species have been reported (Comb et al., 1965; Yang & Comb,
1968; Anderson, 1969; Bagshaw et al., 1970; Sharma et al., 1971; O'Melia &
Villee, 1972) and support the suggestion that protein synthesis in developing
embryos may be limited by transitional alterations and modifications of tRNA
or the process of initiation.

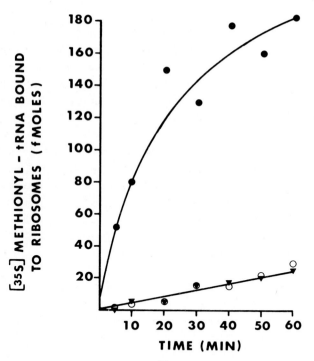

Figure 8: Binding of rabbit embryo ^35S -methionyl-tRNA to rabbit liver
ribosomes as a function of time. Purified methionyl-tRNA from 2-day (▼),
4-day (o) and 6-day (●) rabbit embryos was utilized in ribosome binding experiments. The reaction mixtures contained in a final vol. of 100μ 1: 100 mM Tris-
acetate buffer (pH 7.2), 20 mM magnesium acetate, 50 mM KCl, 1.5 A_{260} units of
rabbit liver ribosomes, 1 A_{260} unit of ApUpG and 0.3 pmol of ^35S methionyl-
tRNA (Reproduced by permission of Academic Press (London) Inc. from Clandinin
& Schultz, 1975).

TABLE 2

METHYLATION OF METHIONYL-tRNA BY A RABBIT LIVER METHYLASE EXTRACT

tRNA	methyl-^{14}C bound per pmol* of methionyl-tRNA (pmol)
2-day embryo	v 46.5
4-day embryo	4.56
6-day embryo	4.04
Rabbit liver	2.76
E. coli met B36	4.18

* Methionyl-tRNA purified by reverse phase chromatography was methylated
using S-adenosyl-L-methionine as the methyl donor and a rabbit liver
methylase extract (Reproduced with permission by Academic Press, (London)
Inc. from Clandinin & Schultz, 1975).

CONCLUDING REMARKS

In this paper we have reviewed both qualitative and quantitative changes
in protein synthesis during preimplantation rabbit development, including some
new information obtained in our laboratory. Using two-dimensional electrophoresis to analyze newly synthesized proteins at various stages of development,
we have demonstrated that a complex array of polypeptides are produced at all
stages. While there are some stage-specific polypeptides, including distinctive
lactate dehydrogenase isozyme patterns, we believe that a large number of polypeptides are common to all stages of preimplantation development. There are,
however, marked variations in the quantitative proportion of many of these
polypeptides through this early development period.

In terms of the overall rate of protein synthesis, there appears to be a
marked increase as the morula transforms into the blastocyst with further
increase in rate on an embryo basis as the blastocyst expands and matures to a
stage just prior to implantation. This quantitative change in protein synthetic
rate does not seem to be totally under transcriptional control. The rate of
synthesis of putative mRNA molecules (poly(A)-containing RNA) decreases during
the transitional period of increase in protein synthetic rate. Hence, greater
availability of template RNA molecules for translation is not an adequate
explanation for the increased rate in protein synthesis. A limiting complement of ribosomes appears to be largely responsible for the low rate of
synthesis during the cleavage period. We have provided evidence that new
ribosomes begin to be produced at 3-days post-coitum, the stage at which the
rate of protein synthesis begins to increase markedly. Additionally, proper
methylation of the initiator tRNA with a concomitant increase in its ability to

form initiation complexes with mRNA and ribosomes may contribute to further
increases in protein synthesis during the stages of blastocyst maturation.

Since mRNA synthesis is more active on a cellular basis in the cleavage
than in the blastocyst period, it is possible that some form of post-
transcriptional regulation is operative in these early embryos.   Messenger
RNA molecules may be actively synthesized and accumulated in the form of
ribonucleoprotein (RNP) particles for translation at a later stage of development.   For example, mRNAs for specific polypeptides required for blastocyst
formation could possibly be produced and accumulated during cleavage.   While
such a phenomenon has not been demonstrated in preimplantation rabbit embryos,
such post-transcriptional regulation has been described during myoblast growth
and differentiation (Buckingham et al., 1976), during the transition of mouse
fibroblasts from a quiescent to growing state (Rudland et al., 1975), as well
as in murine myeloma cells (MacLeod, 1975).   In all these cases, poly(A)-
containing RNA is stored nonpolysomally and is inactive or unable to initiate
protein synthesis until some translational control mechanism leads to its
utilization.   While there are no careful kinetic studies of the fate of poly(A)-
containing RNA in preimplantation rabbit embryos, a significant amount of
poly(A)-RNA has been demonstrated in the post-ribosomal supernatant of both
unfertilized and fertilized eggs (Schultz, 1975).   Additionally, we have shown
that newly synthesized poly(A)-containing RNA has a considerably long half-life
of about 16 h throughout the preimplantation period.   Thus, the existence of
regulation of protein synthesis at the level described above cannot be ruled out
as a possible mechanism in the control of early rabbit development.

The quantitative differences in individual polypeptides at various stages
of preimplantation rabbit development shown by our two-dimensional electrophoretic studies suggest that stage-specific or cell-specific translation factors
might be present.   Message and cell specific initiation factors which lead to
preferential translation of purified mRNAs in reconstituted cell-free systems
have been described in other experimental situations (Heywood, 1970; Fuhr & Natta,
1972; Revel et al., 1973; Thompson et al., 1973).   Additionally, translational
control RNAs which have either positive or negative effects on the translation
of purified mRNAs in vitro have also been described (Bester et al., 1975;
Heywood et al., 1975).   However, considerable controversy exists as to how the
findings above observed in cell-free systems compare with in vivo situations.
When mammalian globin mRNA is injected into fertilized Xenopus eggs and
efficiency of translation of the globin mRNA is later analyzed at stages in which
cell differentiation has occurred, no differences are observed (Woodland et al.,
1974).   It is possible that message or cell specific initiation factors or
translational control RNAs are characteristic only of cells in a terminal state
of differentiation.   If Xenopus embryos are a model, then such factors would not

appear to play any major role in the early development process. Despite this, our review of protein synthesis during preimplantation rabbit development suggests that both transcriptional and translational regulatory mechanisms are operative.

ACKNOWLEDGEMENTS

The work discussed from the authors' laboratory was supported by the Medical Research Council of Canada.

REFERENCES

ANDERSON, W.F. (1969) The effect of tRNA concentration on the rate of protein synthesis. Proc. Nat. Acad. Sci. U.S.A. 62, 566-573.

AUERBACH, S. & BRINSTER, R.L. (1967) Lactate dehydrogenase isozymes in the early mouse embryo. Exp. Cell Res. 46, 89-92.

AVIV, H. & LEDER, P. (1972) Purification of biologically active globin messenger RNA by chromatography on oligo-thymidylic acid-cellulose. Proc. Nat. Acad. Sci. U.S.A. 69, 1408-1412.

BAGSHAW, J.C., FINAMORE, F.J. & NOVELLI, G.D. (1970) Changes in transfer RNA in developing brine shrimp. Develop. Biol. 23, 23-35.

BESTER, A.J., KENNEDY, D.S. & HEYWOOD, S.M. (1975) Two classes of translational control RNA: their role in the regulation of protein synthesis. Proc. Nat. Acad. Sci. U.S.A. 72, 1523-1527.

BONNER, W.M. & LASKEY, R.A. (1974) A film detection method for tritium-labelled proteins and nucleic acids in polyacrylamide gels. Eur. J. Biochem. 46, 83-88.

BRANDHORST, B.P. & HUMPHREYS, T. (1971) Synthesis and decay rates of major classes of deoxyribonucleic acid in sea urchin embryos. Biochemistry 10, 877-881.

BRINSTER, R.L. (1975) Lactate dehydrogenase in the preimplanted mouse embryo. Biochim. Biophys. Acta 110, 439-441.

BRINSTER, R.L. (1973a) Paternal glucose phosphate isomerase activity in three-day mouse embryos. Biochem. Genet. 9, 187-191.

BRINSTER, R.L. (1973b) Lactate dehydrogenase isozymes in the preimplantation rabbit embryo. Biochem. Genet. 9, 229-234.

BROWN, D.D. & LITTNA, E. (1966) Synthesis and accumulation of DNA-like RNA during embryogenesis of Xenopus laevis. J. Mol. Biol. 20, 81-94.

BUCKINGHAM, M.E., COHEN, A. & GROS, F. (1976) Cytoplasmic distribution of pulse-labelled poly(A)-containing RNA, particularly 26S RNA, during myoblast growth and differentiation. J. Mol. Biol. 103, 611-626.

CAPRA, J.D. & PETERKOFSKY, A. (1968) Effect of in vitro methylation on the chromatographic and coding properties of methyl-deficient leucine transfer RNA. J. Mol. Biol. 33, 591-607.

CHEN, D., SCHULTZ, G.A. & KATCHALSKI, E. (1971) Early ribosomal RNA transcription and appearance of cytoplasmic ribosomes during germination of the wheat embryo. Nature New Biol. 231, 69-72.

CLANDININ, M.T. & SCHULTZ, G.A. (1975) Levels and modification of methionyl-transfer RNA in preimplantation rabbit embryos. J. Mol. Biol. 93, 517-528.

COMB, D.G., KATZ, S., BRANDA, R. & PINZANO, C. (1965) Characterization of RNA species synthesized during early development of sea urchins. J. Mol. Biol. 14, 195-213.

94

DIXON, G.H. (1972) The basic proteins of trout testis chromatin: Aspects of their synthesis, post-synthetic modification and binding to DNA. Acta Endocrinol. 5, 130-154.

DUCIBELLA, T., ALBERTINI, D.F., ANDERSON, E. & BIGGERS, J.D. (1975) The pre-implantation mammalian embryo: characterization of intercellular junctions and their appearance during development. Develop. Biol. 45, 231-250.

ECKLUND, P.S. & LEVINE, L. (1975) Mouse sperm basic nuclear protein: electrophoretic characterization and fate after fertilization. J. Cell Biol. 66, 251-262.

EMERSON, C.P. Jr. & HUMPHREYS, T. (1970) Regulation of DNA-like RNA and apparent activation of ribosomal RNA synthesis in sea urchin embryos: quantitative measurements of newly synthesized RNA. Develop. Biol. 23, 86-112.

EMERSON, C.R. Jr. & HUMPHREYS, T. (1971) A simple and sensitive method for quantitative measurement of cellular RNA synthesis. Anal. Biochem. 40, 254-266.

ENGEL, W., FRANK, W. & PETZHOLDT, U. (1975) Isozymes as genetic markers in early mammalian development. In: C.L. Markert (Ed.), Isozymes, Vol. 3, Academic Press, Inc., N.Y., pp. 67-81.

EPSTEIN, C.J. (1975) Gene expression and macromolecular synthesis during pre-implantation embryonic development. Biol. Reprod. 12, 82-105.

EPSTEIN, C.J. & SMITH, S.A. (1973) Amino acid uptake and protein synthesis in preimplantation mouse embryos. Develop. Biol. 33, 171-184.

EPSTEIN, C.J. & SMITH, S.A. (1974) Electrophoretic analysis of proteins synthesized by preimplantation mouse embryos. Develop. Biol. 40, 233-244.

FRITZ, P.J., VESELL, E.S., WHITE, E.L. & PRUITT, K.M. (1969) The roles of synthesis and degradation in determining tissue concentrations of lactate dehydrogenase-5. Proc. Nat. Acad. Sci. U.S.A. 62, 558-565.

FUHR, J.E. & NATTA, C. (1972) Translational control of - and -globin chain synthesis. Nature New Biol. 240, 274-276.

GARDNER, R.L. (1975) Analysis of determination and differentiation in the early mammalian embryo using intra-and inter-specific chimeras. In: C.L. Markert & J. Papaconstantinou (Eds.), The Developmental Biology of Reproduction. 33rd Symposium of the Society for Developmental Biology, Academic Press, N.Y., pp. 207-236.

GEFTER, M.L. & RUSSELL, R.L. (1969) Role of modifications in tyrosine transfer RNA: A modified base affecting ribosome binding. J. Mol. Biol. 39, 145-147.

GREENBERG, J.R. (1975) Messenger RNA metabolism of animal cells. J. Cell Biol. 64, 269-288.

HEYWOOD, S.M. (1970) Specificity of mRNA binding factors in eukaryotes. Proc. Nat. Acad. Sci. U.S.A. 67, 1782-1788.

HEYWOOD, S.M., KENNEDY, D.S. & BESTER, A.J. (1975) Studies concerning the mechanism by which translational-control RNA regulates protein synthesis in embryonic muscle. Eur. J. Biochem. 58, 587-593.

HURLBERT, R.B. (1957) Properties of diphosphates and triphosphates. In: S.P. Colowick and N.O. Kaplan (Eds.), Methods in Enzymology, Vol. 3, Academic Press, N.Y., pp. 801-805.

KARP, G., MANES, C. & HAHN, W.E. (1973) RNA synthesis in the preimplantation rabbit embryo: radioautographic analysis. Develop. Biol. 31, 404-408.

KARP, G.C., MANES, C. & HAHN, W.E. (1974) Ribosome production and protein synthesis in the preimplantation rabbit embryo. Differentiation 2, 65-73.

KIERSZENBAUM, A.L. & TRES, L.L. (1975) Structural and transcriptional features of the mouse spermatid genome. J. Cell Biol. 65, 258-270.

KLEPPE, K., VAN DE SANDE, J.H. & KHORANA, H.G. (1970) Polynucleotide ligase catalyzed joining of deoxyribo-oligonucleotides on ribopolynucleotide templates and of ribo-oligonucleotides on deoxyribopolynucleotide templates. Proc. Nat. Acad. Sci. U.S.A. 67, 68-73.

LAM, D.M.K., FURRER, R. & BRUCE, W.R. (1970) The separation physical characterization and differentiation kinetics of spermatogonial cells of the mouse. Proc. Nat. Acad. Sci. U.S.A. 65, 192-199.

MACLEOD, M.C. (1975) Comparison of the properties of cytoplasmic poly(adenylic acid)-containing RNA from polysomal and non-polysomal fractions of murine myeloma cells. Biochemistry 14, 4011-4018.

MANES, C. (1969) Nucleic acid synthesis in preimplantation rabbit embryos. I. Quantitative aspects, relationship to early morphogenesis and protein synthesis. J. Exp. Zool. 172, 303-310.

MANES, C. (1971) Nucleic acid synthesis in preimplantation rabbit embryos. II. Delayed synthesis of ribosomal RNA. J. Exp. Zool. 176, 87-96.

MANES, C. (1973) The participation of the embryonic genome during early cleavage in the rabbit. Develop. Biol. 32, 453-459.

MANES, C. (1975) Genetic and biochemical activities in preimplantation embryos. In: C.L. Markert & J. Papaconstantinou (Eds.), The Developmental Biology of Reproduction. 33rd Symposium of The Society for Developmental Biology, Academic Press, N.Y., pp. 133-163.

MANES, C. & DANIEL, J.C., Jr. (1969) Quantitative and qualitative aspects of protein synthesis in the preimplantation rabbit embryo. Exptl. Cell Res. 55, 261-268 .

MARMOR, J.B., DICKERMAN, H.W. & PETERKOFSKY, A. (1971) Studies on methyl-deficient methionine transfer ribonucleic acid from Escherichia coli. J. Biol. Chem. 246, 3464-3473.

MILCAREK, C., PRICE, R. & PENMAN, S. (1974) The metabolism of a poly(A) minus mRNA fraction in HeLa cells. Cell 3, 1-10.

MONESI, V. & SALFI, V. (1967) Macromolecular synthesis during early development in the mouse embryo. Exp. Cell Res. 46, 632-635.

MOORE, N.W., ADAMS, C.E. & ROWSON, L.E.A. (1968) Developmental potential of single blastomeres of the rabbit egg. J. Reprod. Fert. 17, 527-531.

NEMER, M., DUBROFF, L.W. & GRAHAM, M. (1975) Properties of sea urchin embryo messenger RNA containing and lacking poly(A). Cell 6, 171-178.

NEUHARD, J., RANDERATH, E. & RANDERATH, K. (1965) Ion-exchange thin-layer chromatography XIII. Resolution of complex nucleoside triphosphate mixtures. Anal. Biochem. 13, 211-222.

O'FARRELL, P.H. (1975) High resolution two-dimensional electrophoresis of proteins. J. Biol. Chem. 250, 4007-4021.

O'MELIA, A. & VILLEE, C.A. (1972) De Novo synthesis of transfer and 5S RNA in cleaving sea urchin embryos. Nature 239, 51-53.

PEDERSEN, R.A. & CLEAVER, J.E. (1975) Repair of UV damage to DNA of implantation stage mouse embryos in vitro. Exptl. Cell Res. 95, 247-253.

RAPOLA, J. & KOSKIMIES, O. (1967) Embryonic enzyme patterns: characterization of the single lactate dehydrogenase isozyme in preimplanted mouse ova. Science 157, 1311-1313.

REVEL, M., GRONER, Y., POLLACK, Y., CNAANI, D., ZELLER, H. & NUDEL, U. (1973) Biochemical mechanism to control protein synthesis by mRNA specific initiation factors. Karolinska Symposium on Research Methods in Reproductive Endocrinology 6th Symposium, pp. 54-71.

RUDLAND, P.S., WEIL, S. & HUNTER, A.R. (1975) Changes in RNA metabolism and accumulation of presumptive messenger RNA during transition from the growing to the quiescent state of cultured mouse fibroblasts. J. Mol. Biol. 96, 745-766.

SCHULTZ, G.A. (1973a) Characterization of polyribosomes containing newly-synthesized messenger RNA in preimplantation rabbit embryos. Exptl. Cell Res. 82, 168-174.

SCHULTZ, G.A. (1973b) Stability of cytoplasmic messenger RNA in stationary and exponentially-growing rabbit lung cells. Can. J. Biochem. 51, 1515-1520.

SCHULTZ, G.A. (1974) The stability of messenger RNA containing polyadenylic acid sequences in rabbit blastocysts. Exptl. Cell Res. 86, 190-193.

SCHULTZ, G.A. (1975) Polyadenylic acid-containing RNA in unfertilized and fertilized eggs of the rabbit. Develop. Biol. 44, 270-277.

SCHULTZ, G.A., MANES, C. & HAHN, W.E. (1973) Synthesis of RNA containing polyadenylic acid sequences in preimplantation rabbit embryos. Develop. Biol. 30, 418-426.

SCHULTZ, G.A. & BROWDER, L.W. (1975) Lactate dehydrogenase in preimplantation rabbit embryos. Biochem. Genet. 13, 663-671.

SCHULTZ, G.A. & CHURCH, R.B. (1975) Transcriptional patterns in early mammalian development. In: R. Weber (Ed.), Biochemistry of Animal Development. Vol. 3, Academic Press, N.Y., pp. 47-90.

SHARMA, O.K., LOEB, L.A. & BOREK, E. (1971) Transfer RNA methylases during sea urchin embryogenesis. Biochim. Biophys. Acta 240, 558-563.

SHUGART, L., CHASTIN, B.H., NOVELLI, G.D. & STULBERG, M.P. (1968) Restoration of aminoacylation activity of undermethylated transfer RNA by in vitro methylation. Biochem. Biophys. Res. Comm. 31, 404-409.

TARKOWSKI, A.K. & WROBLEWSKA, J. (1967) Development of blastomeres of the mouse egg isolated at the 4- to 8-cell stage. J. Embryol. Exp. Morph. 18, 155-180.

TASCA, R.J. & HILLMAN, N. (1970) Effects of actinomycin D and cycloheximide on RNA and protein synthesis in cleavage stage mouse embryos. Nature 225, 1022-1025.

THOMPSON, W.C., BUZASH, E.A. & HEYWOOD, S.M. (1973) Translation of myoglobin mRNA. Biochemistry 12, 4559-4565.

VAN BLERKOM, J., MANES, C. & DANIEL, J.C. Jr. (1973) Development of preimplantation rabbit embryos in vivo and in vitro. I. An ultrastructural comparison. Develop. Biol. 35, 262-282.

VAN BLERKOM, J. & MANES, C. (1974) Development of preimplantation rabbit embryos in vivo and in vitro. II. A comparison of qualitative aspects of protein synthesis. Develop. Biol. 40, 40-51.

VAN BLERKOM, J. & BROCKWAY, G.O. (1975) Qualitative patterns of protein synthesis in the preimplantation mouse embryo. Develop. Biol. 44, 148-157.

VAN BLERKOM, J., BARTON, S.C. & JOHNSON, M.H. (1976) Molecular differentiation in the preimplantation mouse embryo. Nature 259, 319-321.

WALES, R.G. (1975) Maturation of the mammalian embryo: Biochemical Aspects. Biol. Reprod. 12, 66-81.

WASSARMAN, P.M. & LETOURNEAU, G.E. (1976) Meiotic maturation of mouse oocytes in vitro: association of newly synthesized proteins with condensing chromosomes. J. Cell Sci. 20, 549-568.

WOODLAND, H.R., GURDON, J.B. & LINGREL, J.B. (1974) The translation of mammalian globin mRNA injected into fertilized eggs of Xenopus laevis. II. The distribution of globin synthesis in different tissues. Develop. Biol. 39, 134-140.

WUDL, L. & CHAPMAN, V. (1976) The expression of β-Glucuronidase during pre-implantation development of mouse embryos. Develop. Biol. 48, 104-109.

YANG, S.S. & COMB, D.G. (1968) Distribution of multiple forms of lysyl transfer RNA during early embryogenesis of sea urchin, Lytechinus variegatus. J. Mol. Biol. 31, 139-142.

RNA POLYMERASE ACTIVITY IN PREIMPLANTATION
MAMMALIAN EMBRYOS

Carol M. Warner

Department of Biochemistry and Biophysics
Iowa State University
Ames, Iowa 50011

It is generally recognized that the control of early development in mammals
is, at least in part, the consequence of the regulation of gene activity.   Since
each adult cell both of plants (Steward et al., 1958; Vasil & Hildebrandt, 1965)
and animals (Gurdon, 1962) contains the necessary genetic information for the
generation of an entire new organism, it is believed that genetic information is
neither gained nor lost during development.   There are two major modes by which
gene activity may be regulated and these may be broadly classified as transcriptional and post-transcriptional controls.   In Table 1, a number of these
parameters are listed for eukaryotic cells.   Although control mechanisms may
exist for any of the categories listed in Table 1, most experimental data support
the view that transcriptional control is the most significant method of regulating gene activity (Davidson & Britten, 1973).

It is the purpose of this contribution to assess the role of multiple forms
of RNA polymerase in the control of gene activity in early mammalian development.
For multiple forms of RNA polymerase to play some role in the control of development, it is important that certain criteria be met.   First, RNA polymerase
activity must be shown to be present in the tissue being studied.   Second,
multiple forms of RNA polymerase must be demonstrated.   Third, the relative
proportion of the enzyme forms should change in a coordinate fashion with the
types of RNA being synthesized at any particular stage in development.   And
finally, any observed changes in RNA polymerase activity, or RNA synthesis,
should be coordinated with some important physiological role for these processes.

EUKARYOTIC RNA POLYMERASES

It was first demonstrated by Roeder and Rutter (1969) that two chromato-
graphically separable forms of RNA polymerase enzyme exist in rat liver and sea
urchin cells.   This finding precipitated an avalanche of experiments which have
helped elucidate the structure and function of eukaryotic RNA polymerases.   We
now know that all eukaryotic cells contain multiple forms of RNA polymerase, as
opposed to prokaryotic cells which contain only one major enzyme form.   Several

TABLE 1

POSSIBLE MODES FOR THE CONTROL OF GENE ACTIVITY IN EUKARYOTES

Transcriptional Controls

   1.   Chromatin Template

        a. Histones

        b. Nonhistone chromosomal proteins

        c. Conformational parameters, such as
            superhelix formation

   2.   RNA Polymerases

        a. Multiple forms

        b. Specific initiation and termination factors

        c. Modification, such as phosphorylation

   3.   "Other Factors", as yet unidentified

Post-Transcriptional Controls

   1.   RNA processing, transport, and degradation

   2.   Translational controls

   3.   Post-translational controls, such as protein half-life

---

reviews of the structure and function of eukaryotic RNA polymerases have appeared during the past few years (Jacob, 1973; Rutter et al., 1974; Chambon, 1974, 1975; Rutter, 1976). Only the most relevant data about these enzymes will be summarized here.

There are three major forms of RNA polymerase, designated either as I, II, and III or as A, B, and C respectively. The first nomenclature will be used in this chapter. Eukaryotic cells always contain enzyme forms I and II, and may contain either small or large amounts of the form III enzyme. In addition to the three major enzyme forms, many organisms show subforms designated IA, IB or IIA, IIB, etc. Whether these subforms have any functional significance or are degraded or modified enzyme forms is not yet clear.

The general properties of the three major enzyme forms are listed in Table 2, as a compilation of data from a number of laboratories (Chambon, 1975; Rutter, 1976), including data on the mouse from our own laboratory (Versteegh & Warner, 1973, 1975). One of the problems in studying RNA polymerases has been the difficulty encountered in obtaining sufficient quantities of highly purified enzymes for biochemical and biophysical studies. Typically, only a few milligrams of enzyme are obtained from a kilogram of tissue (Chambon, 1975). When enough material has been obtained, usually from lower eukaryotes such as yeast, a striking similarity of the physical properties of the different enzyme forms

TABLE 2

GENERAL PROPERTIES OF EUKARYOTIC RNA POLYMERASES

	Property	Form of Polymerase		
		I	II	III
(1)	Elution from DEAE-Sephadex M $(NH_4)_2SO_4$	0.05-0.15	0.25-0.28	0.30-0.35
(2)	$(NH_4)_2SO_4$ for optimal activity (M)	~0.03	~0.09	~0.04
(3)	Mn^{2+}/Mg^{2+} activity ratio	1	5-50	2-3
(4)	% Inhibition by 0.03μ g/ml α-amanitin	0	100	0
(5)	% Inhibition by 300μ g/ml α-amanitin	0	100	100
(6)	% Inhibition by Rifampin (40μ g/ml)	0	0	0
(7)	Cellular Localization	nucleolus	nucleoplasm	nucleoplasm and/or cytoplasm
(8)	Major Transcription Product	rRNA	HnRNA mRNA	5S RNA 4S RNA

to one another and also among species, has been observed (Chambon, 1975; Rutter, 1976). The enzymes are all quite large (MW 400,000-600,000), have two large subunits (MW >100,000), and a number of smaller subunits, depending on the organism and the laboratory reporting the data. In general, then, the enzymes seem similar in size and physical characteristics to the E. coli or B. subtilis enzymes, which have been purified to homogeneity (Chamberlin, 1974). Recently, studies using antibody prepared to the various enzyme forms have shown that the two large subunits do not cross-react, but there may be some cross-reaction among the smaller subunits (Ingles, 1973; Kedinger et al., 1974; Greenleaf & Bautz, 1975). It is a major thrust of research in this area to identify the function of each of the subunits in each particular enzyme form.

The compartmentalization of the form I enzyme in the nucleolus and the form II enzyme in the cytoplasm immediately suggested that the different enzyme forms might be involved in the synthesis of different classes of RNA (see Table 2). This has been proven by Weinmann and Roeder (1974) with the use of α-amanitin, either administered in vivo, or to isolated nuclei, in vitro. However, in spite of a considerable amount of work, these results have not been proven using isolated and highly purified components in a reconstitution study (Chambon, 1975).

Some of the problems encountered have involved the isolation and purification of
an intact chromatin template, and the isolation of pure RNA polymerases that
have not lost any important initiation or termination factors in the purification
process.   Also, if a reconstitution study does not lead to fidelity of transcription, some other, unknown, factors than simply the chromatin and the enzyme
might be missing.   Nevertheless, it does seem clear that intact cells or intact
nuclei do show selective transcription of the different RNA classes by the
different enzyme forms.

THE CONTROL OF DEVELOPMENT BY MULTIPLE FORMS OF RNA POLYMERASE

Almost as soon as the multiple forms of RNA polymerase were discovered, it
was suggested that they might play some role in the control of development.
The first study of this problem was that of Roeder and Rutter (1970) on sea
urchin development.   They found that there was a correlation of the types of
enzymes present with the types of RNA synthesized at stages of development from
fertilization to the late gastrula.   They concluded that this data suggested
the involvement of the multiple forms of RNA polymerase in the regulation of RNA
synthesis in the sea urchin.   A few years later, two studies on Xenopus laevis
embryos came to exactly the opposite conclusion (Roeder et al., 1970; Roeder,
1974).   In these studies it was shown that control of gene activity was not
simply related to or controlled by the cellular levels of specific RNA polymerases.

In view of the previous discussion of RNA polymerases in eukaryotic cells,
several points become apparent with respect to RNA polymerases in preimplantation
mammalian embryos.   First, due to the small size of the embryos, it clearly
would be impossible to isolate and purify the enzyme forms.   Second, since
multiple forms of RNA polymerase have been found in every eukaryotic organism
studied so far, it seems highly likely that mammalian embryos also would contain
multiple forms of the enzyme.   Third, since in vitro reconstitution experiments
with purified materials have at best been equivocal, some interesting results
should be forthcoming using in vivo culture and assay conditions.   Thus, armed
with the foregoing rationale, studies on multiple forms of RNA polymerases in
preimplantation mammalian embryos have been undertaken by a number of laboratories,
including our own, during the past few years.

There are two general methods which have been used to study the RNA polymerase activity in preimplantation mammalian embryos.   First, the synthesis of
RNA has been measured as an indirect assay of RNA polymerase activity.   Second,
the enzyme activity has been measured directly in homogenized embryos.   The
mouse embryo has been the mammalian developmental system most extensively studied
by both of these methods.   For this reason the bulk of this chapter will deal

with experimental results in preimplantation mouse embryos, with references to other systems, such as the rabbit or the rat, when appropriate.

Indirect Assays of RNA Polymerase Activity in Mammalian Embryos

If a cell or tissue can be shown to be capable of synthesizing RNA, this may be interpreted as presumptive evidence that RNA polymerase is functional in that cell.   Such experiments typically involve the incubation of intact cells with an RNA precursor, such as ^{3}H-uridine, and analysis for the incorporation of the label into high molecular weight product.   In order to compare the rate of RNA synthesis at one cell stage to another, one must know the specific activity of the intracellular pool of labeled precursor.   To date, in spite of a number of attempts, this has not been possible with mouse embryos because of their small size (Daentl & Epstein, 1971; Epstein & Daentl, 1971; Bernstein & Mukherjee, 1973; Borland, this volume).   One way to circumvent this requirement is to saturate the intracellular pool so that the specific activity of the pool is the same as the specific activity of the culture medium.   However, this has generally been possible only with blastocysts (Daentl & Epstein, 1973; Borland, this volume).   Thus, even the best studies must describe "relative" and not absolute rates and amounts of RNA synthesis in preimplantation mouse embryos, at the various preimplantation stages (Schultz & Tucker, this volume).

The first indication that mouse embryos are active in RNA synthesis very early in development was the autoradiographic studies of Mintz (1964a).   She showed that incorporation of ^{3}H-uridine into the nucleus was only into the non-nucleolar portion up to the 4 cell stage.   By the 8 cell stage, nuclear RNA synthesis was easily detectable, and in fact was the predominant type of RNA being synthesized by the embryos.   Subsequent studies of RNA synthesis in pre-implantation mouse embryos have been aimed at answering questions such as the following: Exactly when does the embryonic genome become active?   Is there RNA synthesis in one-cell embryos?   What types of RNA are made at each stage?

Activation of the Mammalian Embryonic Genome

The answer to the question 'when does the embryonic genome first become active?' is still not completely clear.   Using mutations at the T locus (e.g. the t^{12}/t^{12} mutation), it has been shown that the mutation is manifested by the morula stage (Mintz, 1964b).   Thus, the embryonic genome must be active by this stage.   Evidence for even earlier activation of the embryonic genome has been obtained by assaying embryos from XX and XO mice for X-linked enzymes such as glucose-6-phosphate dehydrogenase (Epstein, 1969), hypoxanthine guanine phosphoribosyl transferase (Epstein, 1972) and phosphoglycerate kinase (Kozak et al., 1974).   In all cases the evidence was for the activation of the embryonic genome by the 8-16 cell stage.

Another approach to the determination of the time of embryonic genome activation has been to use enzymes, such as glucose phosphate isomerase-1 (Chapman et al., 1971), or β-glucuronidase (Chapman & Wudl, 1975), which may both exist as different isozymes in different inbred strains of mice. Then, females homozygous for one enzyme form are mated to males homozygous for another and the time at which the male isozyme becomes apparent is measured. The conclusion is that paternal enzymes are detectable by the 8 cell stage of development. Thus, the evidence is in favour of activation of the embryonic genome at least by the 8 cell stage. It must be borne in mind that negative results before the 8 cell stage may simply be a reflection of insensitivity of the assay procedures used, rather than a true lack of embryonic gene activity before this stage.

RNA Synthesis in One-Cell Embryos

RNA synthesis is easily detectable before the 8 cell stage, so the next question becomes just when in development is it first detectable. There is general agreement (reviewed by Epstein, 1975) that 2 cell mouse embryos can synthesize significant amounts of RNA. The lack of detectable synthesis in 1 cell embryos has been attributed either to inability of the radioactive RNA precursor to enter 1 cell embryos (Woodland & Graham, 1969) or else the lack of an enzyme, such as uridine kinase, to convert the ^{3}H-uridine to UTP (Daentl & Epstein, 1971). A recent study by Moore (1975) has circumvented both of these objections by using 1 cell embryos which were treated with an organic solvent which allowed them to take up ^{3}H-UTP. It was shown that even though the RNA precursor was available as the triphosphate, there was no RNA synthesis detectable by autoradiography in 1 cell embryos.

The Amount of RNA Synthesized by Preimplantation Mammalian Embryos

The amount of RNA synthesized by preimplantation mouse embryos has been well studied. It was first shown by Monesi and Salfi (1967), on a quantitative basis, that the total amount of RNA synthesized per embryo increases markedly between the 12-16 cell and blastocyst stages. This study confirmed the earlier autoradiographic data of Mintz (1964a). However, since the rate of uptake of ^{3}H-uridine by embryos can only be separated from the rate of incorporation of ^{3}H-uridine for the blastocyst stage (Daentl & Epstein, 1973), on a per cell basis, it is still not clear whether the total amount of RNA synthesized at each embryonic stage remains constant or increases somewhat. As mentioned previously, this problem would be eliminated if the specific activity of the intracellular precursor pools was known for each stage of development. On a per embryo basis, the total amount of RNA synthesized between the 8 cell and blastocyst (taken as 32 cells) stages has been calculated from incorporation studies to be 750 pg

(Epstein, 1975).   This value is quite close to that of 912 pg, which is the
chemically measured RNA increase between the 9 and 34 cell stages reported by
Olds et al. (1973).

The Types of RNA Synthesized by Preimplantation Mammalian Embryos

The next problem with regard to RNA synthesis studies in mouse embryos is
the types (classes) of RNA which are synthesized at each developmental stage.
This has been described both qualitatively and quantitatively by a number of
laboratories, using a variety of methods.   The cell stages before the 8 cell
stage have been studied by Woodland and Graham (1969), and by Knowland and
Graham (1972), using the techniques of MAK column chromatography, sucrose
gradient centrifugation, Sephadex G-100 chromatography, and polyacrylamide gel
electrophoresis.   In the first study (Woodland & Graham, 1969), it was
reported that both high molecular weight RNA (presumed by the authors to be
HnRNA and mRNA) and low molecular weight RNA (presumed by the authors to be a
4S RNA precursor) synthesis was detectable starting at the 2 cell stage, and
that rRNA and 4S RNA synthesis became detectable starting at the 4 cell stage.
In the second study (Knowland & Graham, 1972), using more embryos (1000 2 cell
embryos) and a more sensitive technique (polyacrylamide gel electrophoresis),
all major RNA classes were detectable in 2 cell embryos, including rRNA and
tRNA.

The stages from 8 cell to blastocyst have been studied by Ellem and Gwatkin
(1968), using MAK column chromatography, by Piko (1970), using sucrose gradient
centrifugation, and by ourselves (Warner & Hearn, unpublished) using polyacrylamide gel electrophoresis.   Qualitatively, all three studies are in
agreement, showing that, starting at the 8 cell stage and continuing through
the blastocyst stage, all classes of RNA are synthesized, the predominant class
being synthesized is rRNA, and the distribution of newly synthesized RNA among
the different nucleic acid classes is similar for 8 cell and blastocyst embryos.
The quantitative data available from these studies, for 8 cell and blastocyst
embryos, is shown in Table 3.   It is seen that the two studies are in excellent
quantitative, as well as qualitative agreement.   In the earlier study (Ellem &
Gwatkin, 1968), the high molecular weight, messenger-like RNA (HnRNA) is included
with the mRNA, whereas in the later study (Warner & Hearn, unpublished), this
class of RNA is observed with the DNA, at the top of the gel.   It is immediately
apparent, that using the labeling and RNA extraction conditions of these two
studies, very little HnRNA is observed.

We sought to further characterize any HnRNA which might be present in mouse
embryos by using binding to poly-U filters to search for poly-A-containing RNA
in the mouse embryo extracts.   A similar study by Schultz et al. (1973) on
rabbit embryos, had shown that poly-A-containing RNA's are synthesized at least

TABLE 3

A COMPARISON OF NUCLEIC ACID SYNTHESIS IN

8 CELL AND BLASTOCYST MOUSE EMBRYOS

Nucleic acid classes	% of each class reported in two studies			
	Ellem & Gwatkin (1968)*		Warner & Hearn (1976)***	
	8 cell	Blastocyst	8 cell	Blastocyst
rRNA	64	69	65	62
mRNA	18**	17**	13	17
tRNA	10	8	9	13
DNA	8	7	12**	9**

* Data is from Ellem and Gwatkin (1968) and is the result of MAK column chromatography after 5 hr of labeling of embryos in ^{3}H-uridine. The maximum reported error was ±7%.

** This value includes HnRNA.

*** Data is from Warner and Hearn (unpublished) and is the result of poly-acrylamide gel electrophoresis after 5 hr of labeling of embryos in ^{3}H-uridine. The maximum error was ±9%.

as early as the 16 cell stage, and that the portion of the heterogeneous RNA containing poly-A sequences does not appear to change markedly between the cleavage and blastocyst stages of development. The total percent of hetero-geneous RNA (hRNA) was found to be 18% for day 2 rabbit embryos, 12% for day 4 rabbit embryos, and 17% for day 6 rabbit embryos. Estimates of the mRNA content of the heterogeneous RNA based on polyadenylic acid content, suggested that about 20% of the hRNA is mRNA in 6-day rabbit embryos (see Schultz & Tucker, this volume).

We have attempted to make similar quantitative estimates of the amount of poly-A-binding RNA at various stages in mouse embryo development, using exactly the same procedure for RNA extraction reported by Schultz et al. (1973). Their approach was based on the methods of Lee et al. (1971) and Perry et al. (1972), and uses an extraction into chloroform:phenol(1:1 by volume) at pH 9.0, which is essential to maintain the integrity of poly-A sequences. These sequences are lost when extractions are performed with phenol and SDS at neutral pH. It is important to note that the data reported in Table 3 from Ellem and Gwatkin (1968) and from Warner and Hearn (unpublished) used the latter conditions.

The data shown in Table 4 is the amount of total RNA, from the 8 cell and blastocyst stages of development, which binds to a poly-U filter, prepared as described by Sheldon et al. (1972). Data is shown for RNA which was synthesized by mouse embryos which were incubated for five hours in the presence of

TABLE 4

PERCENT OF POLY-A-CONTAINING RNA IN 8 CELL AND BLASTOCYST MOUSE EMBRYOS

Cell Stage	Amount of Actinomycin D (μg/ml)	Percent Binding to Poly U Filter ± SEM
8 cell	0	2.2 ± 0.2
8 cell	0.01	3.7 ± 0.3
8 cell	0.1	4.7 ± 1.0
blastocyst	0	2.3 ± 0.3
blastocyst	0.01	5.2 ± 1.0
blastocyst	0.1	8.8 ± 2.8

500μ Ci/ml ^{3}H-uridine.   This compares to a labeling of four hours in 200μ Ci/ml
of ^{3}H-uridine, as reported by Schultz et al. (1973) for the rabbit embryos.
In several experiments, actinomycin D was added to preferentially block rRNA
synthesis, so that the total percent of poly-A-containing RNA would increase.
It is seen that mouse embryos synthesize much less poly-A-containing RNA than
rabbit embryos at comparable stages of development.   However, in agreement with
the data on rabbit embryos, is the fact that the proportion of poly-A-containing
RNA remains constant from the 8 cell to the blastocyst stage of development.
This of course in no way implies that the same amount of poly-A-containing RNA
is made at these cell stages.   Rather, the data shows that the relative
proportion of poly-A-containing RNA, compared to total RNA, is similar at the
various embryonic stages.

The small amount of poly-A-containing RNA from mouse blastocysts was
further characterized by poly-dT-cellulose chromatography and poly-acrylamide
gel electrophoresis.   Figure 1 shows the profile of the RNA from 75 blastocysts
incubated for five hours in the presence of 500μ Ci/ml ^{3}H-uridine, and extracted,
as described previously, to maintain the poly-A sequences.   The pooled poly-A-
containing fractions were then subjected to polyacrylamide gel electrophoresis
as described by Loening (1967), with the results shown in Figure 2.   It is seen
that essentially all of the poly-A-containing RNA is found in the mRNA region,
with virtually none at the top of the gel.   The apparent peak in the 4S region
is due to spillover from the ^{14}C-tRNA internal marker.   This result is again,
qualitatively different from the rabbit data of Schultz et al. (1973), in which
a significant amount (~20%) of poly-A-containing RNA was found, by sucrose
gradient sedimentation analysis, to be in the high molecular weight region.
Since the incubation and extraction conditions were virtually identical in the
rabbit and mouse studies, it must be concluded that rabbit blastocysts and mouse

blastocysts differ in the relative proportion of high molecular weight poly-A-containing RNA synthesized at this stage of development.   Thus, no generalizations about poly-A-containing RNA's from preimplantation mammalian embryos can be made at this time.

With respect to RNA polymerase activity, it has generally been hypothesized, but not proven, that form II RNA polymerase is responsible for HnRNA synthesis, as well as mRNA synthesis (Chambon, 1975).   Since both HnRNA and mRNA contain poly-A at the 3' end and a "cap" at the 5' end, it has been suggested, but not fully proven that HnRNA in the nucleus gives rise to mRNA in the cytoplasm (Darnell et al., 1971; Rottman et al., 1974).   Thus, one might speculate that HnRNA containing poly-A may be degraded more rapidly in the nucleus of the cells of the mouse blastocyst than in the rabbit blastocyst.   The much larger size and complexity of the rabbit blastocyst compared to the mouse blastocyst may account for the smaller proportion of newly synthesized poly-A-containing mRNA found in the mouse blastocyst.

Direct Assays of RNA Polymerase Activity in Mammalian Embryos

So far we have discussed the evidence for RNA synthesis in preimplantation mouse embryos as presumptive evidence of RNA polymerase enzyme activity in the embryos.   Although about two dozen different enzyme activities have been measured in preimplantation mouse embryos at various stages of development (reviewed by Epstein, 1975), no enzyme seems so crucial to the control of gene activity as RNA polymerase itself.   We therefore embarked on a study of this enzyme several years ago.   The first question to answer is whether or not RNA polymerase enzyme activity is directly detectable in mouse embryo homogenates. This proved to be possible, using reasonable numbers of embryos, when a highly sensitive DEAE-paper filter binding assay, originally described by Litman (1968), was employed.   The principle is that ^{3}H-uridine which has been incorporated into high molecular weight nucleic acid (>10 nucleotides) will bind to the DEAE-paper by electrostatic attraction, but unincorporated ^{3}H-uridine will be released by extensive washing with phosphate buffer.   This assay procedure is much more sensitive and reproducible than conventional TCA precipitation.   The background can be reduced to 0.1% of the added radioactivity.   Using this

Figure 1. (upper) Poly-dT-cellulose chromatography of mouse blastocyst RNA. Seventy five blastocysts were incubated for 5 hr in the presence of 500μ Ci/ml ^{3}H-uridine, and the RNA extracted and chromatographed, as described in the text.

Figure 2. (lower) Polyacrylamide gel electrophoresis of mouse blastocyst RNA which had been bound to a poly-dT-cellulose column. Fractions 15, 16, and 17 from the column described in Figure 1 were pooled and analyzed, as described in the text. The arrows indicate the positions of the 28S, 18S, and 4S external markers, and the 4S (^{14}C-tRNA) internal marker.

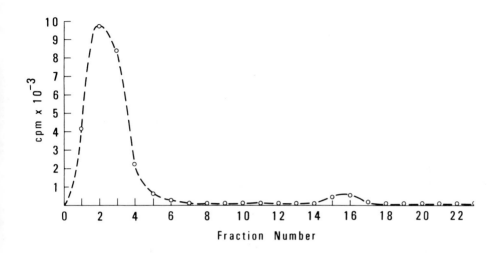

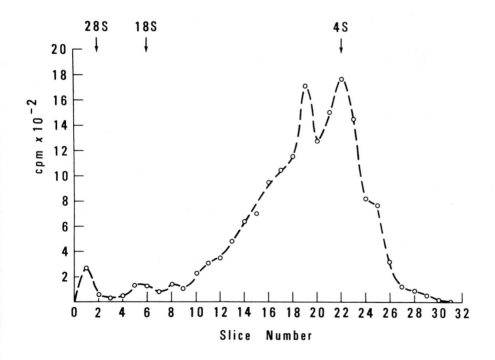

110

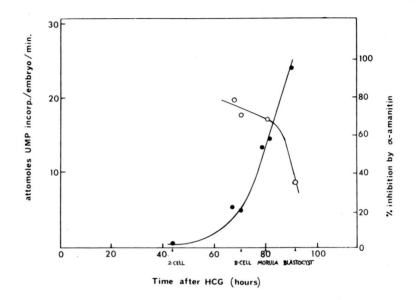

Figure 3: Activity of embryonic RNA polymerase as a function of time of development (●). Percent inhibition of embryonic RNA polymerase activity in the presence of 1.1µ g/ml α-amanitin as a function of time of development (O). All assays were performed at least in duplicate at an ammonium sulfate concentration of 0.09 M. Two cell, 8 cell, morula and blastocyst embryos correspond to 42 hr, 66 hr, 82 hr, and 91 hr post-HCG, respectively.

procedure, groups of 600 2 cell embryos, 150 8 cell embryos, 100 morula stage embryos, and 150 blastocyst embryos were assayed for RNA polymerase activity with the results summarized in Figure 3 (Warner & Versteegh, 1974; Warner & Versteegh, 1975; Versteegh et al., 1975). It is seen that there is very low enzyme activity at the 2 cell stage and that enzyme activity increases markedly by the 8 cell stage. Siracusa (1973), in a similar study, concluded that on a per cell basis, the amount of RNA polymerase activity actually decreases with development. However, using more recent and more accurate values for the cell number at each stage of development (Olds et al., 1973), it can be calculated that the amount of RNA polymerase per cell is actually constant from the 8 cell stage (0.54 attomoles UMP incorporated per minute per cell) to the blastocyst stage (0.60 attomoles UMP incorporated per minute per cell) of development. However, any calculations of RNA polymerase activity (or RNA synthesis) on a per cell basis must be regarded with some caution. Experimentally we can only determine activities per embryo. For instance, in blastocysts, a certain total RNA polymerase activity may reflect an equal distribution of enzyme activity among all cells. Alternatively, a large amount of enzyme activity may be found in the inner cell mass with a concomitant small amount of enzyme activity in the trophoblast cells, or vice versa. A study on rabbit embryos, using autoradiography, has approached this problem (Karp et al., 1973). It was found that in rabbit blastocysts, relatively more nucleoplasmic labeling occurs in the embryoblast, compared to the trophoblast. No such study has been reported for mouse blastocysts.

Changes in RNA Polymerase Enzyme Forms During Mammalian Development

Thus, given the knowledge that total RNA polymerase activity in pre-implantation mouse embryos increases with development on a per embryo basis, and probably remains constant on a per cell basis, the next important question is whether or not the relative proportion of the enzyme forms I, II, and III remains constant or changes with development. As was discussed previously, and summarized in Table 1, the three enzyme forms have a number of different properties which allow them to be distinguished from one another. We have used the inhibitor α-amanitin to distinguish one enzyme form from another. It has been shown, both in vitro (Lindell et al., 1970) and in vivo (Sekeris & Schmid, 1972; Smuckler & Hadjiolov, 1972), and that α-amanitin is a specific inhibitor of form II (nucleoplasmic) RNA polymerase at low concentrations (1μ g/ml). At higher concentrations (100μ g/ml), it is an inhibitor of both form II and form III RNA polymerases (Weinmann & Roeder, 1974). The form I (nucleolar) enzyme is insensitive to inhibition by α-amanitin.

By using various concentrations of α-amanitin we have shown that all three enzyme forms are present from the 8 cell stage on (Warner & Versteegh, 1975; Versteegh et al., 1975). However, the relative proportion of the enzyme forms changes dramatically from the 8 cell to the blastocyst stage. This is shown in Figure 3. It is apparent that the 8 cell stage contains more form II activity than (I+III) activity, whereas the blastocyst stage contains more form (I+III) activity than form II activity. This is demonstrated by the greater total inhibition of enzyme activity, by 1μ g/ml α-amanitin, at the 8 cell stage compared to the blastocyst stage.

A recent report by Siracusa and Vivarelli (1975) has attempted to quantitate the relative amounts of forms I and II enzyme activities by assaying mouse embryo homogenates at low and high ionic strength. Unfortunately, they have failed to take into account the fact that even though the form I enzyme displays

112

optimum activity at low ionic strength, and form II enzyme displays optimum
activity at high ionic strength, there is at least 50% activity of form I at
high ionic strength (Versteegh & Warner, 1973; Versteegh & Warner, 1975).
Therefore, any attempt to quantitate the relative amounts of the different
enzyme forms by using ionic strength differences gives ambiguous results. For
instance, a given decrease in total enzyme activity at high salt could be due
either to a small decrease in the amount of form II enzyme activity or to a
larger decrease in form I activity.   These and other possibilities are not distinguishable on the basis of ionic strength alone.   On the other hand,
α-amanitin is completely specific for form II and form III enzymes, so that
estimates of the relative amounts of the different enzyme forms should be valid
by this method, as discussed above.

The Effect of α-Amanitin on Preimplantation Mammalian Embryos

Of the various inhibitors that have been used to study the inhibition of
macromolecular synthesis in preimplantation mouse embryos, including actinomycin
D, puromycin, cycloheximide, mitomycin C, and fluorophenylalanine (Mintz, 1964a;
Thomson & Biggers, 1966; Monesi & Salfi, 1967; Monesi et al., 1970; Piko, 1970;
Tasca & Hillman, 1970), α-amanitin has yielded some of the most intriguing
results. α-Amanitin has been used to study the control of cleavage in both
mouse (Golbus et al., 1973; Warner & Versteegh, 1974) and rabbit (Manes, 1973)
embryos.   In mouse embryos there is definite inhibition of cleavage by the
2 cell stage.   For example, a concentration of 1μ g/ml α-amanitin prevents the
further cleavage of 71% of 2 cell embryos.   On the other hand, Manes (1973) has
reported that rabbit embryos could undergo the first and a few subsequent
cleavage divisions in the presence of 10^{-4}M (100μ g/ml) α-amanitin.
Unfortunately, it is not clear from these studies if the observed differences
reflect different modes of RNA synthesis in the mouse and rabbit embryos, or
simply differential uptake of α-amanitin.   It would be most revealing to measure
the intracellular concentration of the inhibitor after incubation with embryos
at various preimplantation stages.

In the case of the mouse, inhibition of cleavage at the 2 cell stage can
have several possible interpretations.   First, the α-amanitin may simply bind
to the membrane and physically block further cell division.   However, there is
no precedence for such an effect with α-amanitin.   Second, the α-amanitin may
enter the cell and specifically block form II RNA polymerase.   At first this
would seem to imply that mRNA synthesis would be blocked, but this has proven
to be a complex issue.   In some systems (Bucci et al., 1971; Egyhazi et al.,
1972; Serfling et al., 1972; Hastie & Mahy, 1973), the primary effect of
α-amanitin is to block mRNA synthesis, but in other systems (Stirpe & Fiume,
1967; Jacob et al., 1970; Niessing et al., 1970; Tata et al., 1972), it has been

shown that rRNA synthesis is also inhibited by α-amanitin. When rRNA synthesis
is inhibited by α-amanitin, it has been suggested that the blockage must be by
some indirect mechanism. One possibility is that some crucial mRNA which codes
for a protein necessary for rRNA synthesis may be inhibited by α-amanitin.

The type of RNA synthesis which is blocked by α-amanitin in preimplantation
mouse embryos has been studied in our laboratory. The experiment consisted of
incubating 8 cell or blastocyst embryos with ^{3}H-uridine in the presence and
absence of α-amanitin for various lengths of time. Then the nucleic acids were
extracted and subjected to polyacrylamide gel electrophoresis, with the
resultant typical gel profile shown in Figure 4. The 28S,18S, and 4S peaks
were identified by using both internal and external markers. The high molecular
weight peak at the top of the gel was found to be 98% degradable by DNase, and
was therefore identified as DNA. It is known that ^{3}H-uridine may be converted
to ^{3}H-deoxycytidine and hence the radioactivity is incorporated into DNA
(Comings, 1966; Adams, 1968). An analysis of similar gels to the one depicted
in Figure 4, for embryos which had been incubated for various lengths of time in
α-amanitin, revealed that rRNA is inhibited first, and to the greatest extent
by α-amanitin, mRNA is inhibited next most, and tRNA least. This pattern of
inhibition was found to be identical for 8 cell and blastocyst embryos. The
only difference between 8 cell and blastocyst embryos was that in the blastocysts,
but not in 8 cell embryos, DNA synthesis, as well as RNA synthesis is inhibited
by α-amanitin. Inhibition of DNA synthesis by α-amanitin has been observed in
a number of systems (Hastie et al., 1972; Wanka et al., 1972; Montecuccoli et al.,
1973). Among the many speculative explanations for this phenomenon is the possibility that α-amanitin inhibits the RNA polymerase which produces the RNA for
the priming of DNA synthesis. To date, which of the RNA polymerase enzyme
forms is responsible for this activity is not known. Thus, it is possible that
α-amanitin blocks cleavage of mouse embryos by inhibiting DNA synthesis, as well
as RNA synthesis.

In summary, direct assays of RNA polymerase enzyme activity have been
possible using preimplantation mouse embryo homogenates. Similar studies have
not yet been reported for any other mammalian embryonic system. It is clear,
that in the mouse, on a per embryo basis, a large increase in RNA polymerase
enzyme activity begins at the 8 cell stage, which coincides nicely with the
large increase in RNA synthesis seen shortly afterwards. It is also clear that
the relative amounts of the different enzyme forms changes between the 8 cell
and the blastocyst stages. Thus, the first two criteria for a critical role
for RNA polymerase in development have been met. The next question then becomes
whether or not the changes in the RNA polymerase enzyme forms correspond to
changes in the classes of RNA being made. This data is summarized in Table 5.

114

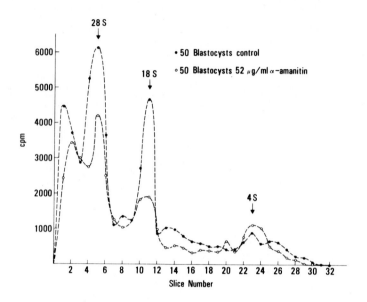

Figure 4: Polyacrylamide gel electrophoresis of phenol-extracted nucleic acids from mouse blastocysts with and without treatment with 52μ g/ml of α-amanitin for 3 hours. The arrows indicate the positions of the 28S, 18S, and 4S markers.

In fact, by again examining Table 3, it is clear that the classes of RNA synthesized by 8 cell and blastocyst embryos are identical. Thus, the conclusion must be that in mouse embryos, as in Xenopus laevis embryos, but not in sea urchin embryos, variations in the relative amounts of form I and form II RNA polymerase do not parallel changes in the types of RNA synthesized. The possible physiological significance of this finding is the subject of the second half of this chapter.

TABLE 5

A COMPARISON OF THE MAJOR TYPE OF RNA POLYMERASE PRESENT
WITH THE MAJOR TYPE OF RNA SYNTHESIZED IN 8 CELL AND
BLASTOCYST MOUSE EMBRYOS

Stage of Development	Major RNA Polymerase Form	Major RNA Class Synthesized
8 cell	II	rRNA
blastocyst	I	rRNA

THE MECHANISM OF ACTION OF STEROID HORMONES

It is clear from the last section that changes in the relative proportion of RNA polymerase enzyme forms I, II, and III do not lead to changes in the overall pattern of RNA synthesis in preimplantation mouse embryos. The question then becomes, what is the physiological significance of the observed changes in the proportion of the different enzyme forms? During the past few years, it has become clear that one of the primary effects of steroid hormones is to cause a transient rise in form II RNA polymerase activity (Glasser et al., 1972). We will hypothesize later that this may be important in early mammalian development, but first we must summarize the current status of the mechanism of action of steroid hormones.

There is a large volume of biochemical evidence that suggests that steroid hormones all act in a similar fashion at the molecular level (reviewed by Yamamoto & Alberts, 1976). The steroid hormone enters the cell and binds tightly to a specific receptor protein in the cell cytoplasm (Raspe, 1971). It is then converted, probably by an allosteric mechanism, to an "activated" state, after which it enters the nucleus and binds to particular chromosomal sites (King & Mainwaring, 1974). It is the steroid-receptor interaction with the genome that then triggers the biological response of a target tissue to a particular steroid hormone. There are three ways in which a tissue may be characterized as being sensitive to a particular hormone. First, the tissue can be shown to bind radioactively labeled hormone. Second, some primary effect, usually apparent within one hour, can be measured. And third of all, secondary parameters, which are usually not apparent for many hours, can be measured.

The overall net effect of steroid hormone action with the target tissue is believed to be the regulation of gene activity (Jensen & De Sombre, 1973; O'Malley et al., 1973; Liao, 1974; Katzenellenbogen & Gorski, 1975; Gorski & Gannon, 1976). However, exactly what the primary effect of the hormone-receptor complex on the nucleus is, remains a matter of some controversy. It has been suggested by Yamamoto and Alberts (1976) that the primary response of a target tissue to a steroid hormone is the transcription of a relatively small number of specific genes, which is only possible when receptor molecules occupy a "patch" of multiple sites in the same genetic region.   However, other workers (Liao, 1973; Jackson & Chalkley, 1974; Szego, 1974; Liang & Liao, 1975; Liao et al., 1973) have suggested a host of alternative mechanisms to account for selective gene activation by steroid hormones.   It is clear that knowledge of the nature of the primary effect of the steroid hormone-receptor complex on the chromatin of the nucleus is essential for the understanding of the start of the cascade of events that often follows the primary effect.   However, since primary effects may be transient, and therefore difficult to analyze, often the more long-term and permanent secondary effects of steroid hormones on their target tissues have been studied.   (See also chapters by Beato and Surani, this volume).

The Effect of Steroid Hormones on RNA Polymerase Activity and RNA Synthesis

One major problem in studying RNA polymerase activity and RNA synthesis in target tissues has been the conciliation of in vivo with in vitro results from different laboratories.   Measurements of total RNA synthesis in target tissues immediately following hormone injection into intact animals must be corrected for the greatly increased transport of nucleotide precursors that often follows (Billing et al., 1969a,b; Miller & Baggett, 1972a,b).   These in vivo results have shown little or no stimulation of total RNA synthesis within the first hour. Billing et al. (1969b) have found that total RNA synthesis, in the immature rat uterus treated with estradiol, does not start to increase until after six hours in contact with the hormone.   These findings suggest that the primary effect of the steroid hormone is the selective stimulation of a small number of genes, and that measurable increases in total RNA (and protein and DNA) synthesis is a secondary effect of the hormone.

However, the above in vivo results conflict with the in vitro results from a number of laboratories.   A typical in vitro protocol, as described by Gorski (1964), might involve injection of hormone (e.g. estradiol) into an immature rat, followed by isolation of nuclei or chromatin from a target tissue (e.g. uterus) at specified time intervals after the hormone injection.   Using this type of protocol, Glasser et al. (1972) measured incorporation of ribonucleoside triphosphates into RNA of isolated rat uterine nuclei, in vitro.   Under conditions of high salt (at 15°C) they observed a three- to four-fold increase in form II RNA polymerase activity within 30 minutes after the injection.   The form II RNA polymerase activity returned to the control level within 2 hours. Following the transient rise of form II RNA polymerase is a five- to eight-fold increase in form I RNA polymerase activity.   The rise in form II activity is insensitive to prior treatment with cycloheximide, but the rise in form I activity is sensitive to both cycloheximide and α-amanitin. These results have been interpreted to mean that the primary effect of the hormone is to stimulate form II RNA polymerase and that the rise in form I activity is a secondary effect which is the result of the selective transcription of certain genes by the form II enzyme (O'Malley & Means, 1974; Borthwick & Smellie, 1975). It has been suggested that these early changes in forms I and II RNA polymerase activity reflect modification of the chromatin template (Borthwick & Smellie, 1975) by the hormone rather than a change in the enzymes themselves. The rationale behind this suggestion is that often hormone interaction with chromatin activates the template for E. coli RNA polymerase which has been added exogenously (Hallick & DeLuca, 1969; Glasser et al., 1972; Schwartz, et al., 1975; Tsai et al., 1975). One suggested mechanism for the modification of the template is histone acetylation (Libby, 1972, 1973), a transient effect which probably precedes RNA polymerase form II activation. Increases in RNA polymerase activity at later times, i.e. concomitant with the burst of RNA synthesis seen by Glasser et al. (1972) after 6 hours, is thought to reflect an actual increase in the number of RNA polymerase molecules (Blatti et al., 1970; Benecke et al., 1973).

The above experiments have all involved treatment of intact cells with hormone. A whole series of experiments have been reported in which attempts have been made to study directly the effects of solubilized hormone-receptor complex on transcription of isolated nuclei or chromatin. These experiments are quite complex because it is never clear whether or not any crucial factors were left out of a particular transcription system. For instance, various studies (Mohla et al., 1972; Jensen et al., 1974) have led to the conclusion that the activated hormone-receptor complex alters uterine chromatin, the RNA polymerases, or both, to stimulate the increased synthesis of rRNA after 30 min. These results are in direct conflict with the experiments of Glasser et al. (1972) described above, which showed that after 30 min, RNA polymerase form II is stimulated. Thus, it is clear that the interpretation and conciliation of data from in vivo and in vitro experiments awaits further progress in this complex area of research. Nevertheless, it is clear that steroid hormones do influence both RNA polymerase activity and RNA synthesis in hormone sensitive target tissues.

THE EFFECT OF STEROID HORMONES ON PREIMPLANTATION MAMMALIAN DEVELOPMENT

Although a great deal of work has been done on maternal tissues, such as the uterus and oviduct, as hormone sensitive tissues, very little work has been done on the preimplantation embryos themselves (Beier, 1974; Smith & Biggers, 1975). There are two crucial aspects of preimplantation mammalian development which are likely candidates for hormonal control: blastocyst formation and implantation. A number of studies have been published describing the morphological effects of hormones on preimplantation mammalian development, but studies aimed at probing the biochemical bases for the observed effects of the hormones have only begun in the last several years.

The effect of Steroid Hormones on the Cleavage of Preimplantation Mammalian Embryos

Perhaps the earliest study in which the in vitro effect of a steroid hormone on preimplantation mammalian development was described is that of Whitten (1957). He studied the effect of progesterone on the ability of 8 cell mouse embryos to form blastocysts when cultured in vitro. Mouse embryos can be cultured in vitro in chemically defined medium, without any added exogenous hormone or serum of unknown hormone concentration (Whitten, 1957; Whitten & Biggers, 1968), so that studies of the effects of hormones on mouse embryos are somewhat simplified. Whitten found that concentrations of progesterone below 2μ g/ml (7×10^{-6}M) produced no observable effect, whereas some toxicity was observed at 4μ g/ml (1.3×10^{-5}M), and few embryos survived higher concentrations. He also found that blastocysts were more sensitive to the action of progesterone than earlier stages, and that the estrogens, estradiol and hexoestrol, afforded no protection against the cytotoxic effects of the progesterone.

Several more recent studies (Kirkpatrick, 1971; Roblero, 1973; Roblero & Izquierdo, 1976) have also measured the effect of steroid hormones on the cleavage of preimplantation mouse embryos. Kirkpatrick (1971) found that concentrations of 25μ g/ml (1×10^{-4}M) estradiol and 8μ g/ml (2.6×10^{-5}M) progesterone completely inhibited the further development of 2 cell embryos. The inhibition became less, the later the stage of development that the embryos were exposed to the hormone. In contrast to the study of Whitten (1957), Kirkpatrick (1971) found that 8 cell embryos seemed to be more sensitive to the action of progesterone than the blastocyst embryos. However, it must be borne in mind that the concentrations of hormone used by Kirkpatrick (1971) in his studies were certainly not physiological, so any observed effects are of dubious meaning.

Roblero (1973) has reported a study in which the effect of progesterone, administered to ovariectomized pregnant mice, was studied. He found that embryos, recovered from ovariectomized, progesterone treated females, had significantly fewer cells when compared to controls, but significantly more cells when compared to embryos of ovariectomized females. These results suggest that ovarian hormones do affect the development of the mammalian embryo. However, experiments such as this one are always open to question since one does not know the concentration of the various steroid hormones in the control, ovariectomized, or hormone-treated animals. An in vitro study aimed at overcoming these objections was recently reported by Roblero and Izquierdo (1976). In direct contrast to the results of Whitten (1957) and Kirkpatrick (1971), they found no effect of 4μ g/ml (1.3×10^{-5}M) progesterone on the cleavage rate of early mouse embryos. However, when they supplemented the culture medium with 15% heat-inactivated rat serum (dialyzed to remove molecules <1000MW), there was a significant increase in the number of blastomeres/blastocyst. They suggested that the rat serum contained a macromolecular carrier for the progesterone, and this led to the stimulatory effect of the hormone.

The effect of hormones on the cleavage rate and development of pre-implantation rabbit embryos has also been studied in a number of laboratories (Daniel, 1964; Daniel & Levy, 1964; El-Banna & Daniel, 1972; Allen & Foote, 1973). Daniel and Levy found that progesterone at a concentration of 10μ g/ml (3×10^{-5}M) blocked cleavage of all stages of rabbit embryos up to the morula, but that the growth of blastocysts was not affected. Estradiol did not block the inhibition of cleavage by progesterone. They concluded that progesterone blocks cleavage by limiting the supply of protein or amino acids to the ovum, thereby blocking protein synthesis. They observed, by autoradiography, ^{14}C-labeled progesterone on the surface of the ovum and zona pellucida, but not inside the ovum. It seems that an alternative explanation for the results of Daniel and Levy (1964) would be simply a blockage of membrane fluidity, and thus cleavage, by the surface-bound hormone. A study by Allen and Foote (1973) has confirmed that progesterone blocks the cleavage of rabbit ova.

Daniel (1964) has reported the effect of twelve different steroid hormones on the rate of cleavage of rabbit ova. Confirming his earlier study (Daniel & Levy, 1964), it was again found that progesterone inhibits cleavage of rabbit ova. At the same concentration (3×10^{-5}M), testosterone proprionate inhibited cleavage somewhat, but not so well as progesterone. The estrogens caused the ova to fragment, but none of the other steroids showed any obvious effect on the rabbit eggs. Another study (El-Banna & Daniel, 1972) has shown that although progesterone alone is detrimental to the development of rabbit embryos, when complexed to a carrier protein from uterine secretions, the hormone actually stimulates blastocyst growth and the uptake of nucleosides and amino acids.

It is clear from the foregoing experiments, in spite of the contradicting results from a number of laboratories, that steroid hormones do have some effect on the development of preimplantation mammalian embryos. This would imply that the embryos are hormone sensitive tissues. However, it must be kept in mind that all the effects, whether stimulatory or inhibitory, were never seen at concentrations below 10^{-5}M hormone. This concentration is certainly higher than physiological, probably by two orders of magnitude, so that the meaning of these results is open to question.

The effect of Steroid Hormones on the Implantation of Mammalian Embryos

Although it is still not clear how steroid hormones affect cleavage of pre-implantation mammalian embryos, it is certainly clear that steroid hormones are necessary for implantation to occur. This has been demonstrated by the fact that implantation can be induced in the ovariectomized pregnant rat or mouse by a single injection of estrogen (Cochrane & Meyer, 1957; Block, 1958; Smith & Biggers, 1968; Hedlund & Nilsson, 1971). In the mouse, by using a radio-immunoassay procedure, McCormack and Greenwald (1974) have shown that a critical rise in estradiol occurs on Day 4 of pregnancy which is essential for implantation to occur. The difficult question to answer is whether the effect of the hormones is only on the maternal tissues, or whether the embryos themselves receive hormonal signals at the time of implantation. Whereas the cleavage stages can be studied in vitro, the implantation reaction always involves both the embryo and the uterus. Thus, any observed effects on the embryos could always be due to a secondary effect by some product secreted by the uterus in response to the hormone.

A useful system for studying the effects of hormones on implantation has been the "delayed implanting" blastocysts which maintain long-term viability, but do not implant in ovariectomized females. It has been shown that estrogen activates delayed implanting mouse blastocysts (McLaren, 1971, 1973). After estrogen treatment, they show increased RNA and protein synthesis (Weitlauf & Greenwald, 1968; Van Blerkom, 1975; Holmes & Dickson, 1975; Weitlauf, 1976), increased CO_2 production (Torbit & Weitlauf, 1974), and surface coat and enzyme changes (Holmes & Dickson, 1973). In delayed implanting rat blastocysts, estrogen enhances RNA synthesis within one hour, protein synthesis within six hours, and DNA synthesis by 42 hours after hormone administration (Dass et al., 1969; Mohla & Prasad, 1971). These dramatic effects are the result of going from the quiescent, metabolically inactive state of the delayed implanting blastocyst, to the status of the normal blastocyst before implantation.

Although the studies with delayed implantation blastocysts are interesting, they don't really tell us much about the hormonal control of implantation of normal blastocysts. RNA synthesis is necessary for implantation to occur. If RNA synthesis is inhibited by actinomycin D, proper development and implantation do not occur in either normal (Psychoyos, 1967; Rowinski et al., 1975) or hormone-treated delayed implanting (Unger & Dickson, 1971; Rollard et al., 1973; Finn & Bredl, 1973; Weitlauf, 1974) mouse blastocysts. Studies with cycloheximide (Unger & Dickson, 1971; Rowinski et al., 1975) also suggest that protein synthesis is necessary for implantation to occur.

The above studies have all utilized either normal or delayed implantation blastocysts for in vivo implantation experiments. Several studies have attempted to allow the implantation of blastocysts to occur in vitro (Grant, 1973; Grant et al., 1975). Grant et al. (1975) have recently reported that when mature and immature uteri from ovariectomized mice were cultured in vitro in chemically defined media, blastocyst invasion occurred in the presence of progesterone, but not in media containing only estradiol. However, the resolution of the problem of separating the effects of steroid hormones on the embryos themselves from effects on the uterus, which are then transmitted to the embryos, may only be circumvented when a good artificial implantation system is developed. So far, all in vitro implantation systems disturb the normal pattern of embryogenesis (Gwatkin, 1966; Hsu, 1973; Wilson & Jenkinson, 1974; Grant et al., 1975).

The Effect of Steroid Hormones on the Uptake of Nucleic Acid and Protein
Precursors by Preimplantation Mammalian Embryos

The effect of estrogen on uptake and incorporation of amino acids has been reported for mouse embryos. In a study by Smith and Smith (1971), it was found that blastocysts cultured in vitro showed increased uptake and incorporation of amino acids within 0.5 hours of exposure to estradiol. The effect of estradiol was biphasic: concentrations of 10^{-10}M, 10^{-9}M, and 10^{-5}M showed significant effects, but 10^{-8}M, 10^{-7}M, and 10^{-6}M concentrations did not. The effect was found to be transient, and disappeared when incubation times were increased to one hour or more. Also, the effect was stage specific, since there was no effect on either uptake or incorporation of amino acids by morulae at any hormone concentration from 10^{-10}M to 10^{-5}M. The results are certainly suggestive that cultured mouse embryos develop a mechanism which is sensitive to estrogen when they advance from morulae to expanded blastocysts (see also, Borland, this volume). However, in a similar study, Weitlauf (1973) has shown that preincubation of blastocysts with estradiol for four hours, followed by exposure to labeled amino acids for 0.5 hours, did not lead to any change in the uptake or incorporation of amino acids either by normal or delayed implanting blastocysts. Based on these results, Weitlauf (1973) has suggested that the uterus regulates the level of amino acid uptake and incorporation in preimplantation mouse blastocysts, and that the ovarian hormones influence the blastocysts indirectly by altering the uterus.

The influence of estrogen on the uptake of nucleic acid precursors has also been reported for mouse embryos. Harrer and Lee (1973) collected embryos at the 2 cell stage and incubated them in vitro for 60-70 hours, to the blastocyst stage. Then they were incubated with 1μ g/ml (4×10^{-6}M) estradiol for one hour and cultured for an additional hour in the presence of the hormone and either ^{3}H-thymidine or ^{3}H-uridine. The embryos were then prepared for autoradiography and both the number of labeled cells, and the number of grains per labeled cell, were counted. With the ^{3}H-thymidine, it was found that the hormone significantly increased the number of labeled cells, as well as increasing the total amount of label per labeled cell. With the ^{3}H-uridine it was found that the number of labeled cells remained constant both with and without hormone treatment. However, the cells which were labeled had twice the amount of radioactive label in the hormone-treated compared to the control sample. The results with the ^{3}H-uridine are especially surprising since it is known that for mouse blastocysts, after a one hour incubation period, 25% of the label presented as ^{3}H-uridine ends up in DNA as a result of conversion to ^{3}H-deoxycytidine (Comings, 1966; Adams, 1968; Warner & Hearn, unpublished). In any event, the results of Harrer and Lee (1973) suggest that estradiol may stimulate both DNA and RNA synthesis in mouse blastocysts.

The Uptake of Steroid Hormones by Preimplantation Mammalian Embryos

We now return again to the question of whether or not preimplantation mammalian embryos are a hormone sensitive tissue. Some of the experiments described in the last several sections have tried to answer this question, but the overall conclusions are not yet clear. First of all, few studies have used physiological concentrations of hormones. Second, studies involving implantation still cannot distinguish between direct effects of the hormones on the embryos themselves, and secondary effects via the uterus. Third, there is conflicting data from a number of laboratories as to whether particular hormones increase or decrease the rate of cleavage, or various biosynthetic processes, in each particular experimental system.

It seems that now that a rational understanding of the mechanism of action of steroid hormones at the molecular level is beginning to emerge, it is time to re-examine the effect of hormones on preimplantation mammalian embryos from this point of view. As mentioned previously, a tissue may be considered to be a hormone sensitive tissue if any of the following criteria can be met: (1) It binds the hormone; (2) It shows a short-term primary effect from the hormone; (3) It shows a long-term secondary effect from the hormone.

Several studies have tried to establish the first criterion. In the mouse, Smith (1968) has shown that blastocysts apparently take up and retain ^{3}H-estradiol. The embryos were treated directly with 2×10^{-8} M ^{3}H-estradiol (specific activity 38.1 Ci/mmole) for one hour, washed extensively, and counted in a liquid scintillation counter. However, the total of 30 cpm above background, which Smith (1968) observed for a sample of about 200 blastocysts, is certainly at the very borderline of significance. To get truly meaningful results either a larger group of embryos, or ^{3}H-estradiol of higher specific activity, would have to be used.

Prasad et al. (1974) have used autoradiography to try to demonstrate directly binding of ^{3}H-estradiol to delayed implanting rat blastocysts from rats which were either injected with the labeled hormone in vivo, or else had the labeled hormone introduced directly into the uterine lumen. The rats were killed 5, 15 and 30 minutes after exposure to the ^{3}H-estradiol (specific activity, 96 Ci/mmole). The results showed that there was no concentration or retention of radioactivity in the cells of the blastocysts. In contrast, the uterine tissue showed significant uptake and retention of the label. In view of these results, it seems questionable, at least in the rat, whether activation of the delayed implanting blastocyst is due to direct action of estradiol on the blastocyst. It seems more likely that in the rat, the activation of the delayed implanting embryo is mediated by some factor elaborated by the uterus after it responds to the estradiol.

In the rabbit, a study by Bhatt and Bullock (1974) offers suggestive evidence, but not proof, for the existence of soluble proteins in rabbit blastocysts capable of binding estradiol. Their study is based on the ability of an anti-estrogenic compound, CI-628, to compete with ^{3}H-estradiol binding to Day 5 or Day 6 rabbit blastocysts. It certainly does seem clear from their data that ^{3}H-estradiol does bind to the blastocysts at significant levels in control experiments, and that the binding is inhibited by CI-628 by 54% in Day 5 blastocysts and by 26% in Day 6 blastocysts. The evidence for the estrogen receptor protein in the blastocysts is, however, very preliminary. Their experimental protocol used scintillation counting, and not autoradiography, so it is not clear whether the hormone bound only to the outside of the embryos, or if it actually penetrated the cells. As mentioned previously, Daniel and Levy (1964), obtained evidence for binding of ^{14}C-progesterone to the surface and zona pellucida of the 1 cell rabbit embryo, but did not observe any labeled hormone inside of the embryos.

Thus, as in many other sets of experiments described in this chapter, the direct binding studies are not clear cut. It is possible that there is true variation among species, but it is also possible that the experimental protocols have not been precise enough to give unequivocal results.

The Effect of Steroid Hormones on RNA Polymerase Activity and RNA Synthesis in Preimplantation Mammalian Embryos

Briefly reviewing, it is known that steroid hormones influence both RNA polymerase activity and RNA synthesis in hormone sensitive target tissues. Both of these parameters are believed to be crucial to the regulation of gene activity, which seems to be the end result of the action of steroid hormones. Thus, the study of both RNA polymerase activity and RNA synthesis is one of the most direct methods for determining whether or not a tissue is a hormone sensitive tissue. Either of these effects could be primary, occurring within one hour of exposure to hormone or secondary, occurring after several hours of exposure to hormone. We have taken the direct approach, of studying RNA synthesis itself, to determine whether or not mouse embryos are a hormone sensitive tissue.

In the first set of experiments, mouse blastocysts were incubated simultaneously with ^{3}H-uridine (250-500μ Ci/ml) and estradiol (10^{-5}M to 10^{-9}) for varying lengths of time. The controls consisted of identical incubation conditions, but without the hormone. It is seen in Figure 5 that there is no apparent effect of estradiol on RNA synthesis, at any of the concentrations studied, for either a 0.5 hour or 5 hour incubation time. It is shown in Figure 6 that there is no apparent effect of 10^{-7}M estradiol on RNA synthesis for times ranging from 0.5 hour to 24 hours of incubation.

In a second set of experiments, both 8 cell and mouse blastocysts were preincubated for 21 hours in the presence of the hormone. Then ^{3}H-uridine (250μ Ci/ml) was added for a three hour additional incubation. As is seen in Table 6, there is no apparent effect of estradiol on RNA synthesis by either blastocyst or 8 cell mouse embryos using this experimental protocol. Thus, the results from both sets of experiments would seem to indicate that 8 cell and blastocyst mouse embryos are insensitive to treatment with physiological concentrations of estradiol. It is, of course, possible that other steroid hormones, such as progesterone, or progesterone and estradiol in combination, would have a direct effect on RNA synthesis in preimplantation mouse embryos. These possibilities have not yet been tested.

Our results contradict several experiments mentioned previously which suggest that estradiol directly affects mouse embryos. First, Smith (1968) has claimed that mouse blastocysts bind estradiol. However, the study was not done using autoradiography, so even if the reported binding is real, it is not clear whether or not the label gets inside the cells, or is merely bound to the zona pellucida or plasma membrane. Second, although Kirkpatrick (1971) found inhibition of cleavage of mouse embryos by estradiol, the concentration used (10^{-4}M) was not physiological, so that the meaning of the experiment is dubious. The study of Harrer and Lee (1973) was based on examining only thirteen blastocysts for ^{3}H-uridine uptake by autoradiography, so their results are at the very borderline of statistical significance with respect to the amount of ^{3}H-uridine found in control and hormone-treated embryos. The study of Harrer and Lee (1973)

Figure 5. (upper) A comparison of RNA synthesis, with and without various concentrations of estradiol, for 0.5 hr ($\bullet$) and 5 hr ($\circ$) incubation times in the presence of hormone and ^{3}H-uridine simultaneously. The bars indicate the standard deviation.

Figure 6. (lower) A comparison of RNA synthesis, with and without 10^{-7}M estradiol, for various times of incubation in the presence of hormone and ^{3}H-uridine simultaneously. The bars indicate the standard deviation.

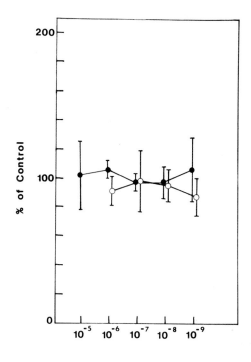

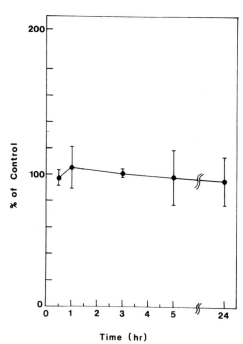

TABLE 6

EFFECT OF ESTRADIOL ON RNA SYNTHESIS IN MOUSE EMBRYOS

| Hormone Concn. (M) | Stage of Development* | | | |
| | Blastocyst | | 8 cell | |
	cpm/embryo	% of Control	cpm/embryo	% of Control
10^{-5}	2,114	96	504	87
10^{-6}	2,657	121	637	110
10^{-7}	2,406	109	601	104
10^{-8}	2,174	99	531	92
10^{-9}	2,096	95	622	107
0**	2,202$\pm$207	–	579$\pm$111	–

* All experimental groups consisted of 25 embryos. The data is corrected for background cpm. Incubations were with hormone only for 21 hrs., followed by a 3 hr. incubation in the presence of hormone and ^{3}H-uridine.

** All controls are the average of at least two determinations. The standard deviation is shown for each value.

measured only uptake, and not incorporation of label. In our studies (see Figures 5 and 6), the precursor pool was saturated, so that hormone could only effect incorporation, and not uptake of the ^{3}H-uridine label. The one study that really does seem to show a true effect of estradiol on mouse blastocysts is that of Smith and Smith (1971). The transient effect of estradiol, namely increased amino acid uptake and incorporation was seen in blastocysts that had been derived from embryos cultured in vitro from the 2 cell stage. Since our embryos were kept in vivo until the stage of development at which they were to be studied (either 8 cell or blastocyst), it is possible that they did not develop the same hormone sensitivity seen by Smith and Smith (1971). It is also possible, but not likely based on the known mechanism of hormone action, that estradiol could cause a transient increase in protein synthesis without causing an increase in RNA synthesis.

On the other hand, there is a body of experimental evidence that supports the notion that mouse embryos themselves are insensitive to steroid hormones (Weitlauf & Greenwald, 1968; Psychoyos, 1967; Weitlauf, 1973, 1974, 1976; Holmes & Dickson, 1975; Van Blerkom, 1975; Roblero & Izquierdo, 1976). These workers believe that the effects of steroid hormones on mouse embryos is mediated by factors secreted by the uterus in response to the hormone. Weitlauf (1976) has recently reported evidence for a factor isolated from the uterine flushings of both normal and delayed implanting mice which inhibits RNA synthesis in mouse blastocysts. The activity was found to be dialyzable and heat resistant, but has not been otherwise identified.

Finally, if preimplantation mouse embryos prove to be truly refractory to exogenously added estradiol (and possibly other steroid hormones), there is an alternate explanation than mediation of hormone effects on the embryos by uterine factors. It has been reported that preimplantation rabbit (Seamark & Lutwak-Mann, 1972; Dickmann et al., 1975), pig (Perry et al., 1973), rat (Dickmann & Dey, 1974), hamster (Dickmann & Gupta, 1974), and mouse (Dey & Dickmann, 1974) embryos may synthesize their own steroid hormones. However, only one of these studies (Seamark & Lutwak-Mann, 1972) has assayed the hormone content of the embryos directly. All the other studies base their conclusions on the presence of enzyme activities in the embryos which would potentially be capable of synthesizing various steroid hormones. It has been hypothesized that production of steroid hormones by the embryos themselves may be necessary for morula-blastocyst transformation, implantation, or as a signaling event to the mother of the presence of the embryo. If it is indeed true that embryos produce their own steroid hormones, one would expect them to be refractory to physiological levels of exogenously added hormones.

CONCLUSIONS AND SPECULATIONS

The control of gene activity and development of preimplantation mammalian embryos is clearly a little understood subject, which is just beginning to be probed at the molecular level. The enzyme RNA polymerase is undoubtedly of some importance in the control of transcription, but the exact mechanism whereby it exerts its control over RNA synthesis is still not clear. In the pre-implantation mouse embryo, multiple forms of RNA polymerase have been demon-strated from the 8 cell stage on, and are probably present earlier, since RNA synthesis is detectable in 2 cell embryos. During the course of development from the 8 cell stage to the blastocyst stage, the relative proportion of the forms I, II and III enzymes changes dramatically. It has been demonstrated that 8 cell embryos contain significantly more form II enzyme activity than do blastocyst embryos.

It has been hypothesized that the different proportion of the RNA polymerase enzyme forms in 8 cell, compared to blastocyst embryos, plays some crucial role in the control of development. One possibility is that the level of the different enzyme forms exerts a coarse control over the classes of RNA synthesized. However, 8 cell and blastocyst embryos synthesize exactly the same proportion of HnRNA, rRNA, and tRNA, in spite of the fact that they contain different proportions of the RNA polymerase enzyme forms I, II, and III.

A second possible reason for different proportions of the enzyme forms in 8 cell and blastocyst embryos was hypothesized to be differential sensitivity of these two stages to steroid hormones. It is known that both RNA polymerase activity and RNA synthesis are affected by both primary and secondary mechanisms

in hormone sensitive target tissues.   We have found no effect of estradiol on
RNA synthesis in either 8 cell or blastocyst mouse embryos.   However, it is
possible that other steroid hormones will be found to affect RNA synthesis in
preimplantation mammalian embryos in future studies.   It is also possible that
preimplantation mammalian embryos are truly refractory to physiological concentrations of exogenously added hormones.   Thus, the known in vivo effects of
hormones on mammalian embryos could either be via some factor secreted by the
uterus in response to hormonal stimulation, or else the embryos might synthesize
their own steroid hormones.   If the latter is the case, the change in RNA
polymerase enzyme forms with development might be due to autostimulation of the
embryos by endogenously produced hormones.

Finally, one might speculate that other, as yet undefined mechanisms are
regulating the switch in the proportion of the RNA polymerase enzyme forms during
development.   For instance, the embryos might develop a mechanism for the
phosphorylation of one enzyme form or another, which leads to a stimulation or
repression of enzyme activity.   Another intriguing possibility is that changes
in the RNA polymerase enzyme forms reflect the fact that the 8 cell stage is the
crucial stage at which the lingering maternally transmitted influence on
development is forsaken, and the embryonic genome becomes fully activated.   Thus,
the embryos may be preprogrammed to undergo a cascade of events which eventually
leads to differences in the relative proportions of the RNA polymerase enzyme
forms, independent of any external influences.   Eventually, if the physiological
significance of the change in the proportion of the RNA polymerase enzyme forms
from the 8 cell to the blastocyst stage of development is understood, we will be
closer to a rational understanding of the control of gene activity and development
of preimplantation mammalian embryos.

ACKNOWLEDGEMENTS

I am indebted to Dr. Larry Versteegh, Mr. Timothy Hearn, Ms. Judi McIvor,
Ms. Ruth Graves, and Ms. Carla Tollefson for making this work possible.   I also
thank my colleagues for the critical reading of this manuscript.   This work was
supported by a grant from The Population Council, New York, and in part by NIH
grant AI-11752.

REFERENCES

ADAMS, R.L.P. (1968) Incorporation of $5\text{-}^3\text{H}$ Uridine into DNA.   FEBS Lett. 2,
91-92.

ALLEN, M.C. & FOOTE, R.H. (1973) The Effect of Progesterone on the Early
Development of the Rabbit Embryo. Fert. Steril. 24, 220-226.

BEIER, H.M. (1974) Ovarian Steroids in Embryonic Development Before Nidation.
In: G. Raspe (Ed.), Advances in the Biosciences, Vol. 13, Pergamon Press, New
York, pp. 199-219.

BENECKE, B.J., FERENCZ, A. & SEIFART, K.H. (1973) Resistance of Hepatic RNA Polymerases to Compounds Affecting RNA and Protein Synthesis in Vivo. FEBS Lett. 31, 53-58.

BERNSTEIN, R.M. & MUKHERJEE, B.B. (1973) Cytoplasmic Control of Nuclear Activity in Preimplantation Mouse Embryos. Develop. Biol. 34, 47-65.

BHATT, B.M. & BULLOCK, D.W. (1974) Binding of Oestradiol to Rabbit Blastocysts and Its Possible Role in Implantation. J. Reprod. Fert. 39, 65-70.

BILLING, R.J., BARBIROLI, B. & SMELLIE, R.M.S. (1969a) The Mode of Action of Oestradiol I. The Transport of RNA Precursors into the Uterus. Biochim. Biophys. Acta 190, 52-59.

BILLING, R.J., BARBIROLI, B. & SMELLIE, R.M.S. (1969b) The Mode of Action of Oestradiol II. The Synthesis of RNA. Biochim. Biophys. Acta 190, 60-65.

BLATTI, S.P., INGLES, C.J., LINDELL, T.J., MORRIS, P.W., WEAVER, R.F., WEINBERG, F. & RUTTER, W.J. (1970) Structure and Regulatory Properties of Eucaryotic RNA Polymerase. Cold Spring Harbor Symp. Quant. Biol. 35, 649-557.

BLOCK, S. (1958) Experimentelle Untersuchungen uber die hormonalen Grundlagen der Implantation des Saugerkeimes. Experientia 14, 447-449.

BORTHWICK, N.M. & SMELLIE, R.M.S. (1975) The Effects of Oestradiol-17 on the Ribonucleic Acid Polymerases of Immature Rabbit Uterus. Biochem. J. 147, 91-101.

BUCCI, S., NARDI, I., MANCINO, G., & FIUME, L. (1971) Incorporation of tritiated uridine in nuclei of Triturus oocytes treated with α-amanitin. Exp. Cell Res. 69, 462-465.

CHAMBERLIN, M.J. (1974) The Selectivity of Transcription. Ann. Rev. Biochem. 43, 721-775.

CHAMBON, P. (1974) Eucaryotic RNA Polymerases. In: P.D. Boyer (Ed.), The Enzymes X, Academic Press, New York, pp. 261-331.

CHAMBON, P. (1975) Eukaryotic Nuclear RNA Polymerases. Ann. Rev. Biochem. 44, 613-638.

CHAPMAN, V.M. & WUDL, L. (1975) The Expression of β-Glucuronidase during Mouse Embryogenesis. In: C.L. Markert (Ed.), Proceedings of the Third International Conference on Isozymes, III, Academic Press, New York, pp. 57-65.

CHAPMAN, V.M., WHITTEN, W.K. & RUDDLE, F.H. (1971) Expression of paternal glucose phosphate isomerase-1 (GPI-1) in preimplantation stages of mouse embryos. Develop. Biol. 26, 153-158.

COCHRANE, R.L. & MEYER, R.K. (1957) Delayed Nidation in the Rat Induced by Progesterone. Proc. Soc. Exp. Biol. Med. 96, 155-159.

COMINGS, D.E. (1966) Incorporation of Tritium of ^{3}H-5-Uridine into DNA. Exp. Cell Res. 41, 677-681.

DAENTL, D.L. & EPSTEIN, C.J. (1971) Developmental interrelationships of uridine uptake, nucleotide formation and incorporation into RNA by early mammalian embryos. Develop. Biol. 24, 428-442.

DAENTL, D.L. & EPSTEIN, C.J. (1973) Uridine Transport by Mouse Blastocysts. Develop. Biol. 31, 316-322.

DANIEL, J.C. (1964) Some effects of steroids on cleavage of rabbit eggs in vitro. Endocrinology 75, 706-710.

DANIEL, J.C. & LEVY, J.D. (1964) Action of progesterone as a cleavage inhibitor of rabbit ova in vitro. J. Reprod. Fert. 7, 323-329.

DARNELL, J.E., PHILIPSON, L., WALL, R., & ADESNIK, M. (1971) Polyadenylic Acid Sequences: Role in Conversion of Nuclear RNA into Messenger RNA. Science 174, 507-510.

DASS, C.M.S., MOHLA, S., & PRASAD, M.R.N. (1969) Time sequence of action of estrogen on nucleic acid and protein synthesis in the uterus and blastocyst during delayed implantation in the rat. Endocrinology 85, 528-537.

DAVIDSON, E.H. & BRITTEN, R.J. (1973) Organization, Transcription, and Regulation in the Animal Genome. Quart. Rev. Biol. 48, 565-613.

DEY, S.K. & DICKMANN, Z. (1974) Steroidogenesis in the Preimplantation Mouse Embryo. Annual Meeting, Society for the Study of Reproduction, Abstr. no. 150.

DICKMANN, Z. & DEY, S.K. (1974) Steroidogenesis in the Preimplantation Rat Embryo and Its Possible Influence on Morula-Blastocyst Transformation and Implantation. J. Reprod. Fert. 37, 91-93.

DICKMANN, Z. & GUPTA, J.S. (1974) ^{5}-3 -Hydroxysteroid Dehydrogenase and Estradiol-17 -hydroxysteroid Dehydrogenase Activity in Preimplantation Hamster Embryos. Develop. Biol. 40, 196-198.

DICKMANN, Z., DEY, S.K. & GUPTA, J.S. (1975) Steroidogenesis in Rabbit Preimplantation Embryos. Proc. Natl. Acad. Sci. U.S. 72, 298-300.

EGYHAZI, E., D'MONTE, B., & EDSTROM, J.E. (1972) Effects of α-amanitin on in vitro labeling of RNA from defined nuclear components in salivary gland cells from Chironomus tentans. J. Cell Biol. 53, 523-531.

EL-BANNA, A.A. & DANIEL, J.C. (1972) Stimulation of rabbit blastocysts in vitro by progesterone and uterine proteins in combination. Fert. Steril. 23, 101-104.

ELLEM, K.A.O., & GWATKIN, R.B.L. (1968) Patterns of nucleic acid synthesis in the early mouse embryo. Develop. Biol. 18, 311-330.

EPSTEIN, C.J. (1969) Mammalian Oocytes: X Chromosome Activity. Science 163, 1078-1079.

EPSTEIN, C.J. (1972) Expression of the Mammalian X Chromosome before and after Fertilization. Science 175, 1467-1468.

EPSTEIN, C.J. (1975) Gene expression and macromolecular synthesis during pre-implantation embryonic development. Biol. Reprod. 12, 82-105.

EPSTEIN, C.J. & DAENTL, D.L. (1971) Precursor pools and RNA synthesis in preimplantation mouse embryos. Develop. Biol. 26, 517-524.

FINN, C.A. & BREDL, J.C.S. (1973) Studies on the Development of the Implantation Reaction in the Mouse Uterus: Influence of Actinomycin D. J. Reprod. Fert. 34, 247-253.

GLASSER, S.R., CHYTIL, F. & SPELSBERG, T.C. (1972) Early Effects of Oestradiol-17β on the Chromatin and Activity of the Deoxyribonucleic Acid-Dependent Ribo-nucleic Acid Polymerases (I and II) of the Rat Uterus. Biochem. J. 130, 947-957.

GOLBUS, M.S., CALARCO, P.G. & EPSTEIN, C.J. (1973) The effects of inhibitors of RNA synthesis (α-amanitin and actinomycin D) on preimplantation mouse embryogenesis. J. Exp. Zool. 186, 207-216.

GORSKI, J. (1964) Early Estrogen Effects on the Activity of Uterine Ribo-nucleic Acid Polymerase. J. Biol. Chem. 239, 889-892.

GORSKI, J. & GANNON, F. (1976) Current Models of Steroid Hormone Action: A Critique. Ann. Rev. Physiol. 38, 425-450.

GRANT, P.S. (1973) The effect of progesterone and oestradiol on blastocysts cultured within the lumina of immature mouse uteri. J. Embryol. exp. Morph. 29, 617-638.

GRANT, P.S., LJUNGKVIST, I. & NILSSON, O. (1975) The hormonal control and morphology of blastocyst invasion in the mouse uterus in vitro. J. Embryol. exp. Morph. 34, 299-310.

GURDON, J.B. (1962) The Developmental Capacity of Nuclei taken from Intestinal Epithelium Cells of Feeding Tadpoles. J. Embryol. exp. Morph. 10, 622-640.

GWATKIN, R.B.L. (1966) Amino Acid Requirements for Attachment and Outgrowth of the Mouse Blastocyst in Vitro. J. Cell. Physiol. 68, 335-343.

GREENLEAF, A. & BAUTZ, E.K.F. (1975) RNA Polymerase B from Drosophila melanogaster Larvae. Purification and Partial Characterization. Eur. J. Biochem. 60, 169-179.

HALLICK, R.B. & DELUCA, H.F. (1969) Vitamin D_3-Stimulated Template Activity of Chromatin from Rat Intestine. Proc. Natl. Acad. Sci. U.S. 63, 528-531.

HARRER, J.A. & LEE, H.H. (1973) Differential Effects of Oestrogen on the Uptake of Nucleic Acid Precursors by Mouse Blastocysts in Vitro. J. Reprod. Fert. 33, 327-330.

HASTIE, N.D. & MAHY, B.W.J. (1973) Effects of α-amanitin in vivo on RNA polymerase activity of cultured chick embryo fibroblast cell nuclei: resistance of ribosomal RNA synthesis to the drug. FEBS Lett. 32, 95-99.

HASTIE, N.D., ARMSTRONG, S.J., & MAHY, B.W.J. (1972) Effect of α-amanitin on protein and nucleic acid synthesis in chick-embryo fibroblast cells. Biochem. J. 130, 28 P.

HEDLUND, K. & NILSSON, O. (1971) Hormonal Requirements for the Uterine Attachment Reaction and Blastocyst Implantation in the Mouse, Hamster and Guinea Pig. J. Reprod. Fert. 26, 267-269.

HOLMES, P.V. & DICKSON, A.D. (1973) Estrogen-induced surface coat and enzyme changes in the implanting mouse blastocyst. J. Embryol. exp. Morph. 29, 639-645.

HOLMES, P.V. & DICKSON, A.D. (1975) Temporal and spatial aspects of oestrogen-induced RNA, protein and DNA synthesis in delayed-implantation mouse blastocysts. J. Anat. 119, 453-459.

HSU, Y. (1973) Differentiation in Vitro of Mouse Embryos to the Stage of Early Somite, Develop. Biol. 33, 403-411.

INGLES, C.J. (1973) Antigenic Homology of Eukaryotic RNA Polymerases. Biochem. Biophys. Res. Commun. 55, 364-371.

JACKSON, V. & CHALKLEY, R. (1974) The Binding of Estradiol-17β to the Bovine Endometrial Nuclear Membrane. J. Biol. Chem. 249, 1615-1626.

JACOB, S.T. (1973) Mammalian RNA Polymerases. Progr. Nucl. Acid Res. Mol. Biol. 13, 93-136.

JACOB, S.T., MUECKE, W., SAJDEL, E.M. & MUNRO, H.N. (1970) Evidence for extra-nucleolar control of RNA synthesis in the nucleolus. Biochem. Biophys. Res. Commun. 40, 334-342.

JENSEN, E.V. & DE SOMBRE, E.R. (1973) Estrogen-Receptor Interaction. Science 182, 126-134.

JENSEN, E.V., MOHLA, S., GORELL, T. & DE SOMBRE, E.R. (1974) The Role of Estrophilin in Estrogen Action. Vitam. Horm. N.Y. 32, 89-127.

KARP, G., MANES, C. & HAHN, W.E. (1973) RNA Synthesis in the Preimplantation Rabbit Embryo: Radioautographic Analysis. Develop. Biol. 31, 404-408.

KATZENELLENBOGEN, B.S. & GORSKI, J. (1975) Estrogen Actions on Synthesis of Macromolecules in Target Cells. In: G. Litwak (Ed.), Biochemical Actions of Hormones 3, Academic Press, New York, pp. 187-243.

KEDINGER, C., GISSINGER, F. & CHAMBON, P. (1974) Animal DNA-Dependent RNA Polymerases, Molecular Structures and Immunological Properties of Calf-Thymus Enzyme AI and of Calf-Thymus and Rat-Liver Enzymes B. Eur. J. Biochem. 44, 421-436.

KING, R.J.B. & MAINWARING, W.I.P. (1974) Steroid-Cell Interactions. University Park Press, Baltimore, pp. 1-440.

KIRKPATRICK, J.F. (1971) Differential Sensitivity of Preimplantation Mouse Embryos in Vitro to Oestradiol and Progesterone. J. Reprod. Fert. 27, 283-285.

KNOWLAND, J. & GRAHAM, C. (1972) RNA synthesis at the two-cell stage of mouse development. J. Embryol. exp. Morphol. 27, 167-176.

KOZAK, L.P., MCLEAN, G.K. & EICHER, E.M. (1974) X Linkage of Phosphoglycerate Kinase in the Mouse. Biochem. Genet. 11, 41-47.

LEE, S.Y., MENDECKI, J. & BRAWERMAN, G. (1971) A Polynucleotide Segment Rich in Adenylic Acid in the Rapidly-Labeled Polyribosomal RNA Component of Mouse sarcoma 180 Ascites Cells. Proc. Natl. Acad. Sci. U.S. 68, 1331-1335.

LIANG, T. & LIAO, S. (1975) A Very Rapid Effect of Androgen on Initiation of Protein Synthesis in Prostate. Proc. Natl. Acad. Sci. U.S. 72, 706-709.

LIAO, S. (1974) Biochemical Studies on the Receptor Mechanisms Involved in Androgen Actions. In: H.L. Kornberg and D.C. Phillips (Eds.), Biochemistry of Hormones, Butterworths, London, pp. 153-185.

LIAO, S., LIANG, T. & TYMOCZKO, J.L. (1973) Ribonucleoprotein Binding of Steroid-"Receptor" Complexes. Nature New Biol. 241, 211-213.

LIBHY, P.R. (1972) Histone Acetylation and Hormone Action, Early Effects of Oestradiol-17β on Histone Acetylation in Rat Uterus. Biochem. J. 130, 663-669.

LIBBY, P.R. (1973) Histone Acetylation and Hormone Action, Early Effects of Aldosterone on Histone Acetylation in Rat Kidney. Biochem. J. 134, 907-912.

LINDELL, T.J., WEINBERG, F., MORRIS, P., ROEDER, R.G. & RUTTER, W.J. (1970) Specific inhibition of nuclear RNA polymerase II by α-amanitin. Science 170, 447-449.

LITMAN, R.M. (1968) A Deoxyribonucleic Acid Polymerase from Micrococcus luteus (Micrococcus lysodeikticus) Isolated on Deoxyribonucleic Acid-Cellulose. J. Biol. Chem. 243, 6222-6233.

LOENING, U.E. (1967) The Fractionation of High-Molecular-Weight Ribonucleic Acid by Polyacrylamide-Gel Electrophoresis. Biochem. J. 102, 251-257.

MANES, C. (1973) The Participation of the Embryonic Genome during Early Cleavage in the Rabbit. Develop. Biol. 32, 453-459.

McCORMACK, J.T. & GREENWALD, G.S. (1974) Evidence for a Preimplantation Rise in Oestradiol-17β Levels on Day 4 of Pregnancy in the Mouse. J. Reprod. Fert. 41, 297-301.

McLAREN, A. (1971) Blastocysts in the Mouse Uterus: The Effect of Ovariectomy, Progesterone and Oestrogen. J. Endocr. 50, 515-526.

McLAREN, A. (1973) Blastocyst activation. In: The Regulation of Mammalian Reproduction. S.J. Segal, R. Crozier, P.A. Corfman and P.G. Condliffe (Eds.), Charles C. Thomas, Springfield, Illinois, pp. 321-388,

MILLER, B.G. & BAGGETT, B. (1972a) The Effects of 17β-Estradiol on the Rate of Synthesis of RNA in the Uterus. Biochim. Biophys. Acta 281, 353-364.

MILLER, B.G. & BAGGETT, B. (1972b) Effects of 17β-Estradiol on the Incorporatic of Pyrimidine Nucleotide Precursors into Nucleotide Pools and RNA in the Mouse Uterus. Endocrinology 90, 645-656.

MINTZ, B. (1964a) Synthetic Processes and early development in the mammalian egg. J. Exp. Zool. 157, 85-100.

MINTZ, B. (1964b) Gene expression in the morula stage of mouse embryos, as observed during development of t^{12}/t^{12} lethal mutants in vitro. J. Exp. Zool. 157, 267-272.

MOHLA, S. & PRASAD, M.R.N. (1971) Early Action of Oestrogen on the Incorporation of ³H Uridine in the Blastocyst and Uterus of Rat during Delayed Implantation. J. Endocr. 49, 87-92.

MOHLA, S., DE SOMBRE, E.R. & JENSEN, E.V. (1972) Tissue-Specific Stimulation of RNA Synthesis by Transformed Estradiol-Receptor Complex. Biochem. Biophys. Res. Commun. 46, 661-667.

MONESI, V. & SALFI, V. (1967) Macromolecular synthesis during early development in the mouse embryo. Exp. Cell Res. 46, 632-635.

MONESI, V., MOLINARO, M., SPALLETTA, E. & DAVOLI, C. (1970) Effect of metabolic inhibitors on macromolecular synthesis and early development in the mouse embryo. Exp. Cell Res. 59, 197-206.

MONTECUCCOLI, G., NOVELLO, F. & STIRPE, F. (1973) Effect of α-amanitin poisoning on the synthesis of deoxyribonucleic acid and of protein in regenerating rat liver. Biochim. Biophys. Acta 319, 199-208.

MOORE, G.P.M. (1975) The RNA polymerase activity of the preimplantation mouse embryo. J. Embryol. exp. Morph. 34, 291-299.

NIESSING, J., SCHNIEDERS, B., KUNZ, W., SEIFART, K.H. & SEKERIS, C.E. (1970) Inhibition of RNA Synthesis by α-amanitin in vivo. Z. Naturforsch. B 25, 1119-1125.

OLDS, P.J., STERN, S. & BIGGERS, J.D. (1973) Chemical estimates of the RNA and DNA contents of the early mouse embryo. J. Exp. Zool. 186, 39-46.

O'MALLEY, B.W. & MEANS, A.R. (1974) Effects of Female Steroid Hormones on Target Cell Nuclei. In: H. Busch (ed.), The Cell Nucleus 3, Academic Press, New York, pp. 379-416.

O'MALLEY, B.W., SCHRADER, W.T. & SPELSBERG, T.C. (1973) Hormone-Receptor Interactions with the Genome of Eucaryotic Target Cells. In: B.W. O'Malley and A.R. Means (Eds.), Adv. Exp. Med. Biol. 36, Plenum Press, New York, pp. 174-196.

PERRY, R.P., LATORRE, J., KELLEY, D.E. & GREENBERG, J.R. (1972) On the lability of poly (A) sequences during extraction of messenger RNA from poly-ribosomes. Biochim. Biophys. Acta 262, 220-226.

PERRY, J.S., HEAP, R.B. & AMOROSO, E.C. (1973) Steroid Hormone Production by Pig Blastocysts. Nature 245, 45-47.

PIKO, L. (1970) Synthesis of macromolecules in early mouse embryos cultured in vitro: RNA, DNA, and a polysaccharide component. Develop. Biol. 21, 257-279.

PRASAD, M.R.N., SAR, M. & STUMPF, W.E. (1974) Autoradiographic Studies on ³H Oestradiol Localization in the Blastocysts and Uterus of Rats During Delayed Implantation. J. Reprod. Fert. 36, 75-81.

PSYCHOYOS, A. (1967) Mechanismes de la Nidation. Archs. Anat. microsc. Morph. exp. 56, 616-623.

RASPE, G., (Ed.) (1971) Schering Workshop on Steroid Hormone 'Receptors'. In: Advances in the Biosciences, Vol. 7, Pergamon-Vieweg, Braunschweig, Germany, pp. 1-417.

ROBLERO, L. (1973) Effect of Progesterone in Vivo Upon the Rate of Cleavage of Mouse Embryos. J. Reprod. Fert. 35, 153-155.

ROBLERO, L. & IZQUIERDO, L. (1976) Effect of progesterone on the cleavage rate of mouse embryos in vitro. J. Reprod. Fert. 46, 475-476.

ROEDER, R.G. (1974) Multiple Forms of Deoxyribonucleic Acid-dependent Ribonucleic Acid Polymerase in Xenopus laevis. J. Biol. Chem. 249, 249-256.

ROEDER, R.G. & RUTTER, W.J. (1969) Multiple Forms of DNA-dependent RNA Polymerase in Eukaryotic Organisms. Nature 224, 234-237.

ROEDER, R.G. & RUTTER, W.J. (1970) Multiple Ribonucleic Acid Polymerases and Ribonucleic Acid Synthesis during Sea Urchin Development. Biochem. 9, 2543-2553.

ROEDER, R.G., REEDER, R.H. & BROWN, D.D. (1970) Multiple Forms of RNA Polymerase in Xenopus laevis: Their Relationship to RNA Synthesis in vivo and Their Fidelity of Transcription in vitro. Cold Spring Harbor Symp. Quant. Biol. 35, 727-735.

ROLLARD, R.M., BREDL, J.C.S. & FINN, C.A. (1973) The Effect of Actinomycin D on the Attachment Reaction of Implantation in Mice. J. Reprod. Fert. 33, 343-345.

ROTTMAN, F., SHATKIN, A.J. & PERRY, R.P. (1974) Sequences Containing Methylated Nucleotides at the 5' Termini of Messenger RNAs: Possible Implications for Processing. Cell 3, 197-199.

ROWINSKI, J., SOLTER, D., & KOPROWSKI, H. (1975) Mouse Embryo Development in Vitro: Effects of Inhibitors of RNA and Protein Synthesis on Blastocyst and Post-blastocyst Embryos. J. Exp. Zool. 192, 133-142.

RUTTER, W.J. (1976) In: ICN-UCLA Winter Conference, Molecular Mechanisms in the Control of Gene Expression, in press.

RUTTER, W.J., GOLDBERG, M.I., & PERRIARD, J.C. (1974) RNA Polymerase and Transcriptional Regulation in Physiological Transitions. In: J. Paul (Ed.), Biochemistry of Cell Differentiation, University Park Press, Baltimore, Maryland, pp. 267-300.

SCULTZ, G., MANES, C. & HAHN, W.E. (1973) Synthesis of RNA Containing Polyadenylic Acid Sequences in Preimplantation Rabbit Embryos. Develop. Biol. 30, 418-426.

SCHWARTZ, R.J., TSAI, M.J., TSAI, S.Y. & O'MALLEY, B.W. (1975) Effect of Estrogen on Gene Expression in the Chick Oviduct V. Changes in the Number of RNA Polymerase Binding and Initiation Sites in Chromatin. J. Biol. Chem. 250, 5175-5182.

SEAMARK, R.F. & LUTWAK-MANN, C. (1972) Progestins in rabbit blastocysts. J. Reprod. Fert. 29, 147-148.

SEKERIS, C.E. & SCHMID, W. (1972) Action of α-amanitin in vivo and in vitro. FEBS Lett. 27, 41-45.

SERFLING, E., WOBUS, U. & PANITZ, R. (1972) Effect of α-amanitin on chromosomal and nucleolar RNA-synthesis in Chironomus thummi polytene chromosomes. FEBS Lett. 20, 148-152.

SHELDON, R., JURALE, C. & KATES, J. (1972) Detection of Polyadenylic Acid Sequences in Viral and Eukaryotic RNA. Proc. Natl. Acad. Sci. U.S. 69, 417-421.

SIRACUSA, G. (1973) RNA polymerase during early development in mouse embryo. Exp. Cell Res. 78, 460-462.

SIRACUSA, G. & VIVARELLI, E. (1975) Low-Salt and High-Salt RNA Polymerase Activity During Preimplantation Development in the Mouse. J. Reprod. Fert. 43, 567-569.

SMITH, D.M. (1968) The Effect on Implantation of Treating Cultured Mouse Blastocysts with Oestrogen In Vitro and the Uptake of ^{3}H Oestradiol by Blastocysts. J. Endocr. 41, 17-29.

SMITH, D.M. & BIGGERS, J.D. (1968) The Oestrogen Requirement for Implantation and the Effect of Its Dose on the Implantation Response in the Mouse. J. Endocr. 41, 1-9.

SMITH, D.M. & SMITH, A.E.S. (1971) Uptake and Incorporation of Amino Acids by Cultured Mouse Embryos: Estrogen Stimulation. Biol. Reprod. 4, 66-73.

SMITH, D.M. & BIGGERS, J.D. (1975) Preparation of Mouse Embryos for the Evaluation of Hormone Effects. In: J.G. Hardman and B.W. O'Malley (Eds.), Methods in Enzymology XXXIX, Part D, Academic Press, New York, pp.297-302.

SMUCKLER, E.A. & HADJIOLOV, A.A. (1972) Inhibition of hepatic deoxynucleic acid-dependent ribonucleic acid polymerases by the exotoxin of Bacillus thuringiensis in comparison with the effects of α-amanitin and cordycepin. Biochem. J. 129, 153-166.

STEWARD, F.C., MAPES, M.O., & MEARS, K. (1958) Growth and Organized Development of Cultured Cells. II. Organization in Cultures Grown from Freely Suspended Cells. Amer. J. Bot. 45, 705-708.

STIRPE, F. & FIUME, L. (1967) Studies on the pathogenesis of liver necrosis by α-amanitin. Biochem. J. 105, 779-782.

SZEGO, C.M. (1974) The Lysosome as a Mediator of Hormone Action. Recent Progr. Horm. Res. 30, 171-233.

TASCA, P.J. & HILLMAN, N. (1970) Effects of actinomycin D and cycloheximide on RNA and protein synthesis in cleavage stage mouse embryos. Nature 225, 1022-1025.

TATA, J.R., HAMILTON, M.J. & SHIELDS, D. (1972) Effects of α-amanitin in vivo on RNA polymerase and nuclear RNA synthesis. Nature New Biol. 238, 161-164.

THOMSON, J.L. & BIGGERS, J.D. (1966) Effect of inhibitors of protein synthesis on the development of preimplantation mouse embryos. Exp. Cell Res. 41, 411-427.

TORBIT, C.A. & WEITLAUF, H.M. (1974) The Effect of Oestrogen and Progesterone on CO_2 Production by 'Delayed Implanting' Mouse Embryos. J. Reprod. Fert. 39, 379-382.

TSAI, M.J., SCHWARTZ, R.J., TSAI, S.Y. & O'MALLEY, B.W. (1975) Effects of Estrogen on Gene Expression in the Chick Oviduct IV. Initiation of RNA Synthesis on DNA and Chromatin. J. Biol. Chem. 250, 5165-5174.

UNGER, B. & DICKSON, A.D. (1971) Effect of cycloheximide and actinomycin D on the mouse blastocyst undergoing the giant cell transformation. J. Anat. 108, 519-525.

VAN BLERKOM, J. (1975) Qualitative Patterns of Protein Synthesis in the Pre-implantation Mouse Embryo II. During Release from Facultative Delayed Implantation. Develop. Biol. 46, 446-451.

VASIL, V. & HILDEBRANDT, A.C. (1965) Differentiation of Tobacco Plants from Single, Isolated Cells in Microcultures. Science 150, 889-892.

VERSTEEGH, L.R. & WARNER, C. (1973) DNA-dependent RNA Polymerases from Normal Mouse Liver. Biochem. Biophys. Res. Comm. 53, 838-844.

VERSTEEGH, L.R. & WARNER, C.M. (1975) A comparative study of mouse liver and mouse blastocyst DNA-dependent RNA polymerases. Arch. Biochem. Biophys. 168, 133-144.

VERSTEEGH, L.R., HEARN, T.F. & WARNER, C.M. (1975) Variations in the Amounts of RNA Polymerase Forms I, II, and III during Preimplantation Development in the Mouse. Develop. Biol. 46, 430-435.

WANKA, F., MOORS, J. & KRIJZER, F.N.C.M. (1972) Dissociation of nuclear DNA replication from concomitant protein synthesis in synchronouse cultures of chlorella. Biochim. Biophys. Acta 269, 153-161.

WARNER, C.M. & VERSTEEGH, L.R. (1974) In vivo and in vitro effect of α-amanitin on preimplantation mouse embryo RNA polymerase. Nature 248, 678-680.

WARNER, C.M. & VERSTEEGH, L.R. (1975) Multiple Forms of RNA Polymerase in Preimplantation Mouse Embryos. In: C.L. Markert (Ed.), Proceedings of the Third International Conference on Isozymes, III, Academic Press, New York, pp. 45-55.

WEINMANN, R. & RODER, R.G. (1974) Role of DNA-Dependent RNA Polymerase III in the transcription of the tRNA and 5S RNA Genes. Proc. Natl. Acad. Sci. U.S. 71, 1790-1794.

WEITLAUF, H.M. (1973) In vitro uptake and incorporation of amino acids by blastocysts from intact and ovariectomized mice. J. Exp. Zool. 183, 303-308.

WEITLAUF, H.M. (1974) Effect of Actinomycin D on protein synthesis by delayed implanting mouse embryos in vitro. J. Exp. Zool. 189, 197-202.

WEITLAUF, H.M. (1976) Effect of Uterine Flushings on RNA Synthesis by 'Implanting' and 'Delayed Implanting' Mouse Blastocysts in vitro. Biol. Reprod. 14, 566-571.

WEITLAUF, H.M. & GREENWALD, G.S. (1968) Influence of estrogen and progesterone on the incorporation of ^{35}S methionine by blastocysts in ovariectomized mice. J. Exp. Zool. 169, 463-470.

WHITTEN, W.K. (1957) The Effect of Progesterone on the Development of Mouse Eggs in vitro. J. Endocr. 16, 80-85.

WHITTEN, W. & BIGGERS, J. (1968) Complete Development in Vitro of the Pre-implantation Stages of the Mouse in a Simple Chemically Defined Medium. J. Reprod. Fert. 17, 399-401.

WILSON, I.B. & JENKINSON, E.J. (1974) Blastocyst Differentiation in Vitro. J. Reprod. Fert. 39, 243-249.

WOODLAND, H.R. & GRAHAM, C.F. (1969) RNA synthesis during early development of the mouse. Nature 221, 327-332.

YAMAMOTO, K.R. & ALBERTS, B.M. (1976) Steroid Receptors: Elements for Modulation of Eukaryotic Transcription. Ann. Rev. Biochem. 45, 721-746.

EFFECT OF ANAESTHETIC AGENTS ON EGGS AND EMBRYOS

Matthew H. Kaufman

Department of Anatomy
Downing Street, Cambridge, U.K.

In this paper the findings of recent in vivo and in vitro experiments designed to investigate the effects of anaesthetics on mammalian eggs and embryos will be considered in detail, as will the biochemical and ultra-structural observations which may shed some light on their possible mode of action during early development.

A great deal of work is being carried out at the present time to investigate how local and general anaesthetics influence the cell surface and plasma membrane, and intracellular cytoplasmic processes. The rapid advances taking place in this area have almost exclusively been aimed at investigating the biochemical and molecular interrelationship between the external environment of the cell, the structure of the plasma membrane, and the microtubules and microfilaments of the cellular cytoskeletal system. While this information may not appear to shed much light on the possible mechanisms of anaesthesia, it provides certain clues which may help to explain recent clinical observations that the reproductive efficiency of operating theatre personnel who are chronically exposed to low levels of anaesthetic gases is impaired.

There is now a considerable body of circumstantial evidence which seems to confirm the notion that anaesthetics, both in clinical and subclinical dosage, may be teratogenic during certain stages of pregnancy. The evidence in favour of this hypothesis has been building up over the last 10 years. Epidemiological studies which have been carried out on operating theatre personnel (eg. Vaisman, 1967; Askrog & Harvald, 1970; Lencz & Nemes, 1970; Cohen et al., 1971; Knill-Jones et al., 1972; Corbett et al., 1974; American Society of Anesthesiologists, 1974; Knill-Jones et al., 1975) have recently been briefly reviewed by Smithells (1976), who comes to the general conclusion that 'there are good reasons for believing that there are teratogenic forces at work in operating theatres'. Up to now no definite cause-effect relation-ship has been established between exposure to trace levels of waste anaesthetic gases and the increased incidence of spontaneous abortions and congenital abnormalities found both in females who work within the operating theatre environment and in the unexposed wives of males who work in operating theatres.

Measurements of the concentration of different anaesthetic gases in operating room air, and specifically within the inhalational zone of the anaesthetist, have been made by various workers (eg. Linde & Bruce, 1969; Corbett & Bell, 1971; Corbett, 1972), and range from barely detectable to very high levels (peak concentration of nitrous oxide of 9700 ppm, Corbett & Bell, 1971).

In addition to the effect of trace anaesthetic concentrations, attention has also been drawn to the possible risk to the human foetus of surgery carried out during pregnancy (Shnider & Webster, 1965). This topic has been discussed in detail by Smith (1968) who considered experimental and clinical evidence for the possible teratogenicity of anaesthetic agents, premedicants and supplemental drugs (eg. narcotics, hypotensive and vasopressor agents, antihistamines and the muscle relaxant drugs). While most of his review is directed towards anaesthesia at later stages of pregnancy than considered in the present discussion, he has directed attention to some of the marked alterations which occur in maternal physiology during anaesthesia (eg. hypoxia, electrolyte disturbances, hypercapnia and increased oxygenation) that may in themselves be foetotoxic either when acting alone or synergistically with one another or with the anaesthetic agent. However, as Wilson (1954) has pointed out, not all alterations in maternal physiology during pregnancy, no matter how severe they may be, are capable of affecting the subsequent offspring, though this may, of course, depend on the duration of the stress and the stage of development of the foetus at the time of exposure.

The results of a large number of teratological studies in which pregnant animals have been subjected to various subanaesthetic and anaesthetic concentrations of different anaesthetics are summarised, and attention drawn to the salient features. The effect of anaesthesia on female germ cells at the 'resting' or dictyate stage of meiosis is also considered in detail, as recent indirect evidence suggests that anaesthesia carried out before conception may, under certain circumstances, affect the outcome of pregnancy.

THE OPERATING THEATRE ENVIRONMENT – CLINICAL STUDIES

In one of the earliest studies concerned with occupational disease in anaesthetists, Vaisman (1967) reported an unusually high incidence of headache, nausea and fatigability, and that 18 out of 31 pregnancies studied ended in spontaneous abortion. A smaller survey of Hungarian anaesthetists of both sexes (Lencz & Nemes, 1970, cited in Corbett, 1972) revealed that only 10 out of 31 deliveries were normal; of the remaining 21 pregnancies, 10 ended in spontaneous abortion, 9 in 'pathologic' pregnancies and 2 in premature delivery.

Cohen et al. (1971) reported a 29.7% incidence of spontaneous abortions (10 out of 36 pregnancies) in 67 operating room nurses compared to a control incidence of 8.6% (3 out of 34 pregnancies) in 92 general duty nurses with no history of exposure to an operating room environment. These authors also compared the incidence of spontaneous abortions in 50 female anaesthetists (38.7%, 14 out of 37 pregnancies) and 81 female physicians (10.3%, 6 out of 58 pregnancies). The differences between the exposed and unexposed groups in both series were highly significant (p 0.045 and p 0.0035, respectively).

Knill-Jones et al. (1972) obtained obstetric information from 563 female anaesthetists and 828 women doctor control subjects. The anaesthetists were divided into two groups depending on whether they had or had not worked during the first and second trimesters of pregnancy. When the ratio of abortion/total pregnancies was examined, the ratio was found to be significantly greater in the "working" anaesthetists (18.2%) compared with the control group (14.7%). The ratio in the anaesthetists "not at work" was 13.7%. The incidence of congenital abnormalities was significantly greater (6.5%) where the mother had worked than when she had not worked (2.5%, p 0.02) but not significantly different from the control frequency (4.9%). Involuntary infertility was twice as frequent among anaesthetists (12%) as in the control group.

In a recent study by Corbett et al. (1974) information on the incidence of birth defects was obtained from 621 nurse anaesthetists. 434 births were recorded in pregnancies during which the mother had worked, and 261 pregnancies in which the mother had not. No significant difference was seen in the incidence of major malformations. However, there appeared to be a significant difference in the incidence of minor defects, 16.4% in the working nurse anaesthetists and 5.7% in those that had not worked during pregnancy. The majority of these malformations consisted of skin defects and inguinal hernias.

In the recent national study in the U.S.A. (American Society of Anesthesiologists, 1974) 49,585 exposed operating room personnel were compared with 23,911 unexposed control individuals. The results indicated that the exposed group of females had an increased risk of spontaneous abortion (17.0 $\pm$ 0.9% to 19.5 $\pm$ 0.9%) compared to unexposed controls (14.4 $\pm$ 1.4% to 15.7 $\pm$ 3.3%). The incidence of congenital abnormalities was also higher in the exposed group (9.6%) compared to the controls (5.9%, p 0.01). The exposed group of females also had a greater risk of developing cancer (1.3 to 2 fold risk), and hepatic disease (1.3 to 2.2 fold risk), and renal disease (1.2 to 1.4 fold). No increased risk of cancer or renal disease was observed in male anaesthetists, but a similar chance of developing hepatic disease was reported. No increased risk of spontaneous abortions was observed in the unexposed wives of male operating room personnel, but an increased risk of bearing children with congenital abnormalities was noted in the unexposed wives of male anaesthetists (p=0.04). The types of abnormalities reported involved most of the major organ systems.

In the latest study of this type, Knill-Jones et al. (1975) surveyed
5507 male doctors (of whom 26% were anaesthetists, 9% surgeons) in the U.K.
The incidence of spontaneous abortion in pregnancies involving the exposed
group was 11.1%, compared to 10.9% in their control series.   However, maternal
exposure alone was associated with a frequency of 15.5% compared to 10.9%
where neither parent was exposed (p 0.01).   A slight increase in the incidence
of minor abnormalities was observed in association with male exposure (4.5%)
compared to controls where neither parent was exposed (3.2%, p 0.01), largely
accounted for by an increase in the reporting of minor congenital abnormalities
by the exposed group.

THE EFFECT OF ANAESTHETICS ON GAMETES

No experimental information is available on the possible effects of
anaesthetics on gametogenesis in males or females, though indirect evidence
from the recent Danish report (Askrog & Harvald, 1970), the American Society
of Anesthesiologists national study (American Society of Anesthesiologists,
1974) and Knill-Jones et al. (1975) that spermatogenesis might be affected by
chronic exposure to subclinical levels of anaesthetics will be discussed later.
Assessment of any effect on gametogenesis in the female is complicated by the
fact that most of oogenesis occurs prenatally.   In this section some recent
observations on the effects of anaesthesia in rodents will be outlined, and an
attempt made to discuss their possible significance.

The current requirements of various government agencies for reproductive
and teratological testing of new drugs have recently been discussed by Berry
and Barlow (1976).   At least three reports of laboratory tests on the effects
of anaesthetics on fertility and reproductive function have recently appeared
in the literature (Bruce, 1973; Kennedy et al., 1976; Kaufman, 1976).   Other
studies of this kind have undoubtedly been carried out by interested drug
companies, but their results are not generally available.   The few reports of
in vivo and in vitro teratological studies on the effect of anaesthetics on
pregnant animals and early postimplantation embryos which are available are
discussed later.

The effect of anaesthetics on ovarian oocytes at the 'resting' or dictyate stage of meiosis

While it has long been known that anaesthetics and other CNS depressants
may impair cell division (for reviews see Andersen, 1966; Fink, 1971), few
attempts have been made to establish whether the germ cells of adult female

mammals, in the 'resting' or dictyate stage of meiotic prophase, are susceptible
to these agents.   In a recent study (Kaufman, 1976) these aspects of the
possible effect of anaesthetics were monitored by assessing the influence of
anaesthesia carried out 21-27 days before conception on the subsequent
fertility of females.   The incidence of pre- and post-implantation embryonic
loss in the experimental groups was compared with that occurring in
unanaesthetised control females (Group 1).   As the effect of prolonged
anaesthesia was being tested, females were anaesthetised with a single i.p.
injection of either 1.5 times (Group 2) or twice (Group 3) the standard dose
of tribromoethanol (Avertin:Winthrop) normally used to anaesthetise an intact
mouse.   The standard dose of a freshly prepared 1.2% solution of Avertin
dissolved in 0.9% NaCl is usually 0.02 ml Avertin/gm body weight.   As
anaesthesia induced by the standard dose of Avertin usually only lasts for
10-20 minutes its effect was not tested.   The twice standard Avertin dose was
close to the LD_{50} for this agent in this series when the body temperature
during anaesthesia was not maintained by external warmth.   In subsequent
studies in which a similar dose of Avertin was given, and the body temperature
maintained during anaesthesia, the mortality was in the region of 25-30%.
The total embryonic losses were significantly higher in Group 3 (p 0.005) than
in Group 1, but the increase in Group 2 was not significantly different from
the control value.   The embryonic loss was most marked in the preimplantation
period, being 13% for Group 2 (p 0.05) and 17% for Group 3 (p 0.005) compared with the level of 8% in Group 1.   The similar corpora lutea counts
which were observed in the 3 groups of animals, suggests that this treatment
did not influence the number of eggs ovulated.

The increased embryonic losses reported by Kaufman (1976) seem to be at
variance with the findings in other studies in which the authors concluded
that halothane given before conception produced no adverse effect on female
fertility in mice (Bruce, 1973) and rats (Kennedy et al., 1976).   However,
because of the different experimental approaches employed in these 3 studies
comparison of results is almost impossible.   In the study by Bruce (1973),
for example, a very low subanaesthetic dose of halothane (16 ppm) was given
to mice for 7 hr/day, 5 days/week for 6 weeks prior to mating, while Kennedy
et al. (1976) exposed rats for 1 hour daily either 1-5, 6-10 or 11-15 days
before mating to an anaesthetic concentration of halothane (1.34-1.48%) in
air.   Obvious differences between these various studies are the use of an
inhalational anaesthetic (halothane) versus an agent given i.p. (Avertin), the
use of the rat as an experimental animal by Kennedy et al. (1976), and the
different type and degree of anaesthesia achieved by these different
anaesthetics.   Thus the level of narcosis obtained using an inhalational
agent from a constant flow gassing apparatus is likely to be much more uniform

than that achieved following an i.p. injection of any anaesthetic agent.
Certainly in the Avertin studies where relatively high doses of this agent
were given i.p., profound narcosis very rapidly ensued.   It was difficult to
assess the level of anaesthesia achieved, but recovery usually occurred
within 1.0 - 1.5 h with the 1.5 times standard dose or 2.0-3.0 h with the
twice standard dose group when body temperature was not maintained by external
warmth.   It can only be assumed that the high circulating levels of Avertin
which were initially achieved following the i.p. injection of this agent had
an adverse effect on a proportion of the oocytes contained within the ovary.
In addition, it seems unlikely that a significant effect on fertility would
have been observed by Kaufman had lower doses of Avertin been used.

Several explanations may be proposed to account for the recent experimental
findings of Kaufman (1976) and the increased incidence of involuntary
infertility and spontaneous abortions seen in operating room personnel.   One
explanation for the observed increase in embryonic loss might be that chronic
exposure to trace levels or acute exposure to high levels of anaesthetics may
be mutagenic either during spermatogenesis or during the dictyate stage of
meiotic prophase.   This might manifest itself in an increase in the incidence
of dominant lethality.   Long-term genetic studies in mice have recently been
initiated to test this possibility (M.F. Lyon, personal communication), but
the results will not be available for some time.   In addition to this possible
mechanism of action, in which it is presumed that damage might be induced at
the molecular level in the genetic material of the germ cells, anaesthetics
may induce the production of gametes with an abnormal chromosome constitution.
Various mechanisms could be envisaged which might produce gametes of this
type.   For example, simple numerical anomalies may be produced as a result
of non-disjunction at meiosis I or II, or more subtle structural anomalies
such as balanced or unbalanced translocations, segment inversions etc. as a
result of more extensive chromosome damage.

The technical and organizational problems involved in carrying out cyto-
genetic analysis of sufficient human abortus material from this source makes
it unlikely that a direct assessment would be feasible on a large enough scale
to produce a meaningful result.   Large numbers would have to be examined
because of the high incidence of chromosome anomalies which are normally present
in spontaneous abortions (Carr, 1965, 1971) where no aetiological factors are
normally found.   In any case, critical material from the pre- and post-
implantation period is likely to be lost before pregnancy is diagnosed and
would therefore not be available for analysis.

Experimental attempts to investigate the possibility that pre- and early post-implantation embryonic losses may, at least in part, be due to numerical anomalies have recently been made by Kaufman (unpublished). Oocytes at the germinal vesicle stage were isolated from the ovaries of randomly-bred CFLP mice (Anglia Laboratories) anaesthetised with 1.5 times or twice the standard dose of Avertin 2 weeks before, and from unanaesthetised controls. Groups of between 30 and 50 eggs/treatment were incubated in vitro for 4 or 13 hours under standard embryo culture conditions (Biggers et al., 1971) in small drops of modified Krebs-Ringer bicarbonate medium (Whittingham, 1971) containing 4 mgm/ml bovine serum albumin. No difference was observed in the proportion of eggs which underwent germinal vesicle breakdown in the 3 groups looked at after 4 hours, or had completed maturation to metaphase II (13 hour groups). However, analysis of the chromosomes of all maturing eggs (method of Tarkowski, 1966) indicated that there was one interesting difference between the morphology of the chromosomes in the experimental and control groups which was probably accentuated by the lengthy (about 20 mins) hypotonic citrate pre-treatment of eggs prior to fixation. Between 30% and 40% of preparations in the experimental groups showed evidence of abnormal chromatid separation (Figure 1a). This feature was not observed in controls (Figure 1b). Attempts to demonstrate a similar effect in eggs isolated at 15-16 hr after the HCG injection for superovulation pretreated for only 10 minutes in hypotonic citrate were not successful.

Because of the technical difficulties involved in accurately estimating the number of chromosomes present in metaphase II preparations a parallel study in which the eggs were activated to take them through to the first cleavage mitosis was undertaken. (C57BL x CBA)F_1 eggs were isolated at about 20 h after HCG and induced to develop parthenogenetically by hyaluronidase treatment in vitro (Kaufman, 1973a). In these preliminary experiments females had previously been anaesthetised with the standard dose of Avertin at intervals between 25 hr before and 2 hr after the HCG injection for superovulation. The chromosome constitution of the activated eggs was determined at the first cleavage mitosis (Kaufman, 1973b). The activation frequency was in the region of 70-80%, and accurate chromosome counts could be made in a very high proportion of preparations. The results of this study are presented in Table 1, and indicate that Avertin anaesthesia carried out during the oestrus cycle leading to ovulation may induce non-disjunction in a small proportion of eggs. Further studies will be required to determine whether this plays any role in the embryonic losses observed when mice are anaesthetised 3 or more weeks before conception. It would in addition be interesting to examine the chromosome constitution of fertilized eggs at the first cleavage mitosis and

144

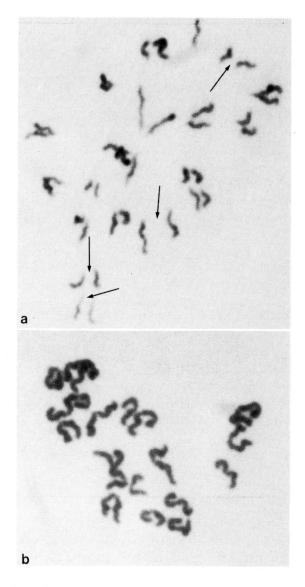

a

b

Figure 1a (upper):   Chromosome complement of mouse oocyte at prometaphase II.
Oocyte isolated from the ovary at germinal vesicle stage and cultured in vitro
for 13 hours. Some sister chromatids have disjoined (arrows). The female had
been anaesthetised with 1.5 times the standard dose of Avertin two weeks before.
Giemsa stain.

Figure 1b (lower):   Chromosome complement of control oocyte at metaphase II.
All chromatid pairs joined at their centromeres. Giemsa stain.

at later preimplantation stages of embryonic development.   It is unfortunate that only gross numerical anomalies can be ascertained by this method of analysis, and that most structural anomalies cannot be detected by this means.

TABLE 1

FIRST CLEAVAGE METAPHASE CHROMOSOME COUNTS IN PARTHENOGENETICALLY ACTIVATED HAPLOID MOUSE EGGS (n = 20).   FEMALES ANAESTHETISED WITH AVERTIN BETWEEN 25 HOURS BEFORE AND 2 HOURS AFTER THE HCG INJECTION FOR SUPEROVULATION

Group Number	Time of anaesthetic HCG $\pm$ hours	Total No. of females	Total No. eggs analysed	Chromosome no.				Percent aneuploid eggs
				18	19	20	21	
1	Nil (control)	9	115		3	112		2.6
2	-25	4	62		1	61		1.6
3	-20	8	80	1	2	74	1	5.0
4	-13	8	121		1	117	3	3.3
5	-4	4	85		4	79	2	7.1
6	-1.5	4	51		4	45	2	11.8
7	+2	12	127		9	113	4	10.2

If anaesthesia does, in fact, increase the incidence of meiotic non-disjunction, then it is likely that it is the interrelationship between the centromere and the meiotic spindle apparatus on which attention should be focused.   Very little is known about the ultrastructure of the centromere (or kinetochore), though slightly more is known about its functional activity as a result of recent studies by Telzer et al. (1975).   These workers have shown that the centromeres of isolated chromosomes of HeLa cells are the only regions on these chromosomes which can act as microtubule assembly sites, and have suggested from this evidence that centromeres are also likely to possess this capacity in vivo.

Many studies have been made on the mitotic spindle apparatus (for review of early work see Mazia, 1961), and numerous models proposed to explain its functioning during mitosis (eg. McIntosh et al., 1969).   The latest models describe an equilibrium between the mitotic spindle and its subunits and the cross-bridges which interconnect spindle microtubules.   It has been postulated that these intertubule links are active mechanochemical units capable of sliding adjacent tubules over one another.   Allison and Nunn (1968) have hypothesised that microtubular protein units are held together by low energy bonds which are easily and reversibly broken, and have suggested that in the presence of general anaesthetics, for example, depolymerisation of these tubules can occur.

This has recently been confirmed by Hinkley and Telser (1974), who demonstrated that halothane caused microfilament breakdown in cultured mouse neuroblastoma cells, and by Haschke et al. (1974) and Nicolson et al. (1976) using local anaesthetics such as lidocaine and tetracaine.

It has been known for many years that anaesthetic agents are capable of interfering with cell division and that metaphase arrest occurs when plant and animals cells are exposed to anaesthetics at concentrations usually within the clinical range (for reviews see Andersen, 1966; Fink, 1971). The inhibitory action on cell division which may be induced by these agents is similar in appearance to the well-defined arrest of mitosis induced by colchicine (c-mitosis, Levan, 1938). It seems likely that anaesthetics in appropriate concentrations would have a similar inhibitory effect on meiotic spindle activity. What is less clear is the effect that these agents may have on 'resting' or dictyate stage ovarian oocytes when the chromosomes are decondensed and the spindle present in a disassembled state. Exposure to trace levels of anaesthetics during the dictyate stage of meiotic prophase may eventually interfere with the normal subunit assembly mechanism which is activated during meiotic maturation. The possibility also exists that exposure to these agents may also produce non-specific chemically-induced changes in the genetic material of oocytes via mutation, the production of major gene defects or chromosome aberrations, or damage to the centromeric region of these chromosomes with the possibility of chromosome imbalance in the resultant products of division. These suggestions go some way towards explaining both the clinical and experimental observations, but should only be considered as a very tentative working hypothesis at this stage.

Delayed ovulation induced by barbiturates

Everett and Sawyer (1950) demonstrated that a range of barbiturates were able to suppress the mid-cycle surge of LH in the rat, delaying ovulation by 24 hours. To produce this effect moderate or profound anaesthesia was usually required, and the sodium salts of the following agents were found to be effective, amytal, barbital, dial, phenobarbital and nembutal.

The chromosome constitution and developmental potential of the oocytes released from delayed follicles (often termed "overripe follicles") have been examined by Butcher and Fugo (1967, see also Butcher, 1975, for summary of recent results). These workers reported a decreased implantation rate when nembutal-treated females were mated to fertile males, and an increased incidence of chromosome anomalies in foetuses at mid-gestation (4.6%) compared to the control level of 1.5% (Butcher & Fugo, 1967). Recent ultrastructural studies on oocytes from delayed follicles by Peluso and Butcher (cited by

Butcher, 1975) demonstrated a range of cytoplasmic changes which included a
reduction in the normal number of cortical granules, and changes in mitochondrial
morphology.   The chromosome constitutions of preimplantation rat embryos
isolated from pentobarbital-treated females have also been studied by Mikamo
and Hamaguchi (1975).   Little effect on meiotic divisions and fertilization was
observed, but the incidence of chromosomally mosaic blastocysts indicated that
non-disjunction or anaphase lagging must have occurred during the early cleavage
divisions.

      These recent findings by Butcher (1975) and Mikamo and Hamaguchi (1975)
suggest that most of the effects on embryonic development previously reported
by these workers was probably due to barbiturate-induced changes in the preovulatory oocyte.   These studies in the rat may well explain the recent
observations by Kaufman (1976, and unpublished) who examined the chromosome
constitution and developmental potential of oocytes from Avertin-treated mice
(see earlier).

Effect of anaesthetics on ovulated unfertilised eggs

      Rodent eggs may be induced to develop parthenogenetically in vivo when
intact animals are anaesthetised shortly after the occurrence of ovulation.
Thus rat eggs have been activated parthenogenetically with ether (Thibault,
1949; Austin & Braden, 1954), chloroform, ethyl chloride, ethyl alcohol,
paraldehyde, nitrous oxide and i.p. Nembutal (Austin & Braden, 1954), and
mouse eggs with ether (Braden & Austin, 1954) and i.p. Avertin (Kaufman, 1975).
Apart from the report by Kaufman (1975), where postimplantation parthenogenetic
development was obtained when pseudopregnant mice were anaesthetised with
Avertin, previous workers had only reported the occasional development of these
eggs as far as the 2 or 4 cell stage.   The incidence of activation and types
of parthenogenones induced is probably related to the post-ovulatory age of
eggs at the time of anaesthesia.   In the mouse, for example, a similar pattern
of response had previously been demonstrated when eggs were isolated at various
times after ovulation and activated in vitro with hyaluronidase (Kaufman, 1973a).
The underlying mechanisms whereby anaesthetics may induce unfertilized rodent
eggs to develop parthenogenetically was initially thought to be due to the
production of tissue anoxia (Austin & Braden, 1954).   It now seems more likely
that anaesthetics initiate a chain of events at the cell membrane which result
in the release of sequestered intracellular calcium ions.   The hypothesis that
similar events may be initiated in eggs of all species, and that this is the
'universal activating mechanism', has recently been proposed by Steinhardt et al.
(1974).   These authors came to this conclusion as a result of studies on
oocyte activation in several species including mammals using the divalent

ionophore A23187. Similar events are also thought to occur at fertilization (Steinhardt & Epel, 1974; Steinhardt et al., 1974).

Extracellular Ca^{2+} ions are also important for second polar body extrusion in mouse parthenogenetic eggs (Surani & Kaufman, unpublished). The presence of an optimal level of extracellular calcium within the ampulla of the mammalian oviduct may facilitate an increase in the intracellular pool of free Ca^{2+} ions within the oocyte during sperm penetration. An optimal ionic concentration may also be a prerequisite for second polar body extrusion following fertilization, being the normal means of reconstituting the diploid state. Cytochalasin B, which disrupts microfilaments, also prevents second polar body extrusion (Balakier & Tarkowski, 1976) probably by displacement of the second metaphase spindle from the periphery of the egg (Johnson et al., 1975). This is relevant in the present context because anaesthetics may competitively affect Ca^{2+}-sensitive functions necessary for microtubule maintenance (Wilson et al., 1970) and might even increase intracellular Ca^{2+} concentrations to levels sufficient to induce microtubule depolymerization (Kirschner & Williams, 1974) and microfilament breakdown (Hinkley & Telser, 1974; Nicolson et al., 1976). Mg^{2+} is required for the polymerization of tubulin, and high levels of Ca^{2+} strongly inhibit this process (Weisenberg, 1972). Intracellular levels of these and other ions could therefore regulate cyclic changes in the formation and breakdown of the cellular cytoskeletal system which are essential in cell division. All the evidence suggests that maintenance of the normal interrelationship between divalent cations and the cellular cytoskeletal system is of fundamental importance, especially in early development, and that this delicate balance may be affected by anaesthetics. Thus even subclinical anaesthetic levels in certain mammalian species may be capable of influencing both meiotic and mitotic chromosome segregation, polar body extrusion, and cytokinesis.

Influence of anaesthetics on spermatogenesis

In three clinical studies (Askrog & Harvald, 1970; American Society of Anesthesiologists, 1974; Knill-Jones et al., 1975) observations have been made on the outcome of pregnancy in unexposed wives of male anaesthetists. The two more recent reports are not in accord with the earlier report of Askrog & Harvald (1970) in which an increase in the incidence of spontaneous abortion and premature delivery had been observed. These findings may, however, have resulted from other factors such as, for example, the effect of maternal age on spontaneous abortion, since they compared obstetric histories before and after starting employment in anaesthetics. A slight increase in the incidence of congenital abnormalities was observed in the American Society of Anesthesiologists (1974) study and by Knill-Jones et al. (1975). Because of the small numbers involved and the high incidence of minor anomalies, caution must be exercised in interpreting these findings. At the most these studies indicate that chronic exposure to trace levels of anaesthetics may have an adverse influence on spermatogenesis, but until further information becomes available the case remains unproven.

Two recent attempts have been made to investigate this problem using rodents (Bruce, 1973; Kennedy et al., 1976). In one study (Bruce, 1973) male mice were exposed to trace levels of halothane 7 hr/day, 5 days/week for 6 weeks prior to mating, while Kennedy et al. (1976) exposed male rats for 1 hr daily to anaesthetic concentrations of this agent for 1-5, 6-10 and 11-15 days prior to mating. Both groups reported negative findings, with no increase in the incidence of pregnancy loss when unexposed females were mated to exposed males. Unfortunately, both studies fall short of the present requirements of various government agencies for testing the effect of drugs on male fertility (see Berry & Barlow, 1976) who recommend that treatment in the male should be carried out at least 60-80 days before mating. Further studies, possibly involving the use of other species (see Jackson, 1970; Baker, 1972), will obviously be required to clarify this issue.

TERATOLOGICAL STUDIES

A number of experiments have been carried out to investigate the effect of halothane, both in subanaesthetic and anaesthetic concentrations and for various exposure durations, on pregnant rats (Basford & Fink, 1968; Katz & Clayton, 1973; Chang et al., 1974; Kennedy et al., 1976; Lansdown et al., 1976), mice (Bruce & Koepke, 1969; Bruce, 1973) and rabbits (Kennedy et al., 1976). Similar studies to test the effect of nitrous oxide on pregnant rats (Parbrook et al., 1965; Fink et al., 1967; Shepard & Fink, 1968; Corbett et al., 1973) and anaesthetic concentrations of sodium pentobarbital on pregnant mice (Setala & Nyyssonen, 1964) have also been reported. In two further studies, the effect of nitrous oxide/halothane mixtures were tested in pregnant rats (Wittman et al., 1974) and hamsters (Bussard et al., 1974).

In these studies three different situations have been tested, namely chronic exposure to subanaesthetic concentrations of these agents, ranging from trace levels up to levels associated with drowsiness and inability to feed, and the effect of a single or repeated exposure to concentrations which induce narcosis. This latter class of experiments where animals have been repeatedly exposed to anaesthetic concentrations of these agents, in some cases for many hours each day, is not particularly close to the clinical situation. In other reports, where subanaesthetic and anaesthetic concentrations have been given, very high levels of skeletal anomalies (eg. Basford & Fink, 1968) and foetal losses (eg. Katz & Clayton, 1973) have been observed in controls, suggesting

that factors other than anaesthetics alone may have contributed to the high
incidence of embryonic mortality and morbidity in these studies.   In attempting
to summarize these findings and present a balanced assessment of their possible
clinical significance, these various considerations have been taken into account.
Very little information is available on the effect of anaesthetics during the
preimplantation period, and only one study has been carried out to investigate
the direct effect of an anaesthetic agent on early postimplantation embryos in
culture (Kaufman & Steele, 1976).

The specific days of pregnancy on which animals were exposed to treatment,
referred to in the following pages, is that found in the original papers.   In
most cases the authors have referred to the day of finding sperm in the vagina
or noted the presence of a vaginal plug as Day 1 of pregnancy.   A few authors
refer to this as Day 0 of pregnancy.   The particular system used is,
unfortunately not always clearly stated.   In addition, it has long been known
that the reaction of the embryo to specific compounds varies not only from
species to species but also within a given species, between each strain and
even between individuals of the same strain (for review, see Tuchmann-
Duplessis, 1969).   The importance of these genetic differences in sensitivity
to different agents and the rationale for choosing a particular experimental
animal in teratological studies has recently been discussed by various authors
(eg. Axelrod, 1970; Berry & Barlow, 1976).   Thus cortisone, for example, is a
potent teratogenic drug in the mouse and rabbit but does not produce malformations in the rat: the same dose of this drug produces cleft palates in
17% of strain C57BL/6 Jax mice and in 100% of strain A/Jax (Fraser et al., 1954).
The use of different strains of experimental animals may partly account for
apparently conflicting findings reported by various workers and reviewed here,
where, for example, increased foetal loss and the presence of malformations
was observed in some studies but not in others.   If further information is
required on these various points, reference should be made to the original
articles.

The effect of halothane

In a recent study by Lansdown et al. (1976) no teratogenic or foetotoxic
effect was observed at autopsy on day 22 when rats were exposed to subanaesthetic levels of halothane ranging from 50-3200 ppm on days 8-12 and
1600 ppm on days 1-21 of pregnancy.   No increase was found in foetal death
and resorption rates, growth retardation or in the frequency of skeletal
anomalies.

No increase in the incidence of embryonic losses or gross congenital
anomalies were reported by Chang et al. (1974) when rats were exposed to a
concentration of 10 ppm halothane throughout pregnancy.   However, the progeny

showed histological changes in the tissue of the central nervous system (vacuoles, some neuronal necrosis, etc.) which these workers hypothesised might later be associated with behavioural changes and learning defects.   The testing of off-spring of exposed females by studies on behavioural and functional development has recently been incorporated into the new British guide-lines for studies on reproduction (Committee on Safety of Medicines, 1974), and the principles and practice of behavioural teratology testing are discussed in detail by Barlow and Sullivan (1975).

In a recent study by Kennedy et al. (1976) rats were exposed to anaesthetic concentrations of halothane (1.35-1.43%) on days 1-5, 6-10 and 11-15 of pregnancy.   No increase was observed in the incidence of skeletal or gross congenital anomalies.   These workers also exposed rabbits to anaesthetic concentrations of halothane (2.15-2.30%) on days 6-9, 10-14 and 15-18 of pregnancy with similar negative results when these animals were autopsied on day 29.

In contrast to these generally negative findings, Basford and Fink (1968) reported an increased incidence of embryonic loss and skeletal anomalies in their experimental group when rats were exposed to anaesthetic doses of halothane (0.8%, which induced light sleep within 1 hour) for 12 hr/day on either day 6, 7, 8, 9 or 10 of pregnancy or daily during this period.   These workers also reported a diurnal variation in response with a higher incidence of resorptions when anaesthetics were administered during the day than at night.

The findings of Katz and Clayton (1973) that low concentrations of halothane were both teratogenic and foetotoxic are particularly difficult to interpret as very high rates of foetal loss were observed in their control series.   In a slightly more complex study, Bruce (1973) exposed various strains of mice to subanaesthetic concentrations of halothane (about 16 ppm) 7 hr/day, 5 days/week for 6 weeks before conception, and daily throughout pregnancy. No adverse effect on the pregnancy outcome was observed in the treatment group compared to the controls.   However, as the results from three strains of mice were pooled, and corpora lutea counts were not given, it is difficult to draw any useful conclusions from this study.   In a previous study by Bruce and Koepke (1969) very few mice mated when exposed daily to 0.1% halothane in air, the reason for this was not established.   An increase in deaths occurred in this group of mice which was neither enhanced nor antagonized by any effects of radiation (1.44 R/hour gamma radiation).

The effect of nitrous oxide

Various groups of workers have examined the effect of nitrous oxide on pregnancy in the rat.   In the majority of these studies, animals were exposed to quite high concentrations of this agent either 8 or 24 hr/day for up to nine successive days during the postimplantation period, test systems far removed from the clinical situation. Only in the study by Corbett et al. (1973), where groups of rats were exposed to 100 ppm and 1000 ppm for 8 hr/day on days 10-13, 14-19 or 10-19 of pregnancy, were subanaesthetic doses given. These workers also exposed rats for 24 hr/day from day 8-13 to levels of 15000 ppm. At this higher dose, an increase in the incidence of foetal death was observed. At the lower doses tested, a diurnal effect was observed in that a significant increase in foetal loss occurred when the anaesthetic was given from 6 a.m. to 2 p.m., but not when animals were exposed from 2 p.m. to 10 p.m. This is similar to the diurnal effect reported by Basford and Fink (1968) noted in the previous section.

In the study by Fink et al. (1967) rats were exposed for 24 hr/day to gas mixtures containing 45-50% nitrous oxide for either 2, 4 or 6 days from day 8 of pregnancy. In the report by Shepard and Fink (1968), further details of this earlier study are given, and additional findings on the effect of exposure of rats to 70% nitrous oxide either 24 hr/day from days 5-11 or for a single exposure period lasting 24 hr on days 5, 6, 7, 8, 9, 10 or 11. Similar findings are reported in both these studies, with an increased incidence of skeletal (100% of foetuses examined in the experimental series) and visceral anomalies, and reduced birth weight in surviving foetuses. A selective loss of male foetuses was thought to have occurred in the early postimplantation period with a concomitant alteration in the sex-ratio at birth. Shepard and Fink (1968) reported that a maximum effect on the incidence of skeletal anomalies occurred when rats were anaesthetised on day 9 of pregnancy. However, a significant increase in skeletal anomalies was found in unanaesthetised females starved for 24 hr on day 7, and others exposed for 48 hr to 10% oxygen on days 8 and 9. A moderate incidence of skeletal anomalies was also found in the control series. The early report by Parbrook et al. (1965) in which rats were exposed to 60% nitrous oxide during early pregnancy is difficult to evaluate due to the almost complete absence of details of their treatment schedules.

Exposure to mixtures of halothane and nitrous oxide

In two studies, experimental animals have been exposed to mixtures of halothane and nitrous oxide, Wittman et al. (1974, cited in Kennedy et al., 1976) exposed rats for 12 hr/day on days 6 and 10 of pregnancy to a gas mixture containing 0.8% halothane in nitrous oxide/oxygen, and reported a 44% incidence of foetal loss compared to a control rate of 15%. The control level was not increased by exposure to nitrous oxide alone. In the second study of this type, Bussard et al. (1974) exposed hamsters for 3 hr/day on days 9, 10 and 11 of pregnancy to anaesthetic concentrations of nitrous oxide/halothane.

At autopsy on day 15 (of a 16 day pregnancy) an increase in foetal loss
(resorptions) was observed, surviving foetuses were reduced in weight and
length compared to controls, with a maximum response on day 11.   No effect
on sex-ratio was found.

Exposure to sodium pentobarbital

In two series of experiments, Setala and Nyyssonen (1964) exposed mice on
days 1-4 and 1-15 of pregnancy to narcotic levels of sodium pentobarbital
(i.p. injection of 1.6 mgm daily), and reported reduced fertility from 80-90%
in controls to about 50% in the experimental series, and a high incidence of
a wide variety of congenital abnormalities.   Very few experimental details
are to be found in this report.

In these last sections the reactions of the foetus to drugs given to the
mother have been outlined.   While several techniques have been described
which could be modified to test the direct action of anaesthetics on the foetus,
either in vivo or in vitro, in only one study has the direct effect of an
anaesthetic agent on the isolated foetus in culture been described.

The effect of Avertin on rat embryos in culture

Kaufman and Steele (1976) using the technique of postimplantation embryo
culture described by New (1971) and New et al. (1973) examined the effect on
headfold stage rat embryos of a single exposure to different levels of Avertin.
The 'standard' dose of Avertin used was similar to the dose normally given to
anaesthetise a rat for a short period, assuming that 1 ml of culture medium be
equivalent to 1 gm body weight of an intact rat.   Groups of embryos were
cultured for 24 hr in 1x, 2x, 4x and 6x the standard dose of Avertin.   Some
embryos were examined histologically at this stage, while others were cultured
for a further 24 hr.   At the end of culture, embryonic development was
assessed, and the protein content of embryos which were not examined histologically was determined.   After 48 hr in culture there was a significant
difference (p 0.0025) in the protein content between controls and embryos
cultured in 1x and 2x Avertin.   Development, assessed by somite number and
other morphological criteria, was unaffected by 1x Avertin, but was greatly
retarded in 2x Avertin (Figure 2a, b, c).   In over half the embryos in this
group which were examined histologically cell debris was present in the amniotic
cavity (Figure 2d).   This study therefore demonstrated that Avertin had a
direct, dose-dependant teratogenic effect on the early postimplantation rat
embryo.   By slightly modifying this technique, it should be possible to
determine the minimal teratogenic dose and exposure duration of a range of
anaesthetics on embryos of different gestational ages and different species.

This test system in which isolated foetuses are exposed for prolonged
periods to anaesthetics in vitro is justified on several grounds.   It provides

154

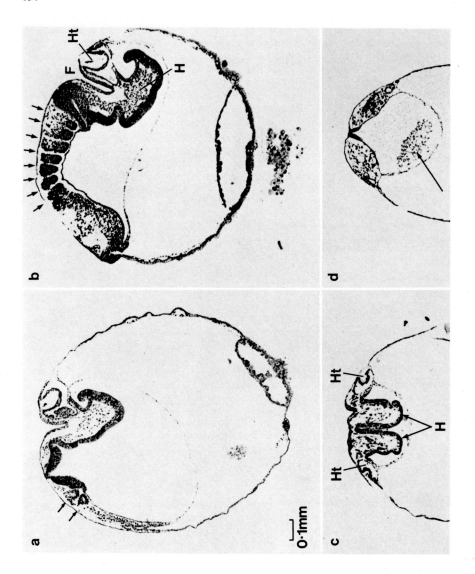

information on the direct effect of known doses of the anaesthetic on the embryo independent of secondary maternal factors such as placental permeability, and the presence of metabolic products which may be more or less teratogenic than the original agent.   Further, the organs of the foetus may be unable to detoxify and excrete many of the anaesthetic substances which he may receive from the mother, and thus be subject to these agents for a much longer period of time than the adult animal with normally functioning detoxification mechanisms (Smith, 1968).

The overall trend in these studies seems to indicate that anaesthetics can act as a positive teratogenic stimulus.   In many of these studies, pregnant animals have been chronically exposed to very high levels of anaesthetics, particularly during the early postimplantation period.   The few studies in which animals have been exposed to trace levels of halothane throughout pregnancy report a less convincing effect on foetal loss and the appearance of gross congenital anomalies in the offspring.   However, at least one study (Chang et al., 1974) indicates that subtle changes in the histo-logical appearance of the central nervous system may be induced under these conditions.   In contrast to these findings, a frank teratogenic and foetotoxic effect has been reported by Katz and Clayton (1973), but the validity of these results has been questioned because of the high rate of foetal loss present in their control series.   When higher doses of halothane were used, the reports are equally conflicting, varying from the totally negative findings of Kennedy et al. (1976) in the rat and rabbit, to the high incidence of foetal losses and skeletal anomalies, reported by Basford and Fink (1968), in the rat. In the majority of studies in which experimental animals were exposed to high levels of nitrous oxide both teratogenicity and foetotoxicity have been reported, though these studies may be criticised on the grounds that they are far removed from the clinical situation.   In the single study where rats were exposed to trace levels of nitrous oxide (Corbett et al., 1973) during the latter half of the organogenetic period, a diurnal response was observed.   No effect on pregnancy occurred when rats were anaesthetised from 2 - 10 p.m., but a significant effect on foetal mortality occurred when animals were

Figure 2 (opposite): (a) top left. Near sagittal section of conceptus cultured for 24 h in twice the standard dose of Avertin. Two poorly formed somites can be seen (arrowed).   (b) top right. Sagittal section of control conceptus cultured for 24 h. Note the larger number of well formed somites (arrowed) compared with (a).   (c and d) bottom left and right. Frontal sections of con-ceptuses cultured for 24 h in twice the standard dose of Avertin. The infused heart primordia (Ht) on either side of the head is shown in (c), and the cellular debris (arrowed) found in the amniotic cavity of experimental embryos is shown in (d). The scale is same for all figures: bar represents 0.1 mm; F, foregut; H, head; Ht, heart.

exposed from 6 a.m. - 2 p.m.   This difference in response is difficult to
explain except in terms of variations in growth of rat foetuses at different
times throughout the day.

   In the experimental models which most closely approximate to the clinical
situation, where operating room personnel are chronically exposed to trace
levels of anaesthetics, the overwhelming trend seems to be in the direction of
a negative or at most a marginally positive teratogenic effect.   However, a
much more consistently positive teratogenic effect is observed when animals are
subjected to a single acute or chronic exposure to high levels of an anaesthetic
agent.   Both situations are not strictly comparable to the conditions pre-
vailing in operating theatres, as in most instances a mixture of different
anaesthetic gases are likely to be present in varied concentrations depending
on the distance from the anaesthetic apparatus and the technique of anaesthesia
being used at any given time.

MECHANISMS OF TERATOGENESIS

   In order to understand some of the possible mechanisms of action of
anaesthetics on the conceptus, it is necessary in the first instance to briefly
consider the sites at which anaesthetics may act.   Most of the studies which
have been carried out to investigate the mode of action of anaesthetics at the
cellular level indicate that nearly all aspects of cell structure and function
are sensitive to these agents (Allison, 1971; Geddes, 1971), for example,
anaesthetics block electron transport at various sites along the respiratory
chain (Snodgrass & Piras, 1966), impair normal functioning of many enzyme
systems within mitochondria (Conney, 1967), and impair liver microsomal activity
which is intimately concerned with drug metabolism (Geddes, 1971).   Exposure
to anaesthetics or their degradation products may induce structural changes in
lipo-protein complexes in mitochondrial membranes which may in turn alter the
activity of enzymes responsible for the coupling of phosphorylation to electron
transport, and lead to depression of cellular respiratory activity (Weinbach &
Garbus, 1969).   Glucose consumption in cultured cells is generally increased
in the presence of anaesthetics, with a resultant rise in lactate levels.
Both these changes are associated with depression of cell multiplication (Fink
& Kenny, 1968).   The effects of these agents on the cell membrane have been
discussed earlier in relation to their possible effect on the microtubules and
microfilaments of the cellular cytoskeletal system (for recent review, see
Nicolson, 1976).   It seems likely that chronic exposure to these agents may
modify maternal metabolism and that this may affect the conceptus either as a
result of the production of toxic metabolites, or by a failure to produce some
basic cell constituent (Tuchmann-Duplessis, 1969; Beck, 1976).

Theoretically anaesthetics act on the mother, the foetus, and the foeto-maternal barrier. The factors which govern whether the embryo will be affected by a particular drug are complex. The means of administration of anaesthetics, whether inhalational, intravenous or intraperitoneal, and their initial and residual plasma and tissue levels may also be critical. To an extent the susceptibility of the foetus may also depend on the general condition of the mother, her dietary, nutritional and hormonal state at the time of drug admini-stration. Thus, for example, the changes which occur in the hormonal state of the mother during pregnancy may serve to increase the risk to the foetus at these times. Exposure during early pregnancy may result in genetic damage to the embryo, or, secondary to maternal factors, prevent implantation or cause early postimplantation mortality, possibly by inducing changes in the concen-tration of certain ions or other essential constituents of the oviduct or uterine luminal fluid. Most drugs cross the placental system to some extent, often reaching the embryo without being modified, and would then be in a position to affect the embryo directly. This is thought to be one of the major routes of drug action. In many instances there may be very little maternal toxicity associated with considerable embryotoxicity. Changes in response to certain drugs during pregnancy in rodents, for example, may depend on the development of the chorio-allantoic placenta (Wilson et al., 1963; Beck & Lloyd, 1966).

Essentially teratogens may act at different structural levels within the developing embryo. Thus disturbances of protein and nucleic acid synthesis, and other metabolic pathways may result when these agents act at the molecular level. Teratogens may induce defects of intra- and extra-cellular differen-tiation with resultant disturbances of cell multiplication. Cell movement and cell contacts may be affected, and this may lead to interference with normal induction processes. Interference at all these levels may, by direct or indirect means, lead to structural and/or functional defects of organogenesis (Menkes et al., 1970; Poswillo, 1976). The type and extent of the damage incurred depends to a large extent on the exposure duration. Exposure during early development may lead to the death of the embryo, whereas exposure at later stages may lead to the production of major or minor malformations, functional anomalies or retarded growth of the foetus. It is well known that differences in response to teratogenic stimuli are observed among individuals of the same strain. This would seem to be the case with operating theatre staff who are exposed to trace levels of anaesthetics, where only a small but significant proportion of those at risk appear to be affected in this way. No overall picture has, as yet, emerged to support the hypothesis that anaesthetics are particularly specific in their teratogenic action. Death usually occurs either during the preimplantation period, or early in the postimplantation period, or foetuses survive and are born with a range of predominantly minor congenital malformations. Detailed analysis of abortus material from this source may eventually demonstrate a common factor which leads to embryonic death at this stage of gestation.

CONCLUSIONS

In this review, I have attempted to put into its proper perspective the small amount of experimental work which has been carried out on ovarian oocytes and preimplantation stages of mammalian embryonic development, as I believe that it is this area rather than in the postimplantation period where more effort should be concentrated in the future. This obviously applies to both in vivo and in vitro studies, where attempts should be made to establish minimal teratogenic levels of the commonly used anaesthetic agents. Most of the clinical studies indicate that the commonest types of reproductive problems encountered in operating room personnel are associated with the significant increase in the incidence of involuntary infertility and spontaneous abortion compared to unexposed controls. Additional problems encountered include premature delivery and the birth of congenitally malformed children. That the majority of the problems encountered by operating room personnel fall into the first group indicates that the pathology, in the case of females, is either already present within the oocyte by the time ovulation occurs, or is induced shortly after conception, possibly during the preimplantation period. In a small proportion of males exposed to trace levels of anaesthetics, adverse changes probably occur during spermatogenesis.

It has generally been considered that teratogenic changes may be induced in response to chronic exposure to trace levels of anaesthetics. However, as indicated by some of the experimental studies reviewed here, a single acute exposure to anaesthetic levels of these agents may also produce similar clinical symptoms if exposure occurs during the critical periods of gametogenesis, embryogenesis or organogenesis. Until minimal teratogenic doses are established for these developmental stages, careful attention should be taken to see that the smallest possible number of people are exposed to any unnecessary risk (for recommendations, see American Society of Anesthesiologists, 1974; Department of Health and Social Security, Circular HC(76)38, 1976). If it can be convincingly demonstrated that the installation of efficient scavenging devices in all operating theatres significantly reduces the incidence of pregnancy loss and the birth of congenitally malformed children in operating theatre staff, then this is surely a small price to pay.

While there still appears to be little human data on the effect of surgery carried out during pregnancy, Smith's (1968) suggestion that, when applicable, 'spinal' (intrathecal) block rather than general anaesthesia should be carried out in early pregnancy seems, at least in theory, to be a reasonable one.

REFERENCES

ALLISON, A.C. (1971) Function and structure of cell components in relation to action of anaesthetics. In: T.C. Gray & J.F. Nunn (Eds.), General Anaesthesia, Vol. 1, Basic Sciences, Butterworths, London, pp. 1-24.

ALLISON, A.C. & NUNN, J.F. (1968) Effects of general anaesthetics on micro-tubules. A possible mechanism of anaesthesia. Lancet, ii, 1326-1329.

AMERICAN SOCIETY OF ANESTHESIOLOGISTS. (1974) Occupational disease among operating room personnel: a national study. Anesthesiology, 41, 321-340.

ANDERSEN, N.B. (1966) The effect of CNS depressants on mitosis. Acta anaesth. scand., Suppl. 22, 1-36.

ASKROG, V. & HARVALD, B. (1970) Teratogen effekt af inhalationsanaestetika. Nord. Med., 83, 498-500 (summary in English).

AUSTIN, C.R. & BRADEN, A.W.H. (1954) Induction and inhibition of the second polar division in the rat egg and subsequent fertilization. Aust. J. biol. Sci., 7, 195-210.

AXELROD, L.R. (1970) Drugs and nonhuman primate teratogenesis. In: D.H.M. Woollam (Ed.), Advances in Teratology, Vol. 4, Logos Press, London, pp. 217-230.

BAKER, T.G. (1972) Gametogenesis. Acta endocr., Copnh., Suppl. 166, 18-41.

BALAKIER, H. & TARKOWSKI, A.K. (1976) Diploid parthenogenetic mouse embryos produced by heat-shock and Cytochalasin B. J. Embryol. exp. Morph,, 35, 25-39.

BARLOW, S.M. & SULLIVAN, F.M. (1975) In: C.L. Berry & D.E. Poswillo (Eds.), Teratology: trends and applications, Springer, Berlin, pp. 103-120.

BASFORD, A.B. & FINK, B.R. (1968) The teratogenicity of halothane in the rat. Anesthesiology, 29, 1167-1173.

BECK, F. (1976) Model systems in teratology. Br. med. Bull. 32, 53-58.

BECK, F. & LLOYD, J.B. (1966) The teratogenic effects of azo dyes. In: D.H.M. Woollam (Ed.), Advances in Teratology, Vol. 1, Logos Press, London, pp. 131-193.

BERRY, C.L. & BARLOW, S. (1976) Some remaining problems in the reproductive toxicity testing of drugs. Br. med. Bull., 32, 34-38.

BIGGERS, J.D., WHITTEN, W.K. & WHITTINGHAM, D.G. (1971) The culture of mouse embryos in vitro. In: J.C. Daniel Jr. (Ed.), Methods in Mammalian Embryology, Freeman, San Francisco, pp. 86-116.

BRADEN, A.W.H. & AUSTIN, C.R. (1954) Reactions of unfertilized mouse eggs to some experimental stimuli. Expl. Cell Res. 7, 277-280.

BRUCE, D.L. (1973) Murine fertility unaffected by traces of halothane. Anesthesiology, 38, 473-477.

BRUCE, D.L. & KOEPKE, J.A. (1969) Interaction of halothane and radiation in mice: possible implications. Anesth. Analg. (Cleve)., 48, 687-694.

BUSSARD, D.A., STOELTING, R.K., PETERSON, C. & ISHAQ, M. (1974) Fetal changes in hamsters anesthetized with nitrous oxide and halothane. Anesthesiology, 41, 275-278.

BUTCHER, R.L. (1975) The role of intrauterine environment and intrafollicular aging of the oocyte on implantation rates and development. In: R.J. Blandau (Ed.), Aging gametes, their biology and pathology, S. Karger, Basel, pp. 72-97.

BUTCHER, R.L. & FUGO, N.W. (1967) Overripeness and the mammalian ova: II Delayed ovulation and chromosome anomalies, Fert. Steril., 18, 297-302.

CARR, D.H. (1965) Chromosome studies in spontaneous abortions. Obstet. Gynec., 26, 308-326.

CARR, D.H. (1971) Chromosomes in abortion. Advances Hum. Genet. 2, 201-257.

CHANG, L.W., DUDLEY, A.W., KATZ, J. & MARTIN, A.H. (1974) Nervous system development following in utero exposure to trace amounts of halothane. Teratology, 9, A-15.

COHEN, E.N., BELLVILLE, J.W. & BROWN, B.W. (1971) Anesthesia, pregnancy, and miscarriage. Anesthesiology, 35, 343-347.

COMMITTEE ON SAFETY OF MEDICINES (1974) Notes for guidance on reproduction studies (MAL 36) Department of Health and Social Security, London.

CORBETT, T.H. (1972) Anesthetics as a cause of abortion. Fert. Steril., 23, 866-869.

CORBETT, T.H. & BALL, G.L. (1971) Chronic exposure to methoxyflurane: a possible occupational hazard to anesthesiologists. Anesthesiology, 34, 532-537.

CORBETT, T.H., CORNELL, R.G., ENDRES, J.L. & MILLARD, R.I. (1973) Effects of low concentrations of nitrous oxide on rat pregnancy. Anesthesiology, 39, 299-301.

CORBETT, T.H., CORNELL, R.G., ENDRES, J.L. & LIEDING, B.A. (1974) Birth defects among children of nurse-anesthetists. Anesthesiology, 41, 341-344.

DEPARTMENT OF HEALTH AND SOCIAL SECURITY (1976) Pollution of operating departments etc. by anaesthetic gases. Health Circular HC(76)38.

EVERETT, J.W. & SAWYER, C.H. (1950) A 24-hour periodicity in the "LH-release apparatus" of female rats, disclosed by barbiturate sedation. Endocrinology, 47, 198-218.

FINK, B.R. (1971) Effect of anaesthesia on cell division. In: T.C. Gray & J.F. Nunn (Eds.), General Anaesthesia, Vol. 1, Basic Sciences, Butterworths, London, pp. 42-54.

FINK, B.R. & KENNY, G.E. (1968) Effect of halothane on cell culture metabolism. In: B.R. Fink (Ed.), Toxicity of Anesthetics, Williams and Wilkins, Baltimore, p. 37.

FINK, B.R., SHEPARD, T.H. & BLANDAU, R.J. (1967) Teratogenic activity of nitrous oxide. Nature, Lond., 214, 146-148.

FRASER, F.C., KALTER, H., WALKER, B.E. & FAINSTAT, T.D. (1954) The experimental production of cleft palate with cortisone and other hormones. J. cell. comp. Physiol., 43, suppl., 237-259.

GEDDES, I.C. (1971) Cellular metabolism in relation to anaesthesia. In: T.C. Gray & J.F. Nunn (Eds.), General Anaesthesia, Vol. 1, Basic Sciences, Butterworths, London, pp. 25-41.

HASCHKE, R.H., BYERS, M.R. & FINK, B.R. (1974) Effects of lidocaine on rabbit brain microtubular protein. J. Neurochem., 22, 837-843.

HINKLEY, R.E. & TELSER, A.G. (1974) The effects of halothane on cultured mouse neuroblastoma cells. I. Inhibition of morphological differentiation. J. Cell Biol., 63, 531-540.

JACKSON, H. (1970) Antispermatogenic agents. Br. med. Bull. 26, 79-86.

JOHNSON, M.H., EAGER, D., MUGGLETON-HARRIS, A. & GRAVE, H.M. (1975) Mosaicism in organization of concanavalin A receptors on surface membrane of mouse egg. Nature, Lond., 257, 321-322.

KATZ, J. & CLAYTON, W. (1973) Fetal mortality in rats chronically exposed to low concentrations of halothane. Proc. Am. Soc. Anesth., Annual Meeting, pp. 57-58.

KAUFMAN, M.H. (1973a) Parthenogenesis in the mouse. Nature, Lond., 242, 475-476.

KAUFMAN, M.H. (1973b) Timing of the first cleavage division of haploid mouse eggs, and the duration of its component stages. J. Cell Sci., 13., 553-566.

KAUFMAN, M.H. (1975) Parthenogenetic activation of mouse oocytes following avertin anaesthesia. J. Embryol. exp. Morph., 33, 941-946.

KAUFMAN, M.H. (1976) Effect of anaesthesia on the outcome of pregnancy in female mice. J. Reprod. Fert. (in press).

KAUFMAN, M.H. & STEELE, C.E. (1976) Deleterious effect of an anaesthetic on cultured mammalian embryos. Nature, Lond., 260, 782-784.

KENNEDY, G.L., SMITH, S.H., KEPLINGER, M.L. & CALANDRA, J.C. (1976) Reproductive and teratologic studies with halothane. Toxicol. and Applied Pharmacol., 35, 467-474.

KIRSCHNER, M.W. & WILLIAMS, R.C. (1974) The mechanism of microtubule assembly in vitro. J. Supramol. Struct., 2, 412-428.

KNILL-JONES, R.P., NEWMAN, B.J. & SPENCE, A.A. (1975) Anaesthetic practice and pregnancy. Controlled survey of male anaesthetists in the United Kingdom, Lancet, ii, 807-809.

KNILL-JONES, R.P., RODRIGUES, L.V., MOIR, D.D. & SPENCE, A.A. (1972) Anaesthetic practice and pregnancy. Controlled survey of women anaesthetists in the United Kingdom. Lancet, i, 1326-1328.

LANSDOWN, A.B.G., POPE, W.D.B., HALSEY, M.J. & BATEMAN, P.E. (1976) Analysis of fetal development in rats following maternal exposure to subanaesthetic concentrations of halothane. Teratology, 13, 299-304.

LENCZ, L., & NEMES, C. (1970) Morbidity of Hungarian anesthesiologists in relation to occupational hazards. In: 3rd European Congress of Anesthesiologists. Prague.

LEVAN, A. (1938) The effect of colchicine on root mitoses in Allium. Hereditas, 24, 471-486.

LINDE, H.W. & BRUCE, D.L. (1969) Occupational exposure of anesthetists to halothane, nitrous oxide and radiation. Anesthesiology, 30, 363-368.

MAZIA, D. (1961) Mitosis and the physiology of cell division. In: J. Brachet & A.E. Mirsky (Eds.), The Cell. Vol. 3., Academic Press, New York, pp. 77-412.

McINTOSH, J.R., HEPLER, P.K. & VAN WIE, D.G. (1969) Model for mitosis. Nature, Lond., 224, 659-663.

MIKAMO, K. & HAMAGUCHI, H. (1975) Chromosomal disorder caused by preovulatory overripeness of oocytes. In: R.J. Blandau (Ed.), Aging gametes, their biology and pathology, S. Karger, Basel, pp. 72-97.

MENKES, B., SANDOR, S. & ILIES, A. (1970) Cell death in teratogenesis. In: D.H.M. Woollam (Ed.)., Advances in Teratology, Vol. 4, Logos Press, London, pp. 169-215.

NEW, D.A.T. (1971) Methods for the culture of post-implantation embryos of rodents. In: J.C. Daniel, Jr. (Ed.), Methods in Mammalian Embryology. Freeman, San Francisco, pp. 305-319.

NEW, D.A.T., COPPOLA, P.T. & TERRY, S. (1973) Culture of explanted rat embryos in rotating tubes. J. Reprod. Fert., 35, 135-138.

NICOLSON, G.L. (1976) Transmembrane control of the receptors on normal and tumor cells. Biochim. biophys. Acta, 457, 57-108.

NICOLSON, G.L., SMITH, J.R. & POSTE, G. (1976) Effects of local anesthetics on cell morphology and membrane-associated cytoskeletal organization in Balb/3T3 cells. J. Cell Biol., 68, 395-402.

PARBROOK, G.D., MOBBS, I. & MACKENZIE, J. (1965) Effects of nitrous oxide on the early chick embryo. Br. J. Anaesth., 37, 990-991.

POSWILLO, D. (1976) Mechanisms and pathogenesis of malformation. Br. med. Bull., 32, 59-64.

SETALA, K. & NYYSSONEN, O. (1964) Hypnotic sodium pentobarbital as a teratogen for mice. Naturwissenschaften, 51, 413.

SHEPARD, T.H. & FINK, B.R. (1968) Teratogenic activity of nitrous oxide in rats. In: B.R. Fink (Ed.), Toxicity of Anesthetics, William & Wilkins, Baltimore, pp. 308-323.

SHNIDER, S.M. & WEBSTER, G.M. (1965) Maternal and fetal hazards of surgery during pregnancy. Am. J. Obstet. Gynec., 92, 891-900.

SMITH, B.E. (1968) Teratogenic capabilities of surgical anaesthesia. In: D.H.M. Woollam (Ed.), Advances in Teratology, Vol. 3, Logos Press, London, pp. 127-180.

SMITHELLS, R.W. (1976) Environmental teratogens of man. Br. med. Bull., 32, 27-33.

SNODGRASS, P.J. & PIRAS, M.M. (1966) The effects of halothane on rat liver mitochondria. Biochemistry, N.Y., 5, 1140-1149.

STEINHARDT, R.A. & EPEL, D. (1974) Activation of sea-urchin eggs by a calcium ionophore. Proc. natn. Acad. Sci. U.S.A., 71, 1915-1919.

STEINHARDT, R.A., EPEL, D., CARROLL, E.J. & YANAGIMACHI, R. (1974) Is calcium ionophore a universal activator for unfertilised eggs? Nature, Lond., 252, 41-43.

TARKOWSKI, A.K. (1966) An air-drying method for chromosome preparations from mouse eggs. Cytogenetics, 5, 394-400.

TELZER, B.R., MOSES, M.J. & ROSENBAUM, J.L. (1975) Assembly of microtubules onto kinetochores of isolated mitotic chromosomes of HeLa cells. Proc. natn. Acad. Sci. U.S.A., 72, 4023-4027.

THIBAULT, C. (1949) L'oeuf des Mammiferes: son developpment parthenogenetique. Annls. Sci. nat. (Zool.), 11, 133-219.

TUCHMANN-DUPLESSIS, H. (1969) Reactions of the foetus to drugs taken by the mother. In: G.E.W. Wolstenholme & M. O'Connor (Eds.), Foetal Autonomy. Churchill, London, pp. 245-270.

VAISMAN, A.I. (1967) Working conditions in surgery and their effect on the health of anesthesiologists. Eksp. Khir Anesteziol.,12 (3), 44-49 (Eng. trans.)

WEINBACH, E.C. & GARBUS, J. (1969) Mechanism of action of reagents that uncouple oxidative phosphorylation. Nature, Lond., 221, 1016-1018.

WEISENBERG, R.C. (1972) Microtubule formation in vitro in solutions containing low calcium concentrations. Science, N.Y., 177, 1104-1105.

WHITTINGHAM, D.G. (1971) Culture of mouse ova. J. Reprod. Fert., Suppl. 14, 7-21.

WILSON, J.G. (1954) Influence on the offspring of altered physiologic states during pregnancy in the rat. Ann. N.Y. Acad. Sci., 57, 517-525.

WILSON, L., BRYAN, J., RUBY, A. & MAZIA, D. (1970) Precipitation of proteins by vinblastine and calcium ions. Proc. natn. Acad. Sci., U.S.A., 66, 807-814.

WILSON, J.G., SHEPARD, T.H. & GENNARO, J.R. (1963) Studies on the site of teratogenic action of C^{14}-labelled trypan blue. Anat. Rec., 145, 300 (Abstract).

WITTMAN, R., DOENICKE, A., HEINRICK, H. & PAUSCH, H. (1974) Abortive effect of halothane. Anesthetist, 23, 30-36.

THE ROLE OF THE ENDOMETRIUM IN BLASTOCYST IMPLANTATION

J.E. O'Grady and S.C. Bell

Department of Biochemistry,
University of Strathclyde,
The Todd Centre,
31 Taylor Street,
Glasgow, G4 ONR, U.K.

The development of the mammalian zygote depends, after a period of independent existence, on the formation of an intimate contact between the cells of the trophoblast and the tissues of the uterus. In order for this association to be initiated, at the time of implantation, and for the foetal and maternal functions of the implant to be satisfied, the blastocyst and uterine endometrium must be preconditioned to be in a state of mutual sensitivity. The cells lining the uterus, the luminal epithelium and their invaginated extension the glandular epithelium, and the stromal cells which form a matrix for the support of the glands, are the major cell types present in the endometrial portion of the rodent uterus (Figure 1). Luminal epithelial differentiation must be such that, at the time of implantation, it may effect blastocyst activation, be sensitive to the stimuli from the blastocyst and in response to such a stimulus be capable of transmitting a decidual stimulus to the underlying stroma. In its turn, the stroma must be competent to respond to the decidual stimulus if the subsequent development of the maternal portion of the placenta is to be secured. Much of the experimental work in this field has been performed in rodents, most frequently the mouse and rat. In these animals implantation is asymmetric, at the anti-mesometrial pole of the uterus, and it is the subepithelial anti-mesometrial region of the stroma (the SEAM cell population) which is initially competent to decidualize. The development of coordinated endometrial sensitivity and blastocyst activation is dependent on the secretion of a specific regimen of ovarian hormones during the pre-implantation period. This consists of the proestrous oestrogen associated with follicular maturation and ovulation, and at least 48 hours of luteal progesterone which is subsequently overlaid by a critical concentration of nidatory oestrogen. In rodents, with short cycles, luteal progesterone falls mainly outside the oestrous cycle, but in man and many other species this sequence of hormones is a normal concomitant of the ovarian cycle. In this review, the effects of hormones secreted during early pregnancy and of exogenous hormone

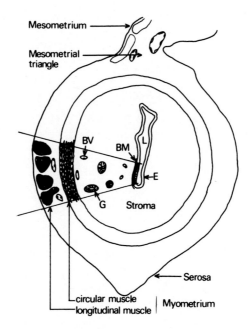

Figure 1: Schematic transverse section of the rodent uterus; L = uterine lumen;
E = luminal epithelium; BM = basement membrane; G = glandular epithelium;
B.V. - blood vessel.

administration will be related to the ultrastructure, cellular division and the
differentiation of the rodent endometrium.   The relation of these phenomena to
biochemical changes in the endometrium will also be discussed.   This data
serves as a background to the delineation of a model for the development of
rodent endometrial sensitivity during the preimplantation period and the subsequent decidualisation of the stroma.   This analysis employs recent concepts
in the fields of cellular proliferation and differentiation.   Finally, the
relevance of this model to implantation in the human uterus will be discussed.

STRUCTURAL CHANGES ASSOCIATED WITH THE PRE- AND POST-IMPLANTATION PERIODS OF PREGNANCY

In this section, the morphological and ultra structural changes taking
place in the endometrium prior to the arrival of the blastocyst in the uterus,
and during its subsequent attachment and invasion of the endometrium will be
described.   An attempt at a comprehensive coverage of these events, drawn as
they are from a number of studies, is hindered due to the different nomenclature adopted by research groups, strain differences and timing reference
used.

Analysis of events will be considered in two sections; those occurring prior to attachment of the blastocyst to the epithelium, and those occurring after attachment. The attachment reaction will be considered to be the first step in the series of events leading to implantation for two reasons. The attachment reaction represents (1) the first observed morphological and physiological change in the luminal epithelium, and (2) the first changes observed after the initiation of implantation by estradiol in rodents under conditions of "delayed implantation", where the development and implantation of the blastocyst is arrested (Nilsson, 1974; Psychoyos, 1973a,b).

Details have been abstracted from studies of the ultrastructure of implantation in rats (Allen, 1931; Enders & Schlafke, 1967, 1969; Jollie & Benscombe, 1965; Krehbiel, 1937; Lobel et al., 1965a,b,c; Ljungkvist, 1972; Nilsson, 1966a,b, 1967, 1970, 1974; Potts, 1966, 1969; Potts & Psychoyos, 1967; Tachi et al., 1970).

Structural changes associated with the preimplantation period of pregnancy

In this survey, the day following mating, which occurs between 24.00 to 02.00 h, was designated day 1 of pregnancy. As many of the observable changes in early pregnancy are similar to those occurring if mating had not taken place and the animal had entered another oestrus cycle, we will equate early pregnancy to the cycle as follows: day 1 = oestrus; day 2 = Dioestrus 1; day 3 = Dioestrus 2; day 4 = Pro-oestrus.

Epithelial tissues

The luminal epithelial cells are columnar, resting on a basement membrane which represents the thin basement lamina and condensed connective tissue fibres, the latter being tropocollagen aggregates (Rowlatt, 1969). The free cell surface contains variable numbers of short microvilli ($1-2\mu$ long and 0.1μ wide). A thin zone of homogeneous, slightly electron dense, finely granular cytoplasmic material, about 1μ deep, is associated with the villi and is free of cytoplasmic residues. An extracellular deposit can be identified on the luminal surface of the microvilli ("fuzz" of Lawn, 1973). Bulbous cytoplasmic projections sometimes appear between the microvilli.

Neighbouring epithelial cells are joined by junctional complexes (Farquahar & Palade, 1963). They probably act as a seal around the apex of the cells preventing direct passage of material between the lumen and the stroma and may be important in intercellular communication (Loewenstein, 1966). At the base of the microvilli are numerous vesicular-appearing structures variously designated caveolae, acanthosomes, micropinocytotic vesicles etc. They are of irregular shape, indented by tongues of cytoplasm and are not characteristic of simple vesicles (Enders & Schlafke, 1967). These are thought to be of

pinocytotic origin (Vokaer, 1952).   In the remaining apical cytoplasm are
abundant cytoplasmic vacuoles (containing secretory products), mitochondria,
prominent Golgi apparatus and a few small lysozomes.   The nucleus occupies the
middle third of the cell, while the basal part is packed with lipid droplets
and mitochondria.

Various changes occurring in the epithelial tissues during the Oestrus
cycle (variations in glycogen, lipid, lysozomes, cellular proliferation, etc)
also occur during the preimplantation period, up to day 4.   Thus, epithelial
cellular division is found on both Dioestrus 1 and the corresponding day 2 of
pregnancy, (Figures 2, 4).   A similar situation pertains to the glandular
epithelial division observed on day 2 (rat - Chaudhury & Sethi, 1970; Clark,
1971; O'Grady et al., 1974; Marcus, 1974a; Tachi et al., 1972; mouse - Finn &
Martin, 1967; Hall, 1969).

Stromal tissues

The stroma cells are fibroblasts which form from undifferentiated
mesenchymal cells.   The stroma immediately around the anti-mesometrial
luminal epithelium is rich in fibroblasts (this zone is referred to by the
authors as the sub-epithelial anti-mesometrial stroma - SEAM stroma and
corresponds to the primary decidual zone of Krehbiel, 1937).   These stromal
cells have an irregular outline, spindly-shaped due to cytoplasmic projections
passing into large areas of extracellular space, have relatively little cytoplasm and rarely make contact with neighbouring cells.   They possess mitochondria, microvilli, some lipid droplets, a small Golgi apparatus, abundant
free ribosomes and a large irregularly shaped nucleus with evenly distributed
chromatin and prominent nucleoli.   Large amounts of extracellular collagen is
found in this area (Enders & Schlafke, 1969).

On day 4 of pregnancy and pseudopregnancy, the stroma undergoes cellular
mitosis (rat - Chaudhury & Sethi, 1970; Marcus, 1974b; O'Grady et al., 1974;
Tachi et al., 1972; mouse - Finn & Martin, 1967; Hall, 1969; Zhinkim &
Samoshkina, 1967) (Figures 3, 4).   In the rat, this mitosis which is first
found on the evening of day 3, is restricted to the crescent-shaped sub-
epithelial anti-mesometrial area of the stroma, i.e. SEAM stroma (Marcus 1974a;
O'Grady et al., 1976; Tachi et al., 1972).   In the mouse, however, Finn and
Martin (1967) report that mitoses were randomly distributed in the stroma with
only slight tendency to cluster around the lumen.   Concomitant with stromal
mitosis, Marcus (1974a) has reported myometrial mitosis.   Stromal and myometrial
mitosis also occurs in Pro-oestrus, but to a more limited extent with respect
to stromal mitosis, and is similar to day 4 of pregnancy.   According to one
estimation (J.E. O'Grady & S.C. Bell - unpublished results) during day 4 the
total SEAM population replicates.

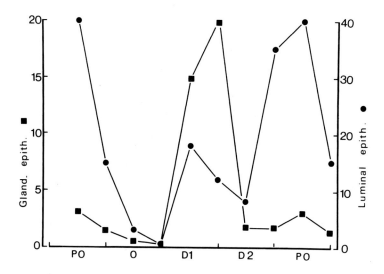

Figure 2: Luminal and glandular epithelial mitosis during the rat oestrous cycle. All values are mitoses per uterine section per 3 h. PO = Proestrus; O = Oestrus; D1,2 = Dioestrus 1 and 2. Data adapted from Marcus (1974).

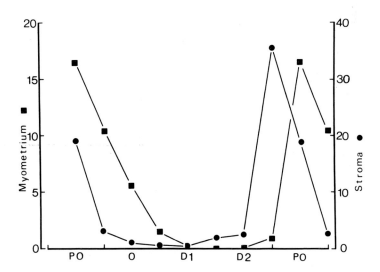

Figure 3: Myometrial and stromal mitosis during the rat oestrous cycle. Values and abbreviations are as indicated in legend to Figure 2.

Jollie and Benscombe (1965) postulated that SEAM stroma contains two stromal populations on day 5,characterized according to their possession of a type of endoplasmic reticulum. The first population possessed narrow somewhat tortuous cisternal tubules which are rough surfaced and appear filled with a homogeneous material of low electron density. The second population contained endoplasmic reticulum of the smooth variety, dilated vascular cisternae which are smooth surfaced and appear optically empty. Leroy and Galand (1969) have also observed two types of stromal cells, classified according to the geometry of their nuclei. One type possessing swollen, round-shaped nuclei, the other containing smaller, irregular "kidney-like" nuclei. The relationship between the populations observed in these two reports is uncertain.

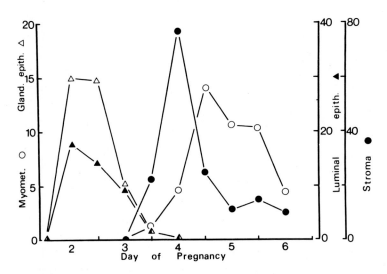

Figure 4: Luminal and glandular epithelial, myometrial and stromal mitosis during the first 6 days of pregnancy in the rat. All values are mitoses per uterine section per 3 h. Data adapted from Marcus (1974).

Blastocyst-epithelial interactions

Egg transport through the oviduct takes 3-4 days in most species (Allen & Doisy, 1923; Blandau, 1961; Hendrickx & Kraemer, 1968; Humphrey, 1968). In the rat, the eggs enter the uterus in the morula or the blastocyst stage late on day 4 of pregnancy at 19.00 h according to Psychoyos (1973b), and morning of day 4 according to Major and Heald (1974). The blastocysts at this stage, still surrounded by the zona pellucida, are probably free-floating in the lumen or tenuously adherent to the uterine epithelium. The uterine lumen is still open and the luminal surface of the epithelial cells is made up of many microvilli bathed in the luminal fluid.   On the morning of day 5, the "first stage of closure" (Finn, 1971) begins, giving rise to the apposition of opposite surfaces of the luminal epithelium to form interlocking villi.   Between noon and 18.00 h on day 5, most of the blastocysts, located in anti-mesometrial crypts, have lost their zona pellucida (Lungkvist & Nilsson, 1974; Surani, 1975a) and this enables the trophoblast to make contact with the luminal epithelium.   The plasma membrane of the trophoblast comes to rest on the microvillous processes and ruffles of the epithelial cells.   The microvilli of the epithelium are shorter, and "bleb-like" (Tachi et al., 1970) or bulbous structures (Potts, 1969; Psychoyos, 1973a,b) formed by the epithelial cells are seen in contact with the blastocyst.   The extracellular space is reduced to 0.5μ or less and contains no visible deposit, and at the points of contact between trophoblast and epithelium a 20 mμ gap intervenes.   The apical cytoplasm of the epithelial cells is vacuolated and rich in mitochondria, and a fibrous core is present in the villi.   Just before noon of day 5, in pregnant and pseudopregnant rats, the nuclei of the epithelial cells begin to change their position and are found several hours later close to the epithelial basal membranes (Allen, 1931; Vokaer, 1952).   Large lipid droplets (Alden, 1947) that were adjacent to the based pole of the nuclei disappear and the cytoplasm is filled with small vacuoles giving it a spongy appearance.

This contact has been termed the "pre-attachment" phase (Larsen, 1975; Potts, 1969; Reinius, 1967) and corresponds to the "first stage of closure" in areas of the uterus not containing blastocysts, or "stage of apposition" (Enders & Schlafke, 1969; Schlafke & Enders, 1975) or "state of pre-sensitivity" and "state of sensitivity" (Pollard & Finn, 1972).   The pre-attachment phase is morphologically similar to the state in which blastocysts and the epithelium are arrested in conditions of delayed implantation (Nilsson, 1974).   The epithelium is covered by numerous regular, short microvilli with occasional thin pseudopodia-like protrusions (cf. "bleb-like" and "bulbous" structure) among them (Psychoyos & Mandon, 1971).   During the delay, the plasma membranes of the trophoblast and epithelial cells are generally separated by a distance of 1μ , and the trophoblast cells occasionally possess microvilli similar to those of the epithelium.

Structural changes associated with the post-implantation period of pregnancy

In this section the ultrastructural events occurring at and after "attachment" which culminate in apposition of the trophoblast against the differentiated stromal or decidual cells are described.   During these periods, the two tissues, the luminal epithelium and the stroma, undergo precise temporally related morphological changes.   The luminal epithelium undergoes changes resulting in its dissolution, initiated by interactions with the trophoblast.

Growth and differentiation occurs in the stroma and are integral parts of the morphogenesis of the deciduoma.

Blastocyst - epithelial interactions

The interactions have been divided into three temporally related stages - attachment, adhesion/fusion and invasion.

(1) Attachment. The transformation of the pre-attachment to the attachment reaction is attended by reorganisation and differentiation of the luminal epithelium. During this stage the microvilli become more squat, bulbous and assume irregular profiles (Nilsson, 1966b, 1970; Potts, 1969) until the villous pattern is lost. Larger bulbous cytoplasmic projections occur and the overall junctional zone assumes a more serpentine contour (Enders & Schlafke, 1969; Mayer et al., 1967; Potts, 1966). An electron microscopic study of the epithelial surface on 15.00 h on day 5 reveals that these projections are well-organized formations, consisting of several flat and irregular tentacles formed by, and projecting from, a basal body with a hollowing centre (Psychoyos & Mandon, 1971), and are particularly abundant on the anti-mesometrial luminal surface (Psychoyos, 1973b). In the afternoon of day 5 of pregnancy, epithelial cell organelles change; the finely granular cytoplasm related to the luminal plasma membrane disappears, vesicles become fewer, lysosomes enlarge and become more numerous and the cell lipid spreads into the supranuclear zone. This membrane "attachment reaction" is suggested to be equivalent to the reaction of the "second stage of closure" (Pollard & Finn, 1972) observed in areas of the uterus not containing blastocysts. Studies by Pollard and Finn in the mouse, have demonstrated that the epithelial differentiation observed, is temporally advanced in implantation areas suggesting that the blastocyst augmented the differentiation of the epithelium. However, it was also observed that the effect was induced by oil. This morphological state is also equivalent to the "state of insensitivity" when the uterus is insensitive to decidual stimuli (blastocyst or oil) and follows the "state of sensitivity" (Pollard & Finn, 1972) (See Figure 27).

(2) Adhesion and fusion. In the adhesion phase, the juxtaposed epithelial and trophoblast membranes run parallel for most of their course, separated only by a 20-25 mμ gap, instead of being in contact at limited sites as previously. Where an extracellular space exists, it is filled with a homogenous deposit which stains heavily with electron dense stains. The epithelial cells contain numerous irregularly shaped, clear vesicles (Tachi et al., 1970). Occasionally the trophoblast cytoplasm entirely surrounds cytoplasmic processes of the uterine epithelium (Tachi et al., 1970). Submicroscopic intercellular contacts have been observed, which Potts (1966) described as 'septate desmosomes' and Tachi et al., as 'tight junctions'. Enders and Schlafke (1969) described these

as 'punctate desmosomes' and 'primitive junctional complexes' in the
implantation sites of the rabbit, the ferret and the bat.   Whereas in the pre-
and attachment stages, no distinction can be made in the ultrastructure of the
uterine epithelium facing the uterine cleft and that adjacent to the blastocyst
(except that of temporal development), the events of adhesion are unique to the
area containing the implanting blastocyst.

Fusion of the epithelial and trophoblast cytoplasms following rupture of
their plasma membranes has been described in the rabbit (Larsen, 1961; Glenister,
1965) and the mouse (Potts, 1966).   However, no conclusive evidence was found
by Tachi et al. (1970) in the rat, who point out that proof of cell fusion by
microscopy is often equivocal, since cell boundaries may appear to vanish where
the phase of sectioning is tangential to the cell membrane.   Although precise
timing of these events is problematic, indeed implantation sites in the same
uterine horn have been found to be at different morphological stages, these
stages are over by the noon of day 6 when invasion proper occurs.

(3) Invasion.   After completion of its attachment to the uterine epithelium,
the blastocyst starts to invade the endometrium.   The uterine epithelial cells
at the invasion site contain large fat droplets, and the clear vesicles normally
seen in the apical area are much reduced in number.   Beneath the epithelial
cells, between the basement membrane and the stromal cells, thick bundles of
fibres, resembling collagen, are present at this stage (Tachi et al., 1970).
The earliest sign of trophoblastic invasion can already be recognized by noon
of day 6.   Prior to invasion, the blastocyst pulls slightly from the uterine
surface, showing the imprint of the area which it previously occupied (Schlafke
& Enders, 1975).   This process may be related to secretion.   The subsequent
invasion exhibited by rodents has been called "displacement implantation", in
which the uterine epithelium is readily released from the underlying basement
membrane (basal lamina), facilitating the spread of the trophoblast through and
beneath the epithelium.   Initially, during day 6, the trophoblastic cytoplasm
insinuates itself between adjacent epithelial cells until it reaches the basement membrane (Enders & Schlafke, 1967; Nilsson, 1967; Potts, 1966, 1969;
Tachi, et al., 1970).   Many large inclusion bodies of high electron density are
observed in the trophoblast cells which have been suggested to represent
engulfed, necrotic epithelial cells.   Finn and Lawn (1968) suggest that individual cells which loosen and slough off are phagocytosed by the trophoblast,
prior to any penetration of the epithelial layer per se.   Hinchliffe and El-
Shershaby (1975) have demonstrated that loss of the epithelium in the mouse is
a result of autolytic changes, thus loss occurs in the presence of oil.   The
trophoblast would then phagocytose the degenerating epithelial cells.

During the displacement the trophoblast which remains cellular sends projections extending mesometrially between the basement membrane and the overlying

epithelial cells at the margin of the advancing embryonic pole of the blastocyst.
In these initial stages of invasion (until the early morning of day 7), the
trophoblastic giant cells or cells forming the lateral lining of the blastocyst
cavity, eliminate the uterine epithelial cells, but leave the basement membrane
temporarily intact.   This presents a transient block to the progress of the
trophoblast.   The continuous finely granular deposit of the basement membrane
then swells, and the associated collagen fibres become more numerous as the
trophoblast advances and penetrates the membrane.   By the noon of day 8, the
basement membrane has vanished, and trophoblast cells appear to be in intimate
contact with stromal elements without any detectable intervening fibrinoid
material (Tachi et al., 1970).

Blastocyst-stroma interaction

The first structural change observed at the early attachment phase in the
stroma, is the oedema in the deeper sub-epithelial layers (oedema zone),
resulting from increased capillary permeability (Tachi et al., 1970; Psychoyos,
1960, 1961, 1969a,b; Finn & McLaren, 1967).   This marked change in permeability
can be visualized by the injection of macromolecular dyes (Evan's blue, Greig's
blue or Pontamine blue), and the presence of macroscopically visible blue areas
in the endometrium signify the presence of an implanting blastocyst.

The stromal cells underlying the epithelium in the lateral and anti-
mesometrial areas (SEAM Stroma), which corresponds to the area of proliferation
on day 4 of pregnancy, now align themselves with the lumen and form a distinct
cup around the luminal epithelium.   The alignment of SEAM stromal cells appears
to be dependent upon the degradation of the intercellular collagen fibres
(Fainstat, 1963; Lobel et al., 1965b,c) which loosen and disintegrate at day 5
and disappear by day 6.   The initial breakdown of collagen and the oedema may
be related to the subsequent morphogenetic processes in the stroma.   Grobstein
and Cohen (1965) suggest that collagen "jackets" certain areas, preventing
morphogenesis.   Removal of collagen may be a prerequisite to morphogenesis in
the uterus.   The collagen is replaced by fine argyrophil fibres (Fainstat, 1963).
The SEAM stromal cells now swell, leaving a minimum of extracellular space, and
at places form extensive intercellular junctional complexes (Potts, 1969; Enders
& Schlafke, 1967), resulting in close apposition of their cellular membranes.
The continuous layer leaves a connective tissue cleft between the basement
membrane of the epithelium and the surface of the decidualizing stroma.   It is
the cells of this layer which are the first to exhibit the characteristic
morphological changes associated with decidualization, at 20 h after increased
capillary permeability, and which eventually form the primary decidual zone
(Krehbiel, 1937), starting on the evening of day 6, and reaching completion when
the trophoblast makes contact.   The process of decidualization in the primary

zone (mouse - Finn & Martin, 1967; rat - Krehbiel, 1937) which occurs between
noon day 5, and the evening of day 6 of pregnancy, will be examined later. The
third zone, in the deep stroma, is the zone of sprouting capillaries which are
directed toward the lumen of the uterus (after the classification of Lobel et
al., 1965b,c).

Another distinct morphological observation is the presence of high mitotic
indices in the stroma, corresponding to the oedema zone, i.e. surrounding the
primary decidual zone.   This region subsequently forms the secondary decidual
zone (Krehbiel, 1967).   No mitoses are, however, observed in the primary
decidual zone at this time (mouse - Finn & Martin, 1967; rat - Tachi & Tachi,
1975; Tachi et al., 1972).

These changes precede distinct and defined morphogenetic events leading to
the full development of the decidua and these events are described in the
excellent papers of Krehbiel (1937); Jollie and Benscombe (1965) and Lobel et
al. (1965c).

Decidualization

Decidualization, the process of differentiation of the stromal cell to a
decidual cell, appears to begin during the attachment stage of implantation, in
the subepithelial anti-mesometrial stroma (SEAM stroma).   During the early
attachment stage, the stromal nuclei assume a smoother, more spherical shape,
have prominent and highly granulated nucleoli often located in close proximity
to the nuclear membrane.   Multiple nucleoli (2-3) are observed.   In late
attachment stage, chromatin is condensed under the nuclear membrane, the nuclei
are often polyploid (tetraploid most common), and at the same time the cisternae
of the rough-surfaced endoplasmic reticulum form arrays.   One of the striking
phenomenon observed in the stromal cells during attachment, is the appearance
of numerous polyribosomes.   Polyribosomes are distinctly coil-shaped or v-
shaped and consist commonly of ten to fifteen ribosomes (Tachi et al., 1972) and
are situated at the periphery of the dilated cisternae of the endoplasmic
reticulum and around the outer nuclear membrane.   From the early attachment
phase, fibriller material appears as loosely gathered fasicles of fine filaments
about 50-90 A in diameter (Jollie & Benscombe, 1965; Enders & Schlafke, 1967;
Tachi et al., 1970) and probably about 1μ in length.   During the early phase,
sparse amounts are concentrated in the peri-nuclear region and the periphery of
the cytoplasm, where they are orientated parallel to the surface of the
implantation chamber.   As decidualization proceeds, synthesis increases, until
they become packed in highly regulated, parallel bundles, and in some cells
displace typical cytoplasmic organelles (up to day 8); this is especially true
of those decidual cells in the primary decidual zone in the invasive stage of

implantation (Tachi et al., 1970). The chemical nature of the decidual
fibrillar material is unknown.

The fully formed decidual cell is characterized by the predominance of the
smooth surfaced cisternae of the endoplasmic reticulum, which are distended and
contain fine flocculent material (Jollie & Benscombe, 1965). It is known that
smooth surfaced endoplasmic reticulum is associated with glycogen synthesis
(Milloning & Porter, 1960; Parker & Bonneville, 1963). Glycogen, in fact,
which first appears in cells on day 7, appears as irregular-shaped granules,
approximately 100 A in diameter, diffusely spread in the cytoplasm;
occasionally, where especially abundant, they appear to be related to the
cisternae of the endoplasmic reticulum. Glycogen synthesis continues concomitant with the process of decidualization. These primary decidual cells are
postulated to arise from the population of stromal cells on day 5, containing
smooth endoplasmic reticulum (Jollie & Benscombe, 1965). Lipid droplets also
make their appearance in native decidual cells and may be found in association
with glycogen granules (Enders & Schlafke, 1967). Mitochondria in decidual
cells on day 6 appear to "balloon" (Jollie & Benscombe, 1965) but return to a
normal appearance later. Lysosomes also first appear on day 6 and become
increasingly numerous in mature decidual cells, suggesting impending autolysis.
The "microbodies" identified in these cells late on day 5 are interpreted by
Jollie and Benscombe (1965) as being their immature forms (Porter & Bonneville,
1963). Their presence may be linked to the ultimate fate of the decidual cells
of the primary zone, which is autolysis and phagocytosis by the expanding
embryo in the implantation chamber.

An important ultrastructural feature of decidual cells is the presence of
junctional complexes between neighbouring cells, which have been described as
desmosomes (Jollie & Benscombe, 1965; Enders & Schlafke, 1967). After approximately 48 hrs of decidual growth, decidual cells with varying degrees of polyploidy are observed, predominantely tetraploid and octaploid (Leroy et al., 1974a;
Zybina & Grishenko, 1972).

The properties described are usually ascribed to the SEAM population of
decidual cells which eventually are phagocytosed. Two other types of decidual
cell with different ultrastructural properties, the mesometrial cells forming
the decidua basalis part of the definitive placenta and the mesometrial triangle;
and the deeper anti-mesometrial decidual cells forming the decidua capsularis,
are described by Krehbiel (1937). All presumptive decidual cells, however,
appear to undergo extensive hyperplasia prior to their differentiation as
decidual cells.

177

HORMONAL PRECONDITIONING OF THE UTERUS FOR BLASTOCYST IMPLANTATION

The preparation of the uterus for implantation and the attainment of
maximal sensitivity to the blastocyst and other deciduomatic agents has been
investigated by several workers using the hormone treatment of ovariectomised
animals (Psychoyos, 1967; DeFeo, 1967; Finn & Martin, 1972). Progesterone and
oestrogen are necessary for blastocyst implantation and the development of
deciduomata in response to intraluminally injected oil, and the sequence,
amount and timing of the administration of these hormones is critical to the
development of the sensitivity of the uterus and its subsequent refractory condition. This condition requires that there has been an initial period of
oestrogen dominance followed by progesterone, accompanied after at least 48 hr
in the rat (Psychoyos, 1966a,b) by the presence of further oestrogen (nidatory
oestrogen). In many animals the release of these hormones occurs within the
period of the oestrous cycle, the latter release of oestrogen being in the
luteal phase of the cycle. By this process, the appropriate environment for the
implantation of the blastocyst is ensured within the length of each cycle. In
the mouse and the rat, the species in which most of the basic work on
implantation has been done, the luteal phase of the cycle has been attenuated
and the secretion of "luteal" progesterone and nidatory oestrogen fall outside
the oestrous cycle, nevertheless the initial release of oestrogen is a normal
concomitant of the cycle. In this section, the levels of plasma progesterone
and oestrogen and the concentrations of their uterine receptors found during the
oestrous cycle and pregnancy in rodents will be described. The hormonal changes
in early pregnancy will then be related to alterations in endometrial structure,
cellular division and sensitivity.

The levels of progesterone and oestradiol and their receptors during the
oestrous cycle and pregnancy

The synchronisation of the events leading to the preparation of the
endometrium and blastocyst for implantation is maintained by the secretion of
ovarian steroid hormones. The pattern of release of these hormones has been
the subject of considerable speculation. In particular, the duration and
amount of nidatory oestrogen secretion in the rat, which is critical for the
initiation of blastocyst implantation, has only recently been established and
the necessity for such a secretion in several other species remains controversial.
The majority of the actions of progesterone and oestradiol are mediated by the
formation and maintenance of an intracellular steroid hormone-receptor complex.
The concentrations of the endometrial receptors and their steroid hormone complexes and the mechanism of their generation during early pregnancy are not
firmly established. No investigation has been made of the distribution of

such receptors between the individual cell populations of the endometrium during early pregnancy.

Plasma progesterone and oestrogen levels during the oestrous cycle of the rat

Barraclough et al. (1971) have determined both the oestrous cycle secretion of progesterone and the levels circulating systemically by the measurement of both ovarian and peripheral plasma concentrations in the rat. These animals had four day cycles, with a single day each for Proestrus, Oestrus, Metoestrus combined with Dioestrus 1, and Dioestrus 2. Ovarian plasma concentrations showed a large preovulatory release of hormone followed by a sharp decline in plasma level during Oestrus, with a limited transient rise in secretion during Dioestrus 1. Secretion did not increase again until the next preovulatory release of progesterone during the day of Proestrus (Figure 5). The relatively small rise in secretion on Dioestrus 1 represents the limited luteal function present during the oestrous cycle of rodents with short cycles. The appearance of progesterone in the peripheral plasma reflects these changes in secretion by the ovary. The sharp fall in concentration present in the ovarian effluent being rather less marked at the periphery which reveals a slow decline throughout the luteal phase of the cycle (Figure 5).

The only significant secretion of oestrogen occurring in the rat oestrous cycle is the peak of release between late Dioestrus and Proestrus. When measured as oestradiol alone (Dupon & Kim, 1973) or as a total bioassayable oestrogen (Hori et al., 1968), there is a rise (Figure 6) in plasma concentration around mid-day Dioestrus 2, with a slowing in the rate of increase until there is a further rise after mid-day of Proestrus to twice the Dioestrus value. By the morning of Oestrus, the equivalent of Day 1 of pregnancy, circulating oestrogen has returned to basal levels. As Finn and Martin (1969) have noted, this proestrous oestrogen provides an initial period of oestrogen dominance necessary for the later development of uterine sensitivity to the implanting blastocyst.

Plasma progesterone and oestrogen levels during early pregnancy in the rat

There is general agreement on the levels of progesterone to be found in the plasma of the rat derived from the now functional corpora lutea in the days following the release of proestrous oestrogen. Progesterone plasma concentration is in the region of 2 μg/100 ml on Days 1 and 2 of pregnancy, rises dramatically through Day 3 and 4 to concentrations of 5-7 μg/100 ml and is maintained at this level during the prenidatory period (O'Grady et al., 1974; McCormack, S.A., Glasser, S.R. and Clark, J.W. as cited by Glasser (1975); Watson et al., 1975) (Figure 7).

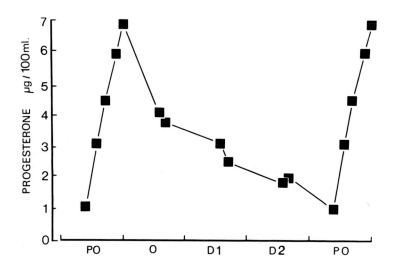

Figure 5: Systemic plasma progesterone concentrations during the rat oestrous
cycle. PO = Proestrus; O = Oestrus; D1,2 = Dioestrus 1 and 2. Data adapted
from Barraclough et al. (1971).

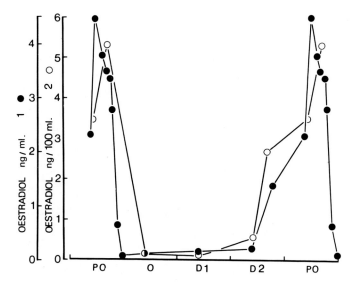

Figure 6: Systemic plasma oestradiol concentrations during the rat oestrous
cycle. Abbreviations as for Figure 5. Data adapted from (1) Dupon and Kim,
1973, and (2) Hori et al. (1968).

The amount, duration and timing of the release of nidatory oestrogen is the subject of considerable controversy but the provision of specific radio-immunoassay techniques for the measurement of oestrogens has allowed some clarification of the position.   The observation by Shelesnyak et al. (1963a) that ovariectomy before 16.00 hr on Day 4 of pregnancy led to a failure of implantation, led these workers to postulate that there was a rise in oestrogen secretion on the afternoon of this day.   Subsequently, Shaikh and Abraham (1969) argued that the results merely indicated that ovarian oestrogen secretion for the induction of implantation cannot be required after this critical time.   Thus, if the preparation of the uterus for implantation requires the attainment of a threshold concentration of intracellular oestrogen, then further exposure to oestrogen will be without effect and previous exposure may have been of either considerable or limited duration.   Attempts have been made to measure oestrogen

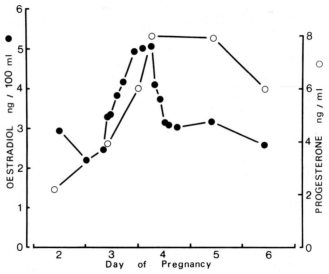

Figure 7: Systemic plasma oestradiol and progesterone concentrations during the first six days of pregnancy. Data adapted from Watson et al. (1974).

concentrations in the ovarian venous plasma and in the peripheral circulation. Pope and Waynforth (1970), when estimating the former during Day 4 of pregnancy, could find no evidence for a peak of oestrogen secretion while Yoshinaga et al. (1969), in a more extended study, noted a small peak of secretion during the afternoon of Day 3, followed by a greater increase on the afternoon of Day 4.   Shaikh (1971) observed two peaks of secretion on Day 4, one early in the morning around 03.00 hr and another between 14.00 hr and 16.00 hr, although in a previous study (Shaikh & Abraham, 1969) there had been a marked decrease in secretion between 10.00 hr and 24.00 hr on this day of pregnancy.

More recently, Nimrod et al. (1972), found that the ovarian venous plasma
concentration rose between Day 2 and 3 of pregnancy and was maintained at this
elevated concentration well beyond the time of implantation.   This study
measured oestrogen by gas-liquid chromatography and, as with the previous studies
of ovarian effluent, are open to the errors induced by surgical intervention
and extended anaesthesia.   Studies of peripheral levels, in spite of the difficulties involved in the measurement of picrogram quantities of steroid (a
problem overcome by the use of specific radioimmunoassay methods) do not suffer
from these complications.   In such a study (Watson et al., 1974) did not find
that estrogen was secreted at an elevated level throughout the preimplantation
period, rather, that on Day 1, the level of secretion was higher than that
observed during the equivalent period of the oestrous cycle, namely Oestrus, and
that this level was maintained on Day 2 and rose to a peak between 20.00 hr on
Day 3 and 04.00 hr on Day 4.   Thereafter, oestrogen levels remained similar to
those observed on the first and second days of pregnancy (Figure 7).   The
corpora lutea, which are presumably the source of this circulating oestrogen,
appear to begin to secrete as soon as they are formed following ovulation and
are presumably stimulated to produce a transient rise in peripheral plasma concentration possibly by FSH (Bindon et al., 1969), between late Day 3 and early
Day 4 and then return to their previous basal level of secretion.   It is of
note that during the oestrous cycle, the proestrous peaks of oestrogen observed
by both Hori et al. (1968) and Dupon and Kim (1973) possess an initial shoulder
with a configuration similar to the prenidatory peak of oestrogen and occurring
at a time equivalent to Days 3 and 4 of pregnancy (Figures 6, 7).   Also the
peak of oestrogen release observed in the prenidatory period is considerably
smaller than that on the day of Proestrus.   The ovariectomy of pregnant mice
at different times on Day 4, followed by progesterone therapy, and the direct
measurement of plasma oestradiol in pregnant mice on that day, indicates that
there is a peak of oestrogen in this species between 08.00 hr and 12.00 hr which
is essential for the initiation of implantation (McCormack & Greenwald, 1974).

The levels of hormone secreted during early pregnancy provide the conditions required for the development of maximal uterine sensitivity suggested
by experiments with ovariectomised rodents (Psychoyos, 1973b; Finn & Martin, 1972)
1972).   That is, the exposure of an oestrogen-primed uterus to a period of pre-
dominately, progestational activity, later accompanied by a transient exposure
to marginally increased levels of oestrogen.

There are several species in which experiments with ovariectomised pregnant
animals have shown that only progesterone is required for blastocyst
implantation(see review DeFeo, 1967 and Surani & Aitkin, this volume).   In
particular, the guinea pig which has been studied both in respect of its
requirement for nidatory oestrogen (Deansley, 1960) and its uterine cellular

mitosis (Mahrota & Finn, 1974; Marcus, 1974b) is reported as having no nidatory oestrogen during normal pregnancy (Challis et al., 1971). Recent studies (G. Auf, personal communication) have shown that there is a transient rise in the secretion of this hormone on Day 4 of pregnancy which immediately precedes stromal mitosis (Figure 8). It is possible that oestrogen derived from the adrenal is circulating in the pregnant ovariectomised guinea pig and is sufficient to support implantation in these animals.

Oestrogen and progesterone receptors in the uterus of the rat during early pregnancy

In order for oestrogen and progesterone to have the majority of their physiological actions on target organs, it is generally accepted that they must first be bound to intracellular receptors. In consequence, the effect of any peripherally circulating hormone will be modulated by the presence or absence of available receptor sites within the uterus. The mechanism of binding of oestradiol in the uterus is well established (Jensen & DeSombre, 1972; O'Malley & Means, 1974) and it is probable that not only the binding of nuclear receptor-estrogen complex to the chromatin but also the maintenance of binding for a sufficient length of time is critical, if there is to be a uterotrophic response (Lan & Katzenellenbogen, 1976; Anderson et al., 1975). Oestradiol is known to promote the formation of its own cytoplasmic receptor (Cidlowski & Muldoon, 1974), the accumulation of its own receptor complex in the nucleus (Clark et al., 1972) and the formation of specific receptors for progesterone (Leavitt et al., 1974; Milgrom et al., 1973; Reel & Shih, 1975).

Tachi et al., (1972) have found that ^{3}H-oestradiol administered to rats in vivo was translocated to the luminal epithelium and stroma in the absence of progesterone, but that previous treatment with this hormone abolished this effect. The observation was not supported by the in vitro experiments of Smith et al. (1970) who found an augmentation of uptake into the epithelium and stroma after treatment with progesterone. Similarly, the treatment of ovariectomised rats with progesterone has been found to increase the ability of the endometrium to bind oestradiol (Mester et al., 1974), but the amount of available cytoplasmic oestrogen receptor in the uterus has also been shown to be negatively modified by treatment with progesterone (Hsueh et al., 1975, 1976). Given the considerable changes in progesterone and oestrogen plasma concentration during early pregnancy, it is reasonable to suppose that there will be alterations in the amounts and distributions of their receptors. As yet, no information is available on the distribution of progesterone and oestrogen receptors between the epithelia and the stroma during early pregnancy. Attempts have been made, however, to measure oestradiol receptor levels in the total uterus during the preimplantation period. Feherty et al. (1970) found no significant variation in the cytoplasmic

183

receptor in the uterus during pregnancy, although they were able to report a
major increase during the Proestrus phase of the cycle corresponding to a period
of high oestrogen secretion.   Clark et al. (1972) have measured the concentration of receptor-oestrogen complex in the nucleus and found peaks of activity
at Proestrus and another on Day 4 of pregnancy, the concentration of receptor
complex increasing from Day 2 to Day 4, and falling abruptly on Day 5.   This
increase in receptor complex may reflect the known peak of oestrogen secretion
in the preimplantation period (Glasser & Clark, 1975; Watson et al., 1975)
(Figure 9).   Measurement of oestrogen receptor concentration has been made in
rat endometrium during the first eight days of pregnancy.   Receptor concentration rose from Day 2 to Day 5, and returned to low levels by Day 7 of
pregnancy.   In these experiments, endometrial concentrations of receptors are

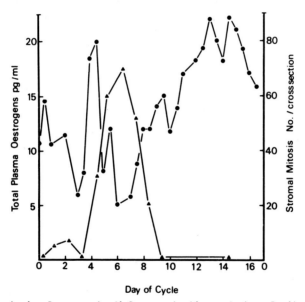

Figure 8: Systemic plasma oestradiol concentration and stromal mitosis in the
guinea pig oestrous cycle. Oestradiol (●) was determined by radioimmunoassay
(personal communication G. Auf, 1976). Stromal mitosis (▲) data adapted from
Mehrotra and Finn (1974).

not directly correlated with the duration of the peak of estrogen release,
indicating that a positive effect of progesterone on the level of receptor may
be involved (Mester et al., 1974) (Figure 10).

      This investigation indicates that oestrogen receptor is available
throughout the prenidatory period, leaving the endometrium receptive to
endogenous oestrogen, including that which may be secreted by the blastocyst
(Dickman et al., 1976).   Milgrom et al. (1972) have attempted to estimate

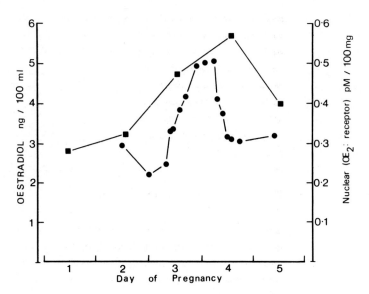

Figure 9: Systemic plasma oestradiol concentration (●) and nuclear oestradiol:
receptor complex in the uterus (■) during early pregnancy in the rat.
Oestradiol data adapted from Watson et al. (1975) and oestradiol:receptor complex data from Glasser and Clark (1975).

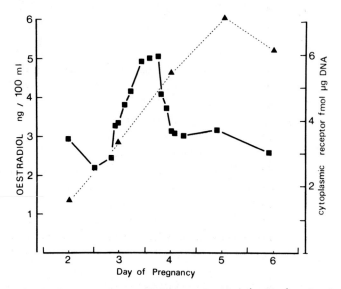

Figure 10: Systemic plasma oestradiol concentration (—■—) and cytoplasmic
oestradiol receptor (---▲---) in the uterus during early pregnancy. Oestradiol
data adapted from Watson et al. 1975 and cytoplasmic receptor data from Mester
et al. (1974).

progesterone receptor levels in the guinea pig uterus and although they found
a peak of receptor concentration at the time of Proestrus, there were only
marginally higher levels of receptor on the day of implantation in the pregnant
uterus when compared with the non-pregnant uterus. This work has subsequently
been criticised on the grounds that glycerol, which aids the stability of the
progesterone-receptor complex, was not present in the preparatory medium (Fiel &
Bardin, 1975). Meaningful interpretation of the results of these studies will
not be possible until the distribution of receptor in the stroma and epithelia
has been established and may then be related to the mitotic and metabolic events
of early pregnancy. This remains an important area of investigation.

Correlation of hormonal changes with endometrial structure, cellular division and sensitivity

As was seen earlier, the morphological changes in the endometrium during
the oestrous cycle and early pregnancy are controlled by the fluctuating levels
of the ovarian hormones, oestrogen and progesterone. The generation of a
receptive uterus, sensitive either to the implanting blastocyst or to any
deciduomatic stimulus for a restricted period, also requires a specific hormonal
interplay. The events rarely depend upon the presence of one hormone, but
rather their sequential or co-operative action. The precise hormonal requirements have been established using ovariectomised animals which have been
subjected to specific regimes. Experimentation in this area is far from comprehensive, and has been mainly restricted to the mouse.

Structural changes

The ultrastructural changes observed in uterine tissue during early pregnancy
are, in part, hormonally conditioned and this is particularly true for those
occurring in the preimplantation period. Correlation of hormonal levels with
cellular proliferation will be considered in a later section.
(1) Epithelial tissues. Oestradiol precipitates many changes in the
epithelial tissues which include hyperplasia (Nilsson, 1958a,b,c) and an
increase in size of nuclei (Schultz et al., 1969; Wilson, 1963), reflecting the
mitogenic action of this hormone on the tissue. In rodents, oestradiol causes
reduction in the number of lipid droplets (Gillman, 1970; Alden, 1947; Elftman,
1963; Fuxe & Nilsson, 1962; Goswami et al., 1963; Bashier & Holloway, 1973).
Luminal epithelial cell surface changes include an increase in length and number
of the microvilli which are surrounded by the deposition of a "Fuzz" (Nilsson,
1958a,b,c, 1959).

The main effect of progesterone upon glandular epithelium is to differentiate
the cells into a "potential secretory condition". A small quantity of oestrogen
given after progesterone treatment in ovariectomized mice will initiate

secretion (Finn & Martin, 1971). Treatment of ovariectomised mice with progesterone, results in an ultrastructural change equivalent to that defined as the "first stage of closure" in early pregnancy (Martin et al., Pollard & Finn, 1972) i.e. produce a simple apposition of opposing luminal walls with inter-digitation of the microvilli on their surface. This morphological state is also like that associated with the period of insensitivity in early pregnancy i.e. prior to nidatory oestrogen, that observed in animals in a condition of delayed implantation, and that seen in the pre-attachment phase of implantation. When oestradiol is injected into a progestational ovariectomised animal, the contracting epithelial cell membranes reorganise so that the cell membranes become intimately apposed, resulting in a morphological condition analogous to the "second stage of closure". Between 8-18 hr after oestradiol administration, the pseudopodia-like protrusions or 'blebs' enlarge, but the microvilli become irregular and flatten (Potts & Psychoyos, 1967). About 24 hr after oestradiol treatment, the plasma membranes become intimately apposed. The morphological transition also results after administration of oestradiol to animals in delayed implantation (Nilsson, 1974; Bergstrom & Nilsson, 1975). The second stage of closure is probably equivalent to the attachment reaction between the trophoblast and luminal epithelium (Nilsson, 1966a,b; Potts, 1969), and to the "refractory state" or "state of insensitivity" immediately following "the state of sensitivity" induced by oestradiol, (Pollard & Finn, 1972, 1974). No morpho-logical correlates have been found for the state of sensitivity (Pollard & Finn, 1972). Priming oestradiol, equivalent to proestrous oestradiol, did not qualitatively effect the responses, but did hasten the onset of both stages (Finn & Pollard, 1973). Although these morphological states would, therefore, appear to be hormonally conditioned, their temporal appearance is affected by the presence of the blastocyst (Pollard & Finn, 1974).

Although no precise hormonal requirements have been established for the sensitivity for epithelial cell death, induced on Day 5 by oil and the blasto-cyst (El-Shershaby & Hinchliffe, 1975; Hinchliffe & El-Shershaby, 1975), the supposition that it is programmed by the progesterone-oestradiol sequence seems probable. This same sequence results in suppression of luminal epithelial proliferation in response to midatory oestradiol, and for the generation of a state in which the tissue is competent to transmit a signal from the implanting blastocyst or oil to the competent stroma. The ability of the epithelium to respond to oil and the blastocyst by cell death may be linked to the latter con-dition.

(2) Stromal tissues. Stroma cells exhibit progesterone-dependent ultra-structural changes, mainly in the nucleus. In ovariectomised animals, stromal cells possess fusiform dense nuclei with clumped chromatin and no nucleoli which after progesterone treatment possess oval nuclei with evenly-distributed

chromatin and prominent nucleoli (Hooker, 1945; Hooker & Forbes, 1947). Once decidualization has been initiated only progesterone is required for decidual growth, and removal of this hormone results in degeneration of the decidual cells (DeFeo, 1967). Progesterone, however, only appears to be permissive for decidual development since even in its presence, decidual growth which continues for 6 days post stimulus, degenerates (Selye et al., 1942; Atkinson, 1944). Oestradiol augments the effect of progesterone and results in an increase in the final decidual growth yield (Glasser & Clark, 1975; Glasser, 1972; DeFeo, 1967).

Intercellular oedema appears to arise as a result of a combination of increased permeability of the blood vessels of the stroma (Bindon, 1969) and a change in the connective tissue matrix of the stroma which results from differentials in collagen biosynthesis. Capillary permeability, which according to Martin et al. (1973c) is due to development of vascular fenestrations and to Majno and Leventhal (1967) and Ham et al., (1970) to active retraction of endometrial cell cytoplasm, is induced by oestradiol. The oedema occurring early on Day 5, thus would result from the nidatory oestradiol secretion. Collagen biosynthesis increases in response to oestradiol (Harkness et al., 1956), although increased synthesis in early pregnancy appears to be due more to the progesterone which suppresses collagenase activity (Jeffrey et al., 1971; Yochim, 1975).

Cellular division

In the preimplantation period of pregnancy in rodents, a marked pattern of cellular proliferation emerges. On Day 2 the luminal and glandular epithelia undergo division, whereas on Day 4 this is restricted to the stroma. Cellular division in the uterus is under hormonal control and has been examined using ovariectomised animals given exogenous hormones and colcemid to arrest cells at metaphase (Clark, 1971; Das & Martin, 1973; Martin & Finn, 1968, 1971; Tachi & Tachi, 1975; Tachi et al., 1972).
(1) Luminal epithelium. A single injection of oestradiol results in an increase in both luminal and glandular mitosis, but the temporal and quantitative response has been found to be dissimilar. Division in the luminal epithelium occurs at 24-30 hr after oestradiol administration and is quantal and very sensitive (rat - Clark, 1971; Tachi et al., 1972; mouse - Allen et al., 1937; Martin & Finn, 1968; Martin et al., 1973b). Das (1972) and Das and Martin (1973) have demonstrated that the luminal epithelia pass through two successive cellular divisions. Continuous injections of oestradiol result in a rapid proliferation for two days, which then abruptly falls (Finn & Martin, 1973). Progesterone inhibits the mitogenic action of oestradiol, when given prior to or within minutes of the oestrogen (Martin & Finn, 1971). However, when given up to 17 hr after oestradiol, it only prevents the population from entering into the second

round of division (Das, 1972; Das & Martin, 1972; Martin et al., 1973a).   It is
suggested by these authors that progesterone blocks luminal epithelial cells in
the early G_1 phase of the cell cycle.   This inhibition by progesterone can,
however, be overcome by high levels of oestradiol (Martin & Finn, 1968).

   Two waves of luminal epithelial division are observed during the oestrous
cycle of the rat (Figure 2), one at Proestrus which is quantitatively more
important, and the second on Dioestrus 1 (Marcus, 1974a).   In the light of the
previous results it is considered that the late Dioestrus 2/proestrous estradiol
peak causes both the immediate proestrous epithelial proliferation and the
secondary wave two days later, on Dioestrus 1.   If mating intervenes at Oestrus,
the oestradiol peak would again produce the proestrous wave and then two days
later, the secondary wave which now falls on Day 2 of pregnancy (Figure 4).
Oestradiol secretion occurring during the night of Day 3 of pregnancy (nidatory
oestradiol) does not induce luminal epithelial proliferation on Day 4, possibly
due to the inhibitory effect of the high progesterone levels over the previous
two days.   Although it has been demonstrated that progesterone pre-treatment
inhibits the mitogenic action of oestradiol upon luminal epithelium (Clark, 1971;
Finn & Martin, 1974; Martin & Finn, 1968; Tachi et al., 1972) the level of
progesterone relative to the oestradiol must be important, since even the
proestrous oestradiol is preceded by three days of progesterone secretion, i.e.
Oestrus to Dioestrus 2 of previous cycle, albeit at lower levels than in early
pregnancy.   Thus, either the levels of proestrous oestradiol are sufficient to
overcome the inhibitory action of progesterone, or the levels of the latter
hormone are insufficient to inhibit the mitogenic action of the former.   The
converse would then apply to the nidatory oestradiol of early pregnancy.
(2) Glandular epithelium.   Glandular epithelial cells respond to oestradiol by
increased proliferation, but the response to a single injection in contrast to
the luminal epithelium, is quantitatively poor and more sustained (rat - Clark,
1971; Tachi et al., 1972; mouse - Finn & Martin, 1973).   Peak response was
obtained after two days.   With continuous oestradiol treatment, however, a
transient response was obtained on the third day after its commencement.   If
the oestradiol treatment lasts for at least two days after the cessation of the
regime, a second wave of proliferation is observed three days after the final
injection.   Thus the glandular epithelium responds to a continuous oestradiol
regime biphasically; a primary wave of proliferation three days after its
initiation, and a secondary wave three days after its cessation.   Progesterone,
as for the luminal epithelium, suppresses the mitogenic action of oestradiol and
inhibits both waves of proliferation when given with, or prior to, the oestradiol.
When injected the day after the cessation of the continuous regime, however, it
inhibits the secondary division (Finn & Martin, 1973).

During the oestrous cycle of the rat, only one wave of glandular epithelial division is observed (Figure 2), occurring on Dioestrus 1 (Marcus, 1974a) which coincides with the second luminal epithelial division.   Experiments with ovariectomised animals point to the induction of this proliferation by Dioestrous 2/proestrous oestradiol, which it precedes by two days.   If mating intervenes, the proestrous oestradiol would again be expected to induce division on Day 2 of pregnancy, which in fact is observed (Figure 4).   The absence of a secondary wave of division observed in hormone replacement experiments, in response to oestradiol, can be explained by the intervention of progesterone.   As for luminal epithelium, nidatory oestradiol is probably an ineffective mitogen due to the high levels of progesterone in early pregnancy.

(3) Stromal cells.   Stromal cells, in contrast to both the epithelial tissues, do not proliferate in response to oestradiol when administered alone, nor are they responsive to progesterone alone.   However, Finn and Martin (1974) elucidated three hormonal parameters which affected the proliferative response of the stroma in the mouse.   Progesterone could only potentiate a response if administered to ovariectomised mice for three days after "priming" of the animals with oestradiol (Martin & Finn, 1968).   Oestradiol acts as a stromal mitogen only when administered as a single injection on the third day of treatment with progesterone.   This response occurred regardless of whether the progesterone treatment was preceded with "priming" oestradiol or not (Martin & Finn, 1968, 1969).   The work of Martin and Finn, however, clearly demonstrates that to achieve maximal and stromal-specific proliferation, all three hormonal parameters are required, i.e. priming oestradiol, progesterone treatment followed by the oestradiol which induces stromal mitosis 24 hr later.   In the rat, oestradiol has also been found to induce stromal cell division after progesterone pre-treatment (Clark, 1971; Tachi et al., 1972).   A minimum of 36 hr progesterone treatment was required in rat (Tachi et al., 1975) compared to 48 hr in the mouse, to potentiate oestradiol action.   Clark (1971) has also noted that the proliferation induced in hormone treated ovariectomised rats appeared to be restricted to the sub-epithelial stromal cell population.

Finn et al. (1969) have examined the effect of multiple injections of oestradiol upon stromal division, using the basic hormonal regime (oestradiol priming and progesterone pre-treatment).   They found that a second injection 48 hr after the first, also induced stromal proliferation 24 hr later.   However, if the second injection was given 12-36 hr after the first, no further division was induced, and in fact when given at 24 hr, the stroma was refractory to an injection given at 48 hr after the first injection.   Thus, it appeared that whilst the first injection of oestradiol induced a stromal population to proliferate (maximum mitotic index observed after 24 hr) a transient period of insensitivity between 12 and 36 hr was produced, regaining competence only after

48 hr.   These authors have suggested that the 48 hr progesterone treatment produces factors essential for stromal cell division, producing a population competent to divide in response to oestradiol.   After the initiation of proliferation, depletion of these factors results in insensitivity, which, in response to a further exposure to 48 hr of progesterone become replenished.

Stromal mitosis which occurs on the morning of proestrus in the oestrous cycle (Figure 3) and on Day 4 of pregnancy (Figure 4) temporally correlates with the pattern of oestradiol secretion (Marcus, 1974a).   The modest proliferation observed on the night of Dioestrus 2 suggests that the levels of progesterone secreted during the cycle are sufficient to potentiate the mitogen action of the oestradiol (rising late on Dioestrus 2).   Division on Day 4 of pregnancy is more intense than the oestrous cycle mitosis, even though the nidatory oestradiol levels are lower than that at Proestrus.   It would, therefore, appear that the level of the prior progesterone secretion, quantitatively greater in early pregnancy, dictates the quantitative stromal response to oestradiol.

Marcus (1974a) observed that myometrial proliferation also occurred during the oestrous cycle and early pregnancy following the pattern of stromal proliferation, in both situations the peaks of myometrial division occurring 12 hr after the peaks of stromal division.   This temporal relationship between the two cell types suggests that the hormonal control of their division is similar, although no experimental evidence pertaining to the hormonal requirements for myometrial division is available.

Cellular proliferation observed in the uterus during the oestrous cycle and the preimplantation period of pregnancy, therefore, is interpretable in terms of the hormonal mileau present, the modifications of the mitogenic action of oestradiol by progesterone and differential sensitivities of the uterine tissue types.   It may be pointed out that although oestradiol is presumed to be the primary mitogen for all uterine tissues, its action being modified by progesterone, this may only be true for epithelial tissues.   An alternative proposal for stromal tissue is that the role of progesterone is as primary mitogen, (progesterone treatment elicits a stromal response, albeit poorly), however a factor whose synthesis could be oestradiol-dependent is required for maximal response.

Endometrial sensitivity

Endometrial sensitivity to the blastocyst is restricted to a short period of early pregnancy in rodents (DeFeo, 1967; Finn & Martin, 1974; Psychoyos, 1973a,b), in the rat occurring on the afternoon of Day 5 of pregnancy.   Outside this period, the uterus is insensitive, and this observation has given rise to names such as "presensitivity" for the prior insensitive period and "refractory state" for the subsequent insensitive condition (Pollard & Finn, 1972).   Using ovariectomized animals, hormonal regimes and transferred blastocysts or eggs, the

hormonal requirements for endometrial sensitivity have been elucidated.   All
studies disclose that a basic 48 hr regime of progesterone is required, followed
by one injection of oestradiol, to produce a period of receptivity to the
blastocyst (Psychoyos, 1960, 1967, 1970).   The oestradiol-induced receptive or
sensitive state is then followed by an endometrial refractory period manifested
by hostility of the uterus towards blastocysts and eggs (Boot & Muhlbock, 1953;
Dickman & Noyes, 1960; Doyle et al., 1963; McLaren & Mitchie, 1956; Noyes &
Dickman, 1960; Psychoyos, 1963a,b).   Thus, the same hormonal sequence that
promotes endometrial sensitivity to the blastocyst is responsible for the subsequent state of non-receptivity (Psychoyos, 1965, 1966,a,b, 1969).   It is not
necessary for implantation to "prime" with oestradiol before the 48 hr sequence
of progesterone (Humphrey, 1969).   However, using the simple progesterone/
oestradiol regime, the uterus responds below its normal capacity, so the "priming"
oestradiol appears to fulfil some role.   Thus the hormonal requirements for
blastocyst implantation are fulfilled in normal pregnancy, i.e. proestrous
oestradiol, progesterone Day 2, 3 and with nidatory oestradiol on Day 4.   We
may note that these are the precise hormonal conditions which induce stromal
mitosis.

One of the major events of implantation is the production of decidual cells
from stromal cells to form the decidua, i.e. the decidual cell reaction
(DeFeo, 1967).   The decidual cell reaction can be produced artificially by
using alternative stimuli, e.g. oil, trauma, beads, etc., and the resulting
tissue mass has been termed a deciduoma.   This reaction has been used to study
the hormonal control of sensitivity in rodents (Corner & Warren, 1919; Brouha,
1928; Finn & Martin, 1972; Shelesnyak, 1933a,b; Rothchild & Meyer, 1942).   The
variety of stimuli used can be subdivided into two groups: a) traumatic, being
those which cause cellular damage e.g. knife scratch and needle; b) non-traumatic,
which act on the surface of the luminal epithelium e.g. air and oil.   The
hormonal requirements for the two groups have been found to be different.
Traumatic stimuli only require the uterus to be pre-treated with progesterone
for approximately 48 hr (DeFeo, 1967; Finn & Martin, 1974; Glasser, 1972) and
thus the period of sensitivity in pseudopregnancy is much wider for trauma
(DeFeo, 1967).   For non-traumatic stimuli, such as oil, the hormonal requirements are more exacting.   In the mouse priming oestradiol is required, followed
by three days of progesterone.   On the third day of such treatment, one injection
of oestradiol results in a transient period of sensitivity 24 hr later, followed
by insensitivity (Finn, 1966; Finn & Hinchliffe, 1964; Finn & Keen, 1962; Finn
& Martin, 1969, 1972, 1974; Meyers, 1970).   This regime is essentially similar
to the regime required for blastocyst implantation and to that occurring in normal
preganancy, and all produce stromal cellular proliferation, implicating its
crucial role in implantation and decidualization.   The relationship between

endometrial sensitivity to implantation or decidualization and stromal mitosis is further implied by the results of Finn (1966) who found the oil-induced decidual cell reaction was inhibited when oestradiol was given to progesterone-treated ovariectomised mice 24 hr before the oestradiol was normally required. This regime would also cause refractoriness of the stroma to oestradiol, and result in no stromal division (Finn et al., 1969). We may note that the time of maximum endometrial sensitivity to oil/blastocysts, which occurs 24 hr after the final injection of oestradiol in the regime corresponds to the time of maximal stromal mitosis after such injection and when the stroma is insensitive to a further injection of the hormone.

Although this data suggests that an important role of the progesterone/oestradiol hormonal sequence lies in its stromal mitogenic action, which in turn appears to be a prerequisite for the decidual cell reaction, other functions must exist. Non-traumatic stimuli, including the natural stimulus of the blastocyst, require the final oestradiol injection in the aforementioned regime, equivalent to nidatory oestradiol in pregnancy, to precipitate action, and since both act on the luminal epithelium, oestradiol may also condition this tissue to transmit a signal to the stroma initiating decidualization. The state of sensitivity of the luminal epithelium may, in fact, only occur in response to oestradiol when the mitogenic action of this hormone on the tissue is suppressed, i.e. by the progesterone in early pregnancy or in the 48 hr of progesterone in the regime. Thus it has been suggested that epithelial cells are hormonally conditioned either to proliferate in response to oestradiol (insensitive to stimuli) or to differentiation in response to a sequence of progesterone and oestradiol (sensitive to stimuli), and such fates are mutually exclusive (Martin & Finn, 1968). Pertinent to this point is the fact that if high doses of oestradiol are injected after a progesterone regime in ovariectomised mice, the inhibition by progesterone is overcome, and the epithelial proliferation induced, but stromal division still occurs. These conditions render the uterus insensitive to non-traumatic stimuli which act on the epithelium and implantation is inhibited (Martin & Finn, 1968; Martin et al., 1961; Martin, 1963).

Another hormonally conditioned event which may be mediated via the epithelium and/or stroma is blastocyst activation. If pregnant rats or mice are ovariectomised prior to the occurrence of significant oestradiol secretion and are maintained on progesterone, blastocysts remain viable but dormant and do not attach or implant (Cochrane & Meyer, 1957; Canivene & Laffargue, 1957; Psychoyos, 1973a,b; DeFeo 1967). This state is known as delayed implantation. Termination of this state is caused by a single injection of oestradiol, the blastocyst attaching to the luminal epithelium one day later (Nilsson, 1974). Several theories have been forwarded to explain the activation by oestradiol but have one common feature in that its action is mediated via the endometrium (McLaren, 1973;

Finn, 1974). One suggestion is that the oestradiol is initiating the synthesis of stimulatory substances by the uterus; while the other is that the hormone inhibits the synthesis of inhibitory substances, e.g. proteins (Psychoyos & Bitton-Casimiri, 1969; Weitlauf, 1973). Recent research in this area has still not clarified the situation (Gore-Langton & Surani, 1976; Pollard et al., 1976; Surani, 1975; see also discussions by Surani and Aitkin, this volume).

UTERINE METABOLISM AND CELL CYCLE RELATED PHENOMENA IN EARLY PREGNANCY

Uterine energetics and the associated metabolism of carbohydrates and lipids during early pregnancy have received considerable attention. These observations have general implications for the development of the endometrium and reveal some metabolic correlates of the differentiation of decidual from stromal cells. The absence of information relating to such metabolism in the individual endometrial cell population precludes interpretation in terms of cell cycle phenomena. Similarly, work on the protein metabolism of the uterus has concentrated on unfractionated tissue and on endometrial preparations. Nevertheless, experiments concerned with the effects of oestrogen and progesterone on the protein metabolism of the stroma and luminal epithelium have been attempted and this work should be extended to the pregnant uterus. The recent development of methods for the preparation of myometrial, stromal and luminal epithelial fractions of the rodent uterus has allowed the nucleic acid metabolism of the fractions to be evaluated and the status of the stroma in the cell cycle immediately prior to decidual transformation to be established. The control of the cell cycle and the expression of the metabolic functions of differentiated cells is effected by cAMP. Variations in the uterine content of this nucleotide during the development of endometrial sensitivity and during decidual transformation can be related to these phenomena.

Carbohydrate and lipid metabolism and energy requirements of the uterus during the preimplantation period

Attempts have been made to relate both carbohydrate metabolism and the availability of oxygen and substrates for energy provision to other metabolic events occurring in the preimplantation uterus. Further, it has been suggested that the availability of energy may be involved in the control and initiation of potential sites of implantation. Uterine oxygen consumption and intra-luminal concentration have been related to total endometrial glucose utilisation, either by glycolysis or the hexose monophosphate shunt, and to the production of lactate and the concentration of endometrial lactate dehydrogenase (LDH), (Yochim, 1971, 1975; Surani & Heald, 1971). Also, these parameters have been related to the synthesis of nucleic acids and proteins during this period and a variety of causal relationships between oxidative pathways and anabolic metabolism

postulated (see review Yochim, 1975). As these observations were all made on
the total endometrium and very little is known of the oxidative pathways and
oxygen concentrations obtaining in the individual cell types, such hypotheses
must await further investigations before they can be substantiated. Measurements of oxygen consumption in both the decidualizing and the non-stimulated
pseudopregnant uterus have shown a marked decline in oxygen utilisation
(Salderni & Yochim, 1967) without an equivalent increase in LDH activity
(Battellino et al., 1971), but with large increases in activity in pyruvate
kinase and phosphofructokinase in the decidualizing tissue (Surani & Heald, 1971).
These latter enzymes are rate limiting and unidirectional in action in the
glycolytic sequence, and their rise in activity is probably responsible for the
rise in glycolytic intermediates occurring in the implant and will be responsible
for an increase in glucose utilisation. Associated with this general stimulation of the glycolytic pathway, there is a marked accumulation of glycogen in
the primary decidual zone during early implantation (Christie, 1966). The
glycogen stored in the anti-mesometrial region disappears as decidualization
proceeds and Christie postulates that it is utilised by the embryo which is
histiotrophic at this time, and that the high level of glucose-6-phosphatase
present in the decidual cells will aid glycogen breakdown (Christie, 1966). As
a result, glucose will be available for trophoblast nutrition and direct utilisation via glycolysis in the proliferating decidual cells. Certainly, the
decidualizing uterus has a higher ATP/ADP ratio on Day 6 than that observed in
the preimplantation tissue (Weber et al., 1972), even though there are increased
levels of adenosine-5'-tri-phosphatase in the primary decidual zone (Christie,
1966).

Both in the rat and rabbit, the concentration of some lipids have been
shown to decline during implantation following a general increase in uterine
lipid during the prenidatory period (Ray & Morin, 1965; Morin & Carrian, 1968;
Beall, 1972) and stainable lipid in the area of decidualization decreases after
implantation in the rat uterus (Williams, 1948). Many authors have supported
the hypothesis that endometrial and decidual lipid serves as an energy source
for the developing blastocyst and for nidation (Gillman, 1941; Ray & Morin, 1965;
Burger, 1966; Krehbiel, 1937) and it has been demonstrated that neutral fat and
triglyceride, in particular, decreases in the implant relative to the inter-
implantation tissues on Day 7 of pregnancy (Beall & Wethessen, 1971). Although
changes in uterine lipids during early pregnancy have been extensively studied
by both histochemical and biochemical means, there is as yet, insufficient
quantitative data to be sure of its general function or its importance in
oxidative metabolism during the prenidatory period. Indeed, none of the data
available is of sufficient detail to be able to directly relate changes in the
proliferation and differentiation of the various cellular populations to the

importance of carbohydrates or lipids to the energy requirements of the
individual cell types.

Protein metabolism

In spite of the heterogeneity of the uterine cell population and their differential division and function, there has been no investigation of changes in
protein synthesis in the individual compartments during early pregnancy; all
studies as yet being confined to the unfractionated uterus or to the total
endometrium.

The effects of progesterone and oestrogen and combinations of those steroid
hormones on protein synthesis in epithelial and stromal fractions of the endometrium have been investigated (Smith et al., 1970).   There have also been
extensive studies of the proteins of the luminal fluid during this period and
alterations in the production and activity of several proteins have been described (Beier, 1974; Surani, 1975) but these are discussed elsewhere in this
volume (Surani; Aitkin; Beato).

Experiments measuring the incorporation of amino acids labelled with single
isotopes have shown that protein synthesis in the rat uterus is elevated on
Day 3, subsides on Day 4, and undergoes a marked increase by the morning of Day 5
(Reid & Heald, 1970, 1971).   Synthesis is maintained on Day 6 in the implant
but declines to low levels in the non-decidualizing uterus (Reid & Heald, 1970)
(Figure 11).   The increase in synthesis on Day 5 occurs in both the myometrium
and endometrium, is suppressed when implantation is inhibited by the administration of the anti-oestrogen taxoxifen (ICI 46, 474) on Day 2, and is not
present in the pseudopregnant uterus (Reid, 1971).   It is feasible that the
presence of the blastocyst in the uterine lumen early on Day 5 has initiated
the decidual reaction and that this increase in protein synthesis is a reflection
of this process.

The nature of the proteins synthesised during early pregnancy has been
examined by techniques previously employed to demonstrate the induction of a
new protein fraction (induced protein) in the uteri of normal cycling and
oestrogen-treated immature rats (Katzenellenbogen, 1975; Iacobelli, 1973;
Notides & Gorski, 1966).   These studies used double-isotope labelling techniques, cytoplasmic proteins being labelled by incubating uteri from rats in media
containing ^{3}H or ^{14}C leucine.   The ^{14}C-labelled proteins synthesised by the
uteri of ovariectomised rats were mixed with the proteins from the uteri of
pregnant rats labelled with ^{3}H leucine and separated by co-electrophoresis on
cellulose acetate strips (Somjen et al., 1973) or the polyacrylamide gel system
of Gorski (Barnea & Gorski, 1970).   'Induced protein' migrates with fewer other
proteins in the cellogel system (Somjen et al., 1973) and in both this and the
gel system examination of the ^{3}H:^{14}C ratios a peak of increased ratio was con-

sistently present from Day 3 to 6 in a position indicating a similar mobility
to 'induced protein' (Bell et al., 1975, 1976).   Measurement of the relative
rate of synthesis of this protein (presumptive induced protein) by the procedure of Mayol (1975) indicates that it is produced biphasically with major
synthesis occurring on Day 4 and Day 6 (Figure 12).   If this protein is
identical with the classical induced protein of Notides and Gorski, a fact yet
to be established, then this work indicates that its production is not solely
oestrogen mediated as the synthesis on Day 6 occurs in the absence of significant amounts of circulating hormone (Watson et al., 1975).   It is possible
that an increased oestradiol receptor population could lead to a greater uptake
of oestradiol, but conclusive proof of this is, as yet, not available.   On
Day 6, decidual tissue is undergoing intense stromal division and as oestradiol-
17β is capable of inducing cellular proliferation and induced protein in all

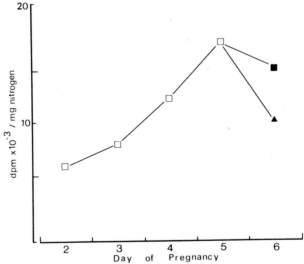

Figure 11: Protein synthesis in the rat uterus during early pregnancy.   The
incorporation of ³H-leucine into the protein of preimplantation (□) implanted
(■) and non-implanted (▲) uterus.   Data adapted from Reid and Heald (1970).

cell types (Dupont-Mairesse & Galand, 1975), it is possible that the increased
synthesis of 'presumptive induced protein' is associated with preparation for
this division or with cell division itself.
     Separations with the gel system have shown a marked increase in synthesis
of proteins in the post transferrin region on Day 5 of pregnancy which is continuous on Day 6 in the implant and decreases in the non-decidualizing uterus
(Bell et al., 1975, 1976).   Synthesis in this fraction appears to be specifically
associated with the onset and continuation of the decidual process and may be
involved in this increased stromal activity.   In addition, the double isotope

technique has shown the sequential production of several protein fractions which can only be demonstrated in implantation tissue on and after Day 6, and which may be characteristic of the decidualizing stroma (S.C. Bell, unpublished observations). The presence of new antigenic components in the mouse and rat deciduomata (Yoshinaga, 1972, 1974) and in the Day 3 pregnant mouse uterus (Sacco & Mintz, 1975) have been demonstrated and it is probable that the increase in the synthesis of protein fractions in the implant and the post-transferrin fraction in the preimplantation uterus may be responsible for some of this alteration in antigenic capacity.

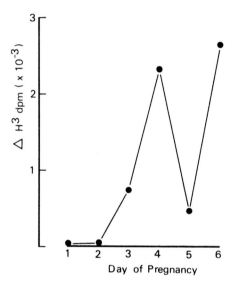

Figure 12: The rate of synthesis of 'presumptive induced protein' (IP) in the rat uterus during early pregnancy. The rate of synthesis of 'presumptive IP' was estimated using the Δ^3 H method of Mayol (1975). Data adapted from Bell et al. (1976b).

These data indicate that during early pregnancy there are specific changes in the protein produced during the latter part of the prenidatory period at the time of stromal mitosis, when the blastocyst is free floating in the uterine lumen, and that the process of decidualization is accompanied by the further synthesis of other specific uterine proteins.

Nucleic acid and phospholipid metabolism

Since alteration in the proliferation, differentiation and function of cells is under the control of nucleic acids, changes in the synthesis of these complex molecular species should also be detected. Studies of their metabolism

in the whole uterus and in the individual cell types by autoradiography have
been related to the known changes in cell division occurring during the
preparation of the uterus for implantation. When measuring the in vivo
incorporation of ^{3}H-thymidine into the total DNA of the uterine horns of
pregnant rats, Tachi et al. (1972) found levels of incorporation which could
be correlated with luminal epithelial and glandular mitosis on Day 2, with a
decline in incorporation on Day 3 associated with a reduction in mitosis in
these cell types, and a return to high levels of incorporation on Day 4,
reflecting the known elevation of stromal mitosis on this day in the pregnant
rat (Marcus, 1974a; Tachi et al., 1972; O'Grady et al., 1975). Following
implantation, when mitosis and the development of polyploidy are occurring
together, the evaluation of data involving the incorporation of ^{3}H-thymidine
into all the DNA contained in a single uterine horn becomes difficult, but it
is clear that once implantation has begun, synthesis proceeds rapidly in the
uterus, whereas the pseudopregnant uterus shows a decline in synthesis (Tachi
& Tachi, 1975; Tachi et al., 1972). This work confirmed earlier histological
studies which showed increased synthesis of nucleic acids in decidualizing
tissue (Shelesnyak & Tic, 1963a,b), the amount of DNA per mg of implant showing
a significant increase relative to the interimplant by Day 7 of pregnancy in
the rat (Heald & O'Grady, 1970). Autoradiographic studies of the in vivo
incorporation of ^{3}H-thymidine into the pregnant rat uterus have also shown that
luminal epithelial and glandular mitosis on Day 2 and mitosis in the stromal and
myometrial cells on Day 4 are accompanied by DNA synthesis (Galassi, 1968; Leroy
& Galand, 1969). These results, which depend on incorporation over relatively
long time periods, do not provide sufficient information for DNA synthesis to
be closely related to changes in the synthesis of RNA in the various cell types
and to alterations in the location and activity of the blastocyst. Techniques
have been developed for the isolation of the cell types of the uterus into
individual, luminal epithelial, stromal (including endometrial glands) and
myometrial fractions (Heald et al., 1975). Fractionation of the uterus, after
the incorporation of a one hour pulse of ^{3}H-thymidine in vivo in the pregnant
rat at six hourly intervals, has allowed a more detailed assessment to be made
of the relationship between preimplantation events and uterine DNA synthesis
(O'Grady & Heald, 1976; O'Grady et al., 1976).

    These experiments were carried out in unilaterally ovariectomised rats so
that a direct comparison could be made between the pregnant and pseudopregnant
uteri. Again, luminal epithelial mitosis on Day 2 and 3 was reflected in
increased synthesis in this fraction in the pregnant and pseudopregnant horns
and was absent at other times during the preimplantation period. Glandular
epithelial mitosis was most probably responsible for the high DNA synthesis
present in the 'stromal' fraction on Day 2, as this fraction contains the

endometrial glands (Heald et al., 1975) (Figures 13, 14, 15). Elevated stromal
incorporation in both pregnant and pseudopregnant uteri at 09.00 hr on Day 4
demonstrates that DNA synthesis rises with stromal mitosis and is maintained at
15.00 hr.   This observation, given that synthesis in other fractions is
relatively low at this time, shows broad agreement with the rise in incorporation
demonstrated by Tachi and Tachi (1975) in the total uterus.   DNA synthesis
declines with stromal mitosis and is only proceeding at relatively low levels
by 03.00 hr on Day 5.   Following this decrease in activity, DNA synthesis in
the stroma remains at low levels in the pseudopregnant uterus throughout the
remainder of Day 5.   It has been suggested that the peak of stromal mitosis on
Day 4 of pregnancy is followed by an immediate further synthesis of DNA in the
area of the primary decidual zone, and that at least a portion of the stromal
cells pass through the G1 and S phases of the cell cycle and are in the tetraploid state, arrested in the G2 phase, before the commencement of decidual DNA
synthesis.   Following the synthesis associated with mitosis, no further DNA
synthesis was observed in the stroma of the pregnant uterus until 15.00 hr on
Day 5, after blastocyst contact (preattachment) with the luminal epithelium at
12.00 hr (Tachi et al., 1970) and the commencement of the decidual response
(Figure 15).   This synthesis probably represents the initiation of the DNA
synthesis associated with the differentiation of decidual cells from the SEAM
population of the uterine stroma.   The observation that DNA synthesis in the
stroma is only slight between 03.00 hr and 09.00 hr on Day 5 of pregnancy,
implies that following mitosis on Day 4 and early Day 5, there is no level of
synthesis on Day 5 comparable to that observed on Day 4, which might place an
appreciable portion of the stromal cells in an arrested G2 phase.   This contention is supported by the work of Leroy et al. (1974a) who examined stromal
cell smears on different days of pregnancy in the rat and found only a few cells
in the tetraploid state during Day 5.   Rather, it is more likely that prior to
decidualization, the cells of the primary decidual zone are in the "A State"
of Smith and Martin (1973).

     The metabolism of uterine RNA during early pregnancy has been extensively
studied both with respect to the time of synthesis and the nature of the RNA
species produced.   Earlier investigations were concerned with the measurement
of total uterine RNA synthesis at relatively infrequent intervals in the pregnant
animal and in ovariectomised rats given hormonal treatments which were capable
of supporting implantation (Emmens et al., 1969; Heald & O'Grady, 1970;
J.E. O'Grady, unpublished observations).   The incorporation of a one hour pulse
of ^{3}H-uridine into the total RNA of the uterus during each day of the preimplantation period of pregnancy established the existence of a biphasic
response in synthesis during this period, with a peak of incorporation on Day 3
and a further rise on Day 5.   The experiments involving hormone replacement

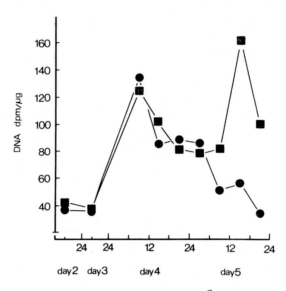

Figure 13: The incorporation of a 1 hour pulse of ^{3}H-thymidine into the DNA of the pregnant (■) and pseudopregnant (●) rat uterus during early prégnancy (O'Grady et al., 1976b).

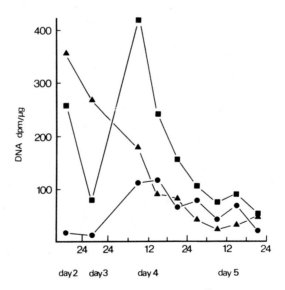

Figure 14: The incorporation of a 1 hour pulse of ^{3}H-thymidine into the DNA of luminal epithelial (▲), myometrial (●), and stromal (■) fractions of the pseudopregnant rat uterus (O'Grady et al., 1976b).

implied that this biphasic response is a necessary concomitant of the normal sequence of uterine cell division. Thus, the biphasic response in RNA synthesis could be demonstrated in rats ovariectomised on Day 2 of pregnancy and then treated with doses of progesterone and oestrogen (2 mg progesterone/ 1 μg oestrone or 2 mg progesterone/0.1 μg oestradiol per day) which will support the decidual reaction and the implantation of transferred blastocysts (DeFeo, 1967; Psychoyos, 1973a,b; O'Grady, J.E., unpublished data) (Figure 16). This regime also supports the patterns of cellular mitosis found in normal rat pregnancy (Clark, 1971; B. Hamidi, unpublished observations). Increasing the oestrogen component of the support system tenfold, i.e. 1.0 μg of oestradiol or 10 μg of oestrone, prevents the induction of the decidual response (Martin et al., 1961; Finn, 1966; DeFeo, 1967) and the biphasic synthesis of uterine RNA (Figure 16).

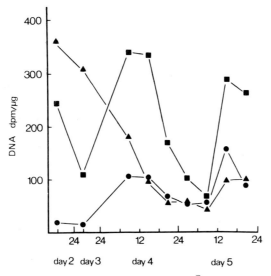

Figure 15: The incorporation of a 1 hour pulse of ^{3}H-thymidine into the DNA of the luminal epithelial (▲), myometrial (●), and stromal (■) fractions of the pregnant rat uterus.

Martin and Finn (1968) have shown that these high doses of oestrogen permit stromal division to proceed but also induce the division of the luminal epithelium.

Analysis of the type and location of RNA synthesis within the uterine cells, indicate that both these parameters alter according to the time of measurement during normal pregnancy. Sucrose density gradient centrifugation was used to fractionate the radioactive RNA synthesised during normal pregnancy and it was found that the heavier species of RNA associated with the ribosome has an increased rate of processing on Day 5 in the uterus and Day 6 in the implant, there being a reduction in the labelling of this fraction in the non-implanted uterus (Heald et al., 1972b) (Figure 17).

Examination of the location of synthesis confirmed this observation. As would be expected with a one hr pulse of radioactive precursor, the majority of the radioactive RNA synthesised in these experiments was located in the nucleus. During Day 5 in the uterus and Day 6 in the implant, radioactivity had been translocated and appeared to a greater extent in the cytoplasmic fraction indicating that more ribosomal species were being produced (Figure 18). The inter-implant tissue on Day 6 had a reduced labelling of RNA in the nuclear and cytoplasmic fractions (Heald et al., 1972a). Examination by electron microscopy of the stromal cells on the morning and afternoon of Day 5 revealed an increased production of polyribosomes during this period which is in accord with the observed increase in the processing of ribosomal RNA (Tachi et al., 1970).

An alternative approach to the problem of examining the RNA produced during early pregnancy has been to prepare total uterine chromatin and to compare the template activity of preparations isolated on different days of pregnancy (O'Grady et al., 1975; Glasser & Clark, 1975); both these studies employed bacterial RNA polymerase. In the former investigation, the template activity of the pregnant uterus mimicked the biphasic pattern of synthesis observed in vivo and showed a further marked increase in the implant on Day 6 associated with a decrease in activity in the interimplant tissue (Figure 19). It is interesting that the subtle differences in hormone secretion between the latter part of the rat oestrous cycle and the first few days of pregnancy are responsible for a relatively lower template activity during Dioestrus than during the equivalent day in early pregnancy, Day 3 (Figure 19). These observations are in broad agreement with those of Glasser and Clark (1975).

The technique of competitive DNA/RNA hybridisation has been used to compare the nature of the RNA synthesised in uterine segments in vitro and templated for by uterine chromatin isolated on differing days of pregnancy. This technique will compare mostly repetitive sequences of RNA, as these are in preponderance in these preparations, but the exclusion of cytoplasmic RNA from the preparations derived from uterine segments increased the possibility of revealing the presence of any new species which may be present on any particular day of pregnancy. This procedure also allows the detection of species which occur on any given day of pregnancy and are otherwise present in only small quantities. Segments from the uterus on Day 2 and 5, and from the implant and interimplant tissue on Day 7 were incubated with ^{3}H uridine and the nuclear RNA isolated. Similar preparations were prepared without prior incubation of the segments with radioactivity. Chromatins prepared from uteri and implants on these days of pregnancy were also employed to support the synthesis of labelled and unlabelled RNA. In a series of experiments competing the labelled and unlabelled preparations derived from

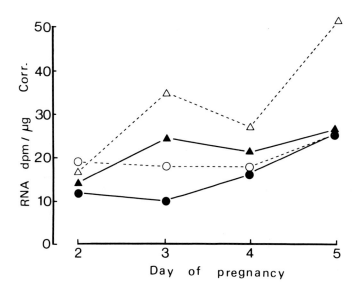

Figure 16: The incorporation of ^{3}H-uridine into the uterine RNA of pregnant rats ovariectomized on Day 2 of pregnancy. These rats received sub-cutaneous daily injection of vehicle only (●), progesterone, 2 mg (▲), progesterone, 2 mg + oestradiol, 0.1 μg (△) or progesterone + oestradiol 1 μg (o).

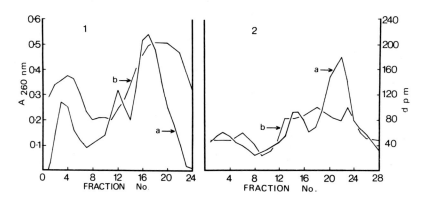

Figure 17: Sucrose density gradients of radioactive RNA extracted from rat uteri on (1) Day 4 and (2) Day 5 of pregnancy. a) is the optical density of each fraction and b) the radioactivity. Data adapted from Heald et al. (1972b).

204

tissues on different and the same days of pregnancy, it was demonstrated that
there are RNA species present on Day 5 which are not present on the other days
of pregnancy investigated (Figure 20). This result was obtained whether the
RNA was synthesised in the uterus in the presence of the endogenous RNA polymerases occurring in the nucleoli and nucleoplasm of this organ (Nicolette &
Babler, 1974) or if the RNA was synthesised in vitro in the presence of bacterial
RNA polymerase prepared from E. coli.

The foregoing outline of the synthesis of RNA in the whole uterus implies
that there are major alterations in nuclear activity in some part of the uterus
on the morning of Day 3 and again Day 5 of pregnancy. A study of the enzyme
ornithine decarboxylase which is known to increase at such times of altered
activity gives support to this view, displaying the typical biphasic response
in total uterine activity (Saunderson & Heald, 1974).

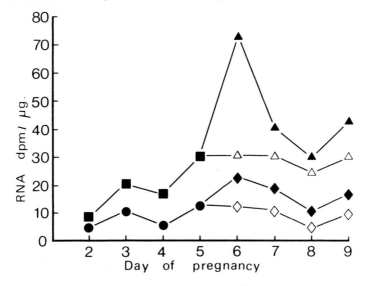

Figure 18: The incorporation of a 1 hr pulse of ^{3}H-uridine into the nuclear and
cytoplasmic fractions of the uterus of the pregnant rat. Measurements were
made in the nuclei of preimplantation (■), implanted (▲), non-implanted
(△) tissue and in cytoplasm of the same tissues designated as (●), (◆)
and (◇) respectively. Data adapted from Heald et al. (1972a).

The question of the location of RNA synthesis during the preimplantation
period in the rat has been investigated by exposure of the uterus to ^{3}H-uridine
in vivo followed by autoradiography (Tachi et al., 1972) or by the separation of
myometrial, stromal and luminal epithelial fractions (Heald et al., 1975a).
The latter experiments were carried out in unilaterally ovariectomised rats
which provide a pregnant and pseudopregnant horn. The autoradiographic approach,

although providing information about localisation in individual cells and regions
of the uterus, was insufficiently quantitative or defined in time to allow a
precise picture of the distribution of activity to be assessed.

The approach, based on the fractionation of the uterus and employing a one
hr pulse of precursor at six hourly intervals, permitted the pinpointing of an
important and early divergence between the metabolism of the pregnant and
pseudopregnant uteri.   When the synthesis of RNA in the unfractionated pregnant
horn is compared to that in the pseudopregnant horn, it is apparent that there
is a marked rise in the pregnant uterus at 09.00 hr on Day 5 (Figure 21) which
is present in all fractions but is greatest in the stromal fraction (Figure 22).

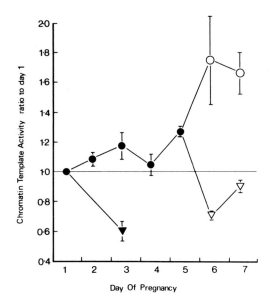

Figure 19:   The template activity of chromatin preparations isolated from the
uterus on Days 2 to 5 of pregnancy (●), from implanted (O) and non-implanted
(∇) tissue of Days 6 and 7 and from the uterus of cycling rats during
Dioestrus (▼).   The activities of the chromatins from pregnant uteri were
expressed in relation to the template activity of Day 1 chromatin.   The activity
of Day 1 Dioestrus chromatin is compared to that of an Oestrus preparation.
The template activity of Day 1 chromatin was 8.61 ± 0.59 (mean $\pm$ S.E.M.)pmol
of UMP incorporated/mg DNA per 10 m (O'Grady et al., 1975).

Comparison of the total amount of stromal RNA being synthesised in the two horns
shows that activity in the pregnant horn exceeds the pseudopregnant as early as
03.00 hr on Day 5 (Heald et al., 1975).   At this time and before 12.00 - 16.00 hr
on Day 5, the blastocyst is free floating in the uterine lumen (Tachi et al.,
1970), but even so is the most likely causative agent for the differences in
synthesis.

Phospholipids are important components in cellular and nuclear membranes and in the structure of polysomes and so their synthesis will be important in the processes of nuclear and cellular replication and the organisation of ribosomal RNA into the structure of the polysome. Incorporation of ^{3}H-choline into cell fractions of pregnant and pseudopregnant uteri have been found to reflect these processes (O'Grady et al., 1976a,b) although no differences were observed between the horns (Figure 23). On Day 4, there were increases in phospholipid synthesis in the stroma and myometrium, which probably reflect mitoses in these fractions and the decline in their frequency with the cessation of this process. Conversely, there is no luminal epithelial division on Day 4 and no increase in synthesis of phospholipid (Figure 24). On Day 5 at 09.00, there was a further rise in synthesis in the stroma which was not associated with

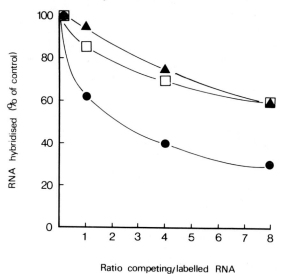

Figure 20: Competition in RNA : DNA hybridisation between ^{3}H-labelled RNA and unlabelled RNAs synthesised in vitro using chromatin templates obtained from uterine nuclei at different stages of pregnancy in the rat. 10 μg of labelled day 5 RNA was competed with 10-80 μg of unlabelled RNA templated by chromatin from day 2 (□) and day 5 (●) uteri and day 7 (▲) implantation site. (O'Grady et al., 1975).

DNA synthesis or mitosis (Figure 24) but was probably related to the formation of polysomes and endoplasmic reticulum in the stroma (Tachi et al., 1970), the former being associated with the increase in RNA synthesis in this fraction during the morning of Day 5 (Heald et al., 1975b).

Evidence provided by comparing nucleic acid synthesis in the cellular fractions of the pregnant and pseudopregnant uteri during the course of Day 5 has thus indicated that the presence of the blastocyst may be responsible for

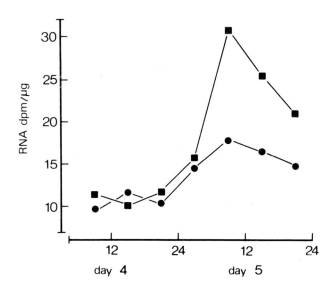

Figure 21: The incorporation of ^{3}H-uridine into the RNA of the pregnant (■) and contralateral pseudopregnant (●) rat uterus on Days 4 and 5 post-coitum. Data adapted from Heald et al. (1975b).

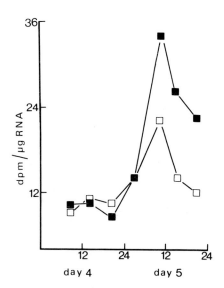

Figure 22: The incorporation of ^{3}H-uridine into the RNA of the stromal fraction of pregnant (■) and contralateral pseudopregnant (□) rat uteri on Days 4 and 5 post-coitum. Data adapted from Heald et al. (1975b).

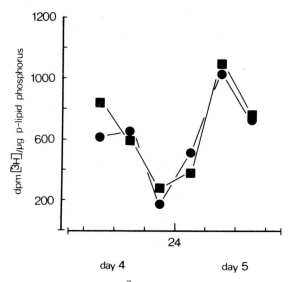

Figure 23: The incorporation of ^{3}H-choline into the phospholipids of the pregnant (●) and pseudopregnant (■) rat uterus.   (O'Grady et al., 1976b).

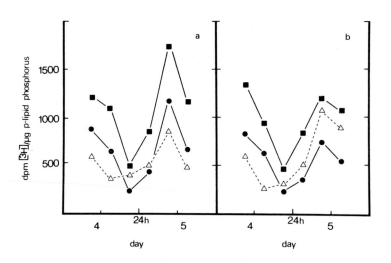

Figure 24: The incorporation of ^{3}H-choline into the phospholipids of the luminal (△) and glandular (●) epithelial and stromal (■) fractions of (a) the pregnant and (b) the pseudopregnant rat uterus.

the relative increase in stromal RNA synthesis in the pregnant horn at 09.00 hr
and for the later stromal DNA synthesis in this horn at 15.00 hr. Nonetheless,
the measurement in RNA synthesis in the pregnant horn could conceivably be
induced by other bodies or agents which are incapable of promoting a later
decidual response.

Experiments have been performed in which RNA synthesis was measured in the
stromal fractions of unilaterally ovariectomised animals mated with vasectomised
males. These rats have then a pseudopregnant horn containing unfertilised ova
which are similar in size to a blastocyst but have no zona pellucida and
another having no ovum or zygote and being identical to the pseudopregnant horns
of the previous experiments (see Heald et al., 1975b). In these animals, there
was no difference in the rates of incorporation of ^{3}H-uridine into the stromal
RNA in the two horns (Table 1). The presence of a body similar in size to the
blastocyst but incapable of inducing decidualization was unable to induce a
similar degree of RNA synthesis in the stroma. Moreover, it is possible that
the blastocyst may release or induce the release of some blood-borne agent which
is capable of stimulating RNA synthesis in the contralateral pseudopregnant
horn. There was a significant reduction in synthesis between both horns from
animals mated with vasectomised males, when compared with the pseudopregnant
horn from the normally mated animal (Table 1).

Intra-luminal injections of arachis oil on Day 5 will induce the
decidualization of the pseudopregnant uterus of the intact rat and of the
ovariectomised rat which has been treated with an appropriate regime of progesterone and oestrogen, but is ineffective when such animals are treated with
similar doses of progesterone alone (Finn & Martin, 1972). When 50 μl of
arachis oil was injected at 06.00 hr on Day 5 into the uterine horn without an
ovary in unilaterally ovariectomised rats mated with vasectomised males, the
RNA synthesis in the stroma of the treated horn at 09.00 hr was greater than
that in the horn containing only ova. The level of synthesis in the untreated
horn was similar to that in the pseudopregnant uterus in the presence of a contralateral pregnant horn (Table 1). Again, an agent which is capable of inducing
decidualization also stimulated the synthesis of stromal RNA on the morning
before decidualization and partially elevated synthesis in a contralateral
pseudopregnant horn. These experiments provide further support for the contention that the presence of the blastocyst in the uterine lumen before attachment provides a stimulus for stromal RNA synthesis which is essential for the
later formation of decidual cells.

Cyclic AMP and prostaglandin metabolism

The activity and metabolism of cyclic adenosine 3'5' monophosphate (cAMP)
are implicated in many cellular processes. In association with cGMP this

TABLE 1

THE INCORPORATION OF ^{3}H-URIDINE INTO THE STROMAL RNA OF PREGNANT, PSEUDO-
PREGNANT AND OIL-TREATED PSEUDOPREGNANT UNILATERALLY-OVARIECTOMIZED RATS*

	right horn	Stromal RNA specific activity dpm/µgm DNA (mean ± S.E.M.) left horn	Number of Experiments
Day 5, 09.00 h			
Pregnant	33.48 ± 3.80	22.83 ± 3.84	5
Pseudopregnant	16.32 ± 4.29	14.40 ± 4.78	4
Pseudopregnant + 50 µl Oil	8.77 ± 0.69	11.81 ± 0.72	4
Day 5, 15.00 h			
Pregnant	25.54 ± 3.48	14.68 ± 1.00	5
Pseudopregnant	4.43 ± 0.76	4.17 ± 0.67	4
Pseudopregnant + 50 µl Oil	14.86 ± 1.09	35.96 ± 7.4	4

* The pregnant animals have a blastocyst in the right horn only. The pseudopregnant animals have unfertilised ova (mean 7/horn) in the right horn
only. The oil-treated animals have additionally 50 µl of arachis oil in
the left horn administered at 06.00 h on Day 5.

nucleotide is important in the control of cell cycle events and differentiation,
particularly in the control of mitosis and in the precipitation of cells into
DNA synthesis which are in the "A state", arrested in G^o. In vitro, exogenous
cAMP may either inhibit DNA synthesis or cause its stimulation but many workers
have shown that the physiological effect of intracellular concentrations of
cAMP is to inhibit processes associated with cell division, cGMP having the
opposite effect (see reviews, Abell & Monahan, 1973; Singhal et al., 1976).
McManus and Whitfield (1974) have pointed out that the response to cAMP differs
according to cell type and that fibroblasts, such as the uterine stroma, divide
in the presence of low levels of cAMP. Hormone action, and in particular in
the context of the uterus, some of the actions of estradiol has been shown to be
mediated via membrane-bound adenyl cyclase and increases in intracellular cAMP
concentrations (Szego & Davis 1967; Rosenfield & O'Malley, 1970; Sinhal &
Lafreniere 1972). In recognition of this multiple action, several workers have
studied the involvement of cAMP in the processes of blastocyst activation and
uterine decidualization. These experiments have taken the form of intra-

peritoneal or intraluminal administration of cAMP in solution or the measurement
of cAMP or its controlling enzymes adenyl cyclase and phosphodiesterase in the
pregnant uterus.   cAMP has been administered intraluminally to mice (Holmes &
Bergstrom, 1975) and rats (Webb & Surani, 1975; Webb, 1975) in delay and to
ovariectomized mice treated with a hormonal regime sufficient to support uterine
sensitivity (Leroy et al., 1974b).

The latter experiments show that even after massive administration of cAMP
either by the intraluminal or intraperitoneal route, the uterus was not stimulated to decidualize and that the mere raising of cAMP levels is not responsible
for the production of substances by luminal epithelium or glands which induce
stromal decidualization, or for the direct induction of that process.   The
possibility remains, however, that while epithelial differentiation may require
increased cAMP, the process of decidual transformation may be dependent on the
opposite condition: further interpretation is complicated by the results of
experiment performed with animals in diapause.   These have met with greater
success, both mice and rats are found to develop implantation sites when
oestrogen-deprived animals in diapause are treated with single doses of intraluminal cAMP.   Attempts to carry the mouse implants through to term met with
failure (Holmes & Bergstrom, 1975).   The site of action of cAMP in these experiments is difficult to assess.   The blastocysts have clearly been activated to
leave diapause and implant, either directly by the luminal cAMP or indirectly by
an oestrogen-mimicking effect on the luminal epithelium which has, through the
cessation of inhibitor production, allowed blastocyst development and implantation
to progress (Surani, 1975; Das et al., 1969).   Similarly, stromal mitosis, a
necessary prerequisite for decidualization, may have been induced directly by
cAMP or by a mechanism similar to that involved in histamine action which may
involve epithelial damage (Tachi et al., 1970 and later).   If the uterine
epithelium of the mouse is damaged, transferred blastocysts have been reported
to attach at all stages of the oestrous cycle (Cowell, 1969) and epithelial
degeneration is normally found in blastocyst implantation (Finn & Hinchliffe,
1964).   A direct action is unlikely as elevated cAMP levels are generally
associated with the inhibition of mitosis (Abell & Monahan, 1973).   Also, the
stimulus to decidualize is unlikely to be a direct effect as no deciduomata were
noted in animals possessing a competent stroma after intraluminal treatment with
cAMP (Leroy et al., 1974b).

The concentrations of cAMP, adenyl cyclase and phosphodiesterase in the rat
uterus have been measured in the pregnant uterus and in horns containing a Nylon
I.U.D. (Sim, 1974).   Measurements were made on a daily basis for the first seven
days of pregnancy.   After an initial drop from Day 1 to Day 2 which probably
reflected the withdrawal of proestrous oestrogen, cAMP regained its Day 1 concentration by Day 4 and showed a slight decline thereafter.   Enzyme concentrations

followed a similar pattern.   The presence of the I.U.D. induced a general increase in all these parameters and it was suggested that the elevated cAMP concentration was responsible for the contraceptive effect of the I.U.D. Unfortunately observations were too infrequent to be open to detailed inter- pretation except to imply that the low levels of cAMP on Day 2 and later on Day 6 and 7 would be compatible with DNA synthesis and cell division and that the higher levels on Day 4 and 5 may stimulate the synthetic functions of the uterus.

When a more frequent measurement of uterine cAMP is made, two peaks of nucleotide concentration are found.   The first at 09.00 hr on Day 4 and the second at 21.00 on Day 5; between these peaks there is a reduction in concen- tration at 15.00 and 21.00 hr on Day 4 followed by a slow recovery through Day 5 until the peak concentration is reached at 21.00 hr.   There is a rapid decline in concentration in both the implant and interimplant tissue to low values throughout the early morning of Day 6 (Swift & O'Grady, 1976) (Figure 25).   All these observations are made in the total uterus and they will be derived from the combination of the low levels of cAMP associated with cell division in one popu- lation of cells and the higher concentrations required for the expression of synthetic processes in others.   Nevertheless, periods of active mitosis either in the epithelia, stroma or the decidualizing tissue, may be associated with a lowering in the level of cAMP.   In particular on Day 4, stromal mitosis is pro- ceeding rapidly and is accompanied by falling levels of uterine cAMP followed by a marked rise over six hours between 21.00 and 06.00 hr on Day 5.   Cells are known to be in the "A state" under the influence of such elevations in cAMP con- centration (MacManus & Whitfield, 1974; Otten et al., 1972; Short et al., 1974) and the rise in uterine cAMP may reflect a change in the status of the SEAM population of the stroma from the B phase to the A state of Smith and Martin (1973) (Figure 25).   It is clear, given the heterogeneity of the uterus that such interpretations are, at best, conjectural.   These studies indicate that cAMP is probably involved in the control of uterine cell function and division during early pregnancy, although the experiments of Leroy et al. (1974b) indicate that elevated levels of cAMP are unlikely to be directly involved in the decidual transformation of the stroma.   The position may be clarified by the measurement of cGMP in the total uterus and the levels of adenyl cyclase and phosphodiesterase in isolated luminal epithelial and stromal fractions of the pregnant uterus.

The intracellular levels of cAMP and cGMP are known to be controlled by Prostaglandin E and $F_{2\alpha}$ respectively.   Changes in the ratio between PGE and $F_{2\alpha}$ produce a similar change in the ratio of cAMP and cGMP (Singhal et al., 1976). Recently, Tobert (1976) has shown that indomethicin, an inhibitor of Prostaglandin synthetase, is capable of suppressing the formation of deciduomata when pseudo- pregnant rats were treated subcutaneously 2 hr before or 8 hr after the intra- luminal administration of olive oil on Day 5 of pregnancy.   This result implicates

the prostaglandins in the induction of the decidual stimulus. Attempts to
induce decidualization in pseudopregnant rats by the instillation of PGE_2 and
$PGF_{2\alpha}$ on Day 5 were without success (Tobert, 1976)

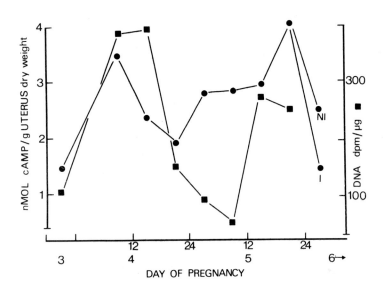

Figure 25: DNA synthesis in the stromal fraction of the uterus (■) compared
with the uterine concentration of cAMP (●) during early pregnancy in the rat.
NI = non-implanted and I = implanted uterine tissue. DNA synthesis data from
O'Grady et al., (1976) and cAMP concentrations from Swift and O'Grady (1976).

A MODEL FOR THE DEVELOPMENT OF ENDOMETRIAL SENSITIVITY
AND DECIDUALIZATION - THE INVOLVEMENT OF THE CELL CYCLE IN DIFFERENTIATION

Ova-implantation and decidualization, integral parts of implantation in
rodents, are complex phenomena resulting from the hormonally directed and
temporally defined operation of three systems: maturation of the blastocyst,
maturation of the endometrium and events arising from the interaction of these
two systems.

The complexity and interdependance of the processes involved has hindered
rigorous analysis and the formulation of models upon which further progress in
this field will be based. Before proceeding we will consider the key systems
the authors believe constitute the process of implantation and will function as
a framework for the conceptual model.

1. Transient production of a metastable, "sensitive" luminal epithelium
characterised by its competence to:

   a) activate the blastocyst.

   b) receive and transmit a stimulus/signal from the blastocyst to the
       stromal tissue.

   c) activate "programmed cell death" in response to the stimulus from the
       blastocyst.

2. Transient production of a metastable "sensitive" stromal cell population
(hereafter designated as presumptive decidual cells, (PD cells) competent to
decidualize (from decidual cells - D cells) in response to the stimulus/signal
from the luminal epithelium.

3. Structurally and temporally defined growth and regression of the
deciduoma.

For successful implantation, a precise synchrony and interdependence must
exist between the constituent systems.   This temporal interlock is achieved by
the dependence of the production of these two "sensitive" tissues upon the same
hormonal regimen proestrous oestradiol, a minimum of 48 hr of progesterone,
followed by nidatory oestradiol, and signals operating between the two tissues
and the blastocyst.   The receptive uterus, comprising as it does the two tissues
involved in receptivity each fulfilling distinct functions, is generated by the
hormone dependent events.   The blastocyst, when present in a mature state, will
be activated by the "sensitive" epithelium (this will involve either secretion
of activators or regression of the synthesis of inhibitory factors) which in
turn will activate "programs" in the "sensitive" epithelium and stroma ensuring
coordination at a multiplicity of organisational levels.

Inherent in the model (Figure 26) is the existence of "sensitive" populations
of epithelium and stromal cells, which we consider to be in "determined" states,
i.e. committed to a specific program of gene expression, but expression of this
program at the transcriptional/translational level in differentiation requires
induction in response to signals.   In the "sensitive" luminal epithelium,
expression of the specific genetic program results in cell death and also possibly
synthesis of a signal to the stroma in response to activation of the luminal cell
membrane by the blastocyst or oil.   In the "sensitive" stroma (PD cells),
expression of the new genetic program results in transformation of the stromal
cell to a decidual cell in response to activation by the signal from the luminal
epithelium.   Both "sensitive" cell types have a transitory existence; if the
activation of the programs does not occur within a limited time-period they pass
into an insensitive state (Figures, 26, 27).

Although both "sensitive" populations are produced by identical hormonal
regimes, the hormone dependent events and the mechanism by which these arise from
the stem cells are diametrically opposed.   The differentiated "sensitive"

epithelial cell population arises via mechanisms not apparently involving prior cellular proliferation. In fact, epithelial proliferation and the production of "sensitive" cells appear to be on mutually exclusive paths of development; thus oestradiol alone induces proliferation, whereas the hormonal regime producing "sensitive" cells inhibits proliferation. In contrast, production of "sensitive" stromal cells (PD cells) is intimately linked with proliferation of the stem stromal cells (S cells) in a progestational environment. It is our thesis that the PD cells are formed from, and are inherently dependent upon, this cellular division, during which reprogramming occurs, and the division is thus essentially a quantal mitotic division.

Few workers in the field of implantation and decidualization have fully appreciated the role of the stromal cell division (defined spatially and temporally) in early pregnancy and this has resulted in a lack of systematic investigations into the kinetic analysis and role of this phenomenon (Finn & Martin, 1974; Glasser, 1972; Marcus, 1970, 1974a). Marcus (1970) has proposed

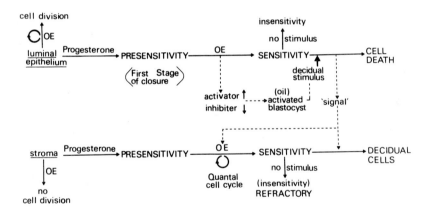

Figure 26: A model for the development of endometrial sensitivity and decidualization in the pregnant rodent uterus.

a model for the differentiation of the stromal cell, similar to our model only in its cytological basis. In the model of Marcus, an initial sub-population of stromal cells conditioned by proestrous oestrogen differentiate to give rise to a second sub-population, the "committed predecidual cells" which in response to minimal stimuli and in the presence of oestrogen and progesterone, differentiate to decidual cells. Marcus then postulates the existence of a third population of metastable cells responsive to inductive stimuli of greater magnitude (traumatic) which can differentiate in a progestational environment to become decidual cells.

In the following sections, the cytological matrix underlying the model and other pertinent concepts will be elaborated.

The role of the stroma
Stromal-decidual differentiation

The fate and role of stromal tissue in ovo-implantation is decidualization, the process initiated at implantation by which the stromal cell is reprogrammed and structurally and functionally differentiates into a decidual cell. Any model for decidualisation must explain the four major biological features of the process: production of a transient period of sensitivity to decidual stimuli, induction of decidualization, differentiation, growth and regression of the

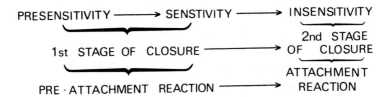

Figure 27: The relationship between the development of endometrial sensitivity and insensitivity and change in luminal morphology and blastocyst attachments.

decidua, and the hormonal dependence of the foregoing features. The process of differentiation has been defined as "those mechanisms that make information not readily available in a particular mother cell readily available in its daughter cells" (Beermann et al., 1975). This has been alternatively described as 'cellular reprogramming' (Tsanev, 1975). In the context of the differentiation of a stromal to a decidual cell, it is necessary to consider not only that new information is made "readily available", but that this information is expressed in the altered structure and function of the cells of the decidua. Galassi (1968) has shown that cells of the SEAM stroma labelled on Day 4 with ^{3}H-thymidine are destined to form labelled decidual cells in the primary decidual zone on Day 7 of pregnancy. That the transformation from a stromal to a decidual cell represents a differentiation is made manifest in the many structural and metabolic differences between the two cell types (Table 2).

TABLE 2

A STRUCTURAL AND METABOLIC COMPARISON OF PRESUMPTIVE STROMAL (PD) CELLS AND DECIDUAL CELLS

	Stromal cells	Decidual cells
Fibriller material	-	+
Smooth cisternae in the endoplasmic reticulum	-	+
Rapid polyribosome production	-	+
Glycogen granules	-	+
Alkaline phosphatase	-	+
Mitosis and DNA synthesis	-	+
Amitotic cycles and polyploidy	-	+

Stromal mitosis and decidualization

Much circumstantial evidence has accumulated suggesting a central role of SEAM stromal mitosis in decidualization, and it is our contention that its occurrence is a pre-requisite for decidualization. O'Grady et al. (1974) examined the mode of action of the anti-oestrogen Tamoxifen (ICI 46,474) in rats during early pregnancy. Treatment with an implantation inhibiting dose also abolished any stromal mitosis in the preimplantation period, whereas doses delaying implantation by 20 hr also resulted in a delay in mitosis, which was observed in the morning of Day 5, in contrast to noon Day 4 during normal preg-nancy (Figure 28). A similar relationship has been observed using the anti-oestrogens U11 100A and U11 55A which, when administered to rats in implantation-inhibiting doses, also appeared to abolish stromal mitosis on Day 4. Doses which reduced the number of implantation sites were found to dramatically reduce the degree of stromal mitosis observed (U. Siddiqui, personal communication). Several workers have shown that stromal mitosis on Day 4 of pregnancy in the rat, occurs in a definitive region of the stroma, i.e. a crescent-shaped area of the subepithelial anti-mesometrial stroma reminiscent of SEAM stroma. When combined with the observation of Galassi (1968) that decidual cells of the primary decidual zone originated from a similarly shaped region of the stroma, these phenomena lend support to the original premise that stromal mitosis is a necessary pre-requisite for decidualization. Another relevant observation is that, in the mouse and rat, concomitant with decidualization of the primary decidual zone, the deeper

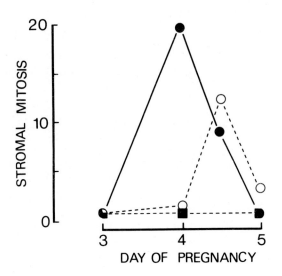

Figure 28: The effect of the anti oestrogen, Tamoxifen (ICI 46,474) on stromal mitosis in the pregnant rat uterus. Rats on Day 2 of pregnancy were dosed orally with vehicle alone (●), 0.1 mg/Kg of Tamoxifen (○) and 0.2 mg/Kg (■). Mitoses are expressed as the number of mitotic figures/100 stromal nuclei. Adapted from O'Grady et al. (1974).

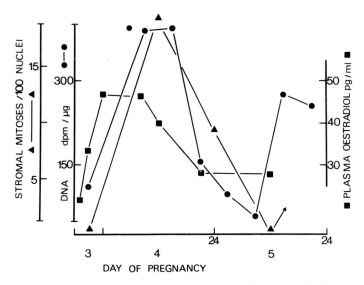

Figure 29: The relationship between plasma oestradiol concentration and DNA synthesis and mitosis in the uterine stroma of the rat. (Oestrogen - Watson et al., (1975); Mitosis - O'Grady et al., (1974); DNA synthesis - Heald et al., (1975b).

stroma undergoes a hyperplasia prior to its decidualization.   The hormonal conditioning required for stromal mitosis is identical to that for decidualization induced by the blastocyst, i.e. proestrous oestradiol, progesterone and nidatory oestradiol (Figure 29).   Further supportive evidence is that all physiological and experimental systems for producing decidualization also produce stromal mitosis.   Thus, in ovariectomized, progesterone-maintained delayed-implantation rats, where no stromal mitosis has occurred, injection of oestradiol sufficient to induce ovo-implantation induces an extensive stromal mitotic response prior to decidualization (Tachi et al., 1972).   In trauma-induced decidualization, rapid proliferation ensues, and it is possible that this in part represents stromal mitosis as well as decidual proliferation.   Glasser (1972) has also observed that decidualization may be induced in cycling rats, although no reference was made to the precise period of sensitivity in the cycle.   This observation is not incompatible with the role of stromal mitosis in decidualization, since it is found on the day of Proestrus (Marcus, 1974a) (Figure 3).

The stromal cell cycle and endometrial sensitivity

The mitosis occurring in the SEAM stroma on Day 4 of pregnancy is part of an entire cell cycle, from G1 late on Day 3 to G1 early on Day 5, from the A state into the B phase and back into the A state.   There is considerable autoradiographic evidence that prior to the onset of stromal DNA synthesis and mitosis on Day 4, there has been little or no DNA synthesis in this compartment since stromal division at Proestrus (Galassi, 1968; Leroy & Galand, 1969; Zhinkin & Samoshkina, 1967).   Experiments measuring the DNA synthesis in isolated stromal fractions also show that there is a small amount of synthesis on Day 2 and 3 of pregnancy; the little synthesis observed probably being due to glandular contamination (O'Grady et al., 1976).   The workers followed the course of DNA synthesis in the stroma at intervals of six hours throughout Days 4 and 5.   DNA synthesis was found to be closely correlated with mitosis on Day 4 (Figure 29) and no further synthesis occurred in the stroma on Day 5 until after the attachment reaction, at 15.00 hr (Figure 15).   This latter synthesis represents the onset of the decidual reaction being absent in the pseudopregnant horn (Figure 14). The measurement of RNA synthesis in stromal fractions of the endometrium at 09.00 hr on Day 5, has shown that, relative to the pseudopregnant uterus, there is an increment in synthetic activity induced by the presence of the blastocyst, free-floating in the uterine lumen (Figures 30, 22).   The increase is associated with a rise in chromatin template activity in the uterus and the production of RNA species which are not found in the uterus at other times during early pregnancy (O'Grady et al., 1975; Heald & O'Hare, 1974).   The increment in stromal RNA production observed in the pregnant uterus is immediately followed by the decidual DNA synthesis on Day 5 and is only induced by agents which are capable

220

of inducing stromal decidualization. The morning of day 5 is also characterised
by the synthesis of stromal phospholipid and ribosomal RNA in the uterus prior to
the formation of polysomes in the endometrial stroma. The production of a new
species of RNA, ribosomal RNA and a general increase in RNA synthesis has been
shown to occur in hepatocytes shortly after partial hepatectomy and is said to be
a G1 synthesis of RNA immediately preceding the DNA synthesis of S phase (Tsanev,
1975). These changes in RNA synthesis in the stroma on the morning of day 5
provide further evidence that a portion of the stroma has passed from mitosis on
day 4 and become arrested in G1, i.e. is in the A state, and is induced to pass
into the B phase by some phenomena associated with the activity of the blastocyst
early on day 5. The fall in cAMP in the uterus during day 4, followed by a rise

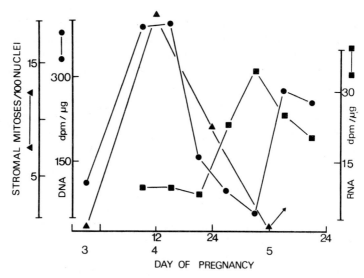

Figure 30: A comparison of mitosis, DNA and RNA synthesis in the stroma of rat
uterus from days 3 to 5 of pregnancy. (Mitosis - O'Grady et al., (1974);
DNA synthesis - O'Grady et al., (1976); RNA synthesis - Heald et al., 1975b).

in concentration throughout day 5 is consistent with this contention. Mitosis
in fibroblastic cells has been shown to be associated with low levels of cAMP
while resting cells contain elevated concentrations (MacManus & Whitfield, 1974,
Figure 25).

Several workers have proposed that the stroma is arrested in G2 on the
morning of day 5. Galassi (1968) observed that stromal DNA synthesis was
elevated on day 4 and on day 5 before the occurrence of any mitosis, and being
unaware that there had been mitosis in the stroma associated with the day 4
synthesis, assumed that many cells were arrested in the G2 state before decidual
induction. Tachi et al. (1972), who like Galassi made only daily measurements

of DNA synthesis in the total uterine horn, observed what appeared to be continous synthesis from day 4 to day 5, the former being associated with stromal mitosis. Only measurements of synthesis during the first 12 hours of day 5 could have revealed that there is a break in synthesis and that synthesis on day 5 is associated with processes in decidualization. The autoradiographic evidence of Leroy and Galand (1969) and the observations in separated stromal fractions provides considerable evidence that on day 4 the presence of oestradiol increases the transition probability of the SEAM stroma passing from the A state to the B phase, and that after a period in the A state on the morning of day 5, this transition probability is again elevated in the absence of oestradiol by the decidual stimulus (Figure 26).

The fundamental hypothesis we now advance is that the SEAM stromal cell population on the morning of day 5 of pregnancy(03.00 - 09.00 h)having traversed the cell cycle is functionally different from the maternal stromal cells (S cells) on day 3. This state will be referred to as A state (being in the A state of the cell cycle) and the cell type as a "presumptive decidual cell" (PD cell). The functional characteristics of this cellular population is its sensitivity to a decidual stimulus and its metastable nature; after receiving such a stimulus it will undergo direct transformation to a decidual cell. The transient nature of the metastable A thus would offer an explanation for the transient sensitivity to decidual stimuli observed on day 5 of pregnancy. Evidence for this correlation lies in the link between the oestradiol-induced stromal cell proliferation on day 4 and "sensitivity", which are both dependent on the same hormonal regimen.

That the post-mitotic stromal cells are different functionally is reflected by the refractory period observed after oestradiol-induced stromal proliferation. For a period of 12-36 hr post-induction, the stroma will not respond to another injection of oestradiol, suggesting that during this period the stroma is functionally altered. Marcus (1970) and Finn and Martin (1970) have previously suggested that the magnitude of "sensitivity" is a function of the number of stromal cells that participate in decidualization. Yochim and De Feo (1963) have reported that the development of maximal sensitivity was dependent primarily on the absolute level of oestrogen, thus if the role of oestrogen was to generate presumptive decidual cells (A* cells or PD cells), then the degree of sensitivity could be correlated with the number of presumptive decidual cells produced. This would provide a basis for the effects of the low doses of the anti-oestrogen U11 100A and U11 55A, where the low level of stromal proliferation would result in a reduction in the number of presumptive decidual cells thus lowering sensitivity for decidualization and the success of blastocyst implantation.

The model

The process by which stromal proliferation is involved in the differentiation of a stromal (S) to a decidual (D) cell is analogous to the systems of differentiation in which a role for a "quantal cell cycle" (Holtzer et al., 1972) or "critical mitosis" (Owens et al., 1973) is postulated. Coupling of prior cell division to differentiation is thought to be mechanistically due to critical events that are linked to the DNA synthesis (S) or the mitotic phase (M) of the cell cycle, which result in developmentally competent daughter cells qualitatively different from their progenitors. These events having occurred, altered phenotypic expression (i.e. differentiation) can take place. These concepts have been applied to several systems, including the oestrogen-primed immature chick oviduct (Oka & Schimke, 1969a,b,c), erythropoietin-sensitive haemoglobin-producing cells (Gross & Goldwasser, 1970; Paul & Hunter, 1968), cultured pancreatic exocrine cells (Wessells, 1968; Wessells & Rutter, 1969), myogenesis (Dienstman & Holtzer, 1975) and neurogenesis (Phelps & Pfeiffer, 1975).

Cell differentiation, within the concept and the model of Phelps and Pfeiffer (1975), can be envisaged as a two step process. In the first step, a stem cell A is changed to a cell B which is "determined", i.e. committed to a specific program of gene expression, but this program is not expressed. This A to B transition may be considered a true phenotypic development, is accompanied by cellular proliferation and is irreversible. Within our system (Figures 26, 34) the stem cell A would be the stromal cell (S cell) preconditioned by exposure to progesterone (State of presensitivity, Figures 26, 27) and the B cell would be the "presumptive decidual cell" (PD cell) possessing the unexpressed or unactivated decidual cell program (state of sensitivity, Figures 21, 27, 34).

In the second step, cell type B is induced either internally or externally, to actually express its organ specific function(s) and thus become cell type C. The B to C transition represents a modulation or further expression of an inherent capacity and may or may not be irreversible. In our system, the presumptive decidual cell (i.e. B cell) is induced externally by a signal from the luminal epithelium to express the decidual program. The cell type C being the decidual cell is only in a stable condition in the presence of the hormone progesterone. The B cell in our system, the presumptive decidual cell, is metastable, possessing the potential for the B to C transition for a short period, although its fate in the absence of this transition is uncertain. This metastable condition could be produced by a short-lived messenger RNA or protein stabilized only by events of the B to C transition.

Differentiation of the stroma is hormonally directed and conditioned and the relationship to the model will now be considered. Production of the pre-mitotic stromal cell (state of sensitivity) is conditioned by progesterone, enabling the cell to respond to the mitogenic action of oestradiol (oestradiol in this context may alternatively facilitate the mitogenic action of progesterone, since the latter hormone will eventually induce stromal mitosis, albeit poorly). Progesterone, in our view, is the determinant hormone responsible for reprogramming the stromal cell during the cell cycle. Mechanistically we may envisage many alternatives for the action of progesterone, but we will only consider two. It is possible that conditioning induces, or inhibits, the synthesis of a cyto-plasmic/nuclear regulatory protein(s), directly or mediated by other proteins, which have the potential to alter genetic expression but which during interphase cannot gain access to the genes. The loci of action, for example, could be a higher order genetic unit, masked by histone/acidic protein combinations and which would be deblocked at unique points in the cell cycle, e.g. S or M phase, allowing the regulator protein to deblock the loci, resulting in a daughter cell of altered genetic potential. This is essentially similar to the model of Gurdon (1975) and Gurdon and Woodland (1970). Secondly, the loci of action of progesterone-dependent regulators may be linked with events which occur at any stage of the B state (Smith & Martin, 1973, 1974), e.g. pre-DNA synthesis G1, G2 or post-mitotic G1 phases (Vonderhaar & Topper, 1974). Although we have considered oestradiol as the stromal mitogen, we must consider the alternative that the decidual signal from the epithelium or from mature decidual cells may act as such since mitosis occurs in the deep stroma at implantation and in response to trauma.

The B to C transition in our system requires a signal from the stimulated luminal epithelium, no direct stimulation will suffice since according to Fainstat (1963) insertion of a thread into the endometrium failed to induce a decidual response when the thread did not reach the luminal epithelium.

Leroy et al. (1974a) have reported that after initiation of decidualization the generation of various endoploidic states characteristic of this proliferating tissue (4n-32n) was delayed by 24-48 hr. The thesis forwarded by these authors postulates that the decidualizing stimulus acts at a locus during G1 phase resulting in shifting the stromal cells into another cycle of higher proliferation rate. From this accelerated circuit, three new cycles will successively be derived as decidualization progresses. At each of these levels, a number of G2 cells will bypass the mitotic phase to enter G1 and S and become polyploid. Another fraction of G2 cells will divide and therefore would act as feedback to the corresponding G1 compartment. Leroy et al. (1974a) account for the lag in generation of polyploidy by suggesting that cells which have progressed in the initial cycle beyond the stage of G1 phase suitable for decidual induction will have to traverse the entire cycle before returning to the G1 loci.

Our model will, however, also account for this observed lag when one considers the spatial and temporal aspects of decidualization. The SEAM stroma

undergoes direct transformation to decidual cells to form the primary decidual
zone and enters amitotic cycles but represents only a small quantitative contribution to the total of endometrial stromal cells which Leroy et al. (1974a) were
examining.   In the deeper stromal areas which rapidly proliferate prior to
decidual transformation and entrance into amitotic cycle, a delay would be
expected in the generation of polyploidy equivalent to the time taken to traverse
the critical quantal cell cycle.   Thus, the observation of Leroy et al. (1974a)
may be interpreted in terms of the masking of the changes in the SEAM stroma
population by changes occurring in the larger deep stromal population.

Another property of the decidual cell is its maintenance of its differentiated function during growth (mitotic and amitotic cycles) and the dependence of
this upon progesterone.   This dependence may be related to the necessity for
the continued synthesis of the regulatory factors (proteins), induced in our
system by progesterone (c.f. the model of Gurdon, 1975).   The finite life-span
of the decidual cell, to account for the regression of decidua in rodents which
spatially and temporally follows previous growth starting in the antimesometrial
cells and spreading outwards, may be linked to cell death programs initiated at
certain points in the mitotic cycle.

The role of the luminal epithelium

Central to our concept of the role of the epithelium, is the production of
a differentiated population of "sensitive" cells.   This population is postulated
to be metastable and if the stimulus is not received during its period of sensitivity, will pass into a refractory state (Figure 26).   The stimuli, whether
they be activated blastocysts or oil, probably act via subtle physiochemical
modifications of the apical cell membrane, facilitating the expression of a
genetic programme resulting in both transmission of a "signal" to the stroma
and programmed cell death.   The concept of programmed cell death has been
recently reviewed by Lockshin and Beulaton (1975).   The cellular reprogramming
involved in the production of the "sensitive" epithelium is analogous to the
"functional-state reprogramming" of Tsanev (1975), not involving mitotic-
dependent alterations in the genome.   In fact, it appears that production of
this "sensitive" state and proliferation is mutually exclusive.

It is postulated that progesterone conditions the cell so that instead of a
mitogenic response to oestradiol, cells in a state of "presensitivity" as a
result of progesterone treatment respond by conversion to the "sensitive" condition.   Contained in this transformation are the events which result in blastocyst activation, i.e. either induction of a blastocyst activator or inhibition
of the synthesis of an inhibitor (synthesis induced by progesterone?).

Both the production of the "signal" and epithelial cell death are postulated
to be linked with activation of the genetic programme in the "sensitive"

epithelium.   Evidence supporting this, includes the observation that implantation/deciualization is always associated with luminal epithelial cell death.   In this context, it is interesting to note that Fainstat (1963) failed to induce decidualization using a needle and thread when the thread failed to reach the luminal epithelium.

Two possible alternatives may explain the linkage of this synthesis of the stromal "signal" to epithelial cell death and the aforementioned states.   It is possible that its synthesis is linked causally with the process and events involved in the autolysis of the epithelial cells.   That is, in the "sensitive" state, stimuli induce programmed cell death and subsequently as a result synthesise and release the signal.   In trauma induced decidualization (no requirement for oestradiol) disruption and necrosis of the epithelium would also result in synthesis and release of the "signal".   In the second case, the process of cell death is delegated to a permissive role.   Here it is suggested that synthesis of the signal occurs at a low level in the progesterone-preconditioned state of presensitivity and that synthesis is augmented by the action of oestradiol in the conversion to the sensitive state.   Initiation of the process of cell death would, in the "sensitive" cells, result in release of high levels of the "signal".   Traumatic induction could arise from destruction of a large number of epithelial cells in the "presensitive" states, releasing enough "signal" to induce decidualization.

In general terms, this model represents a possible cytological matrix for the understanding of the process involved in implantation and decidualization, and may provide a basis for further experimentation in the field.

APPLICATION OF THE RODENT MODEL TO HUMAN OVO-IMPLANTATION

The human menstrual cycle has a fully competent luteal phase ("secretory phase"), a functional corpus luteum being present for half of the cycle. Implantation occurs within the secretory phase of the cycle, first contact being made by the blastocyst on the fifth to sixth day of development (De Feo, 1967). The luteal phase of the human cycle is endocrinologically equivalent to rat pseudopregnancy, the hormone secretion by the ovary being analogous in the two states.   There is a peak of oestrogen secretion on the day of ovulation which is equivalent to proestrous oestrogen in the rat.   Ovulatory secretion is followed by a second and less prominent peak of secretion, eight days later, which is analogous to the day 4 nidatory oestrogen in the rat.   Similarly, the period following ovulation is accompanied by increasing secretion of progesterone, reaching a peak of activity mid-way through the luteal phase.   This secretion is analogous to the increasing levels of progesterone secreted during the preimplantation period in the rat (Dodson et al., 1975) (Figure 31).   During the first 14 days of human gestation, equivalent to the luteal phase of the menstrual

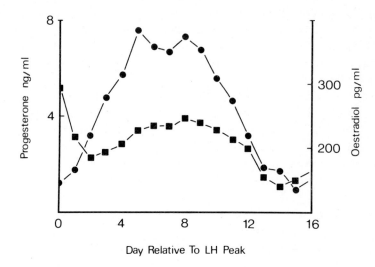

Figure 31: Systemic plasma progesterone (●) and oestradiol (■) during the secretory phase of the human menstrual cycle. Data adapted from Dodson et al. (1975).

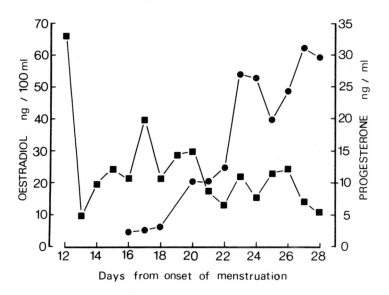

Figure 32: Systemic plasma progesterone (●) and oestradiol (■) during the early stages of human gestation. Data adapted from Sommerville, (1971).

cycle, plasma progesterone concentrations rise and are maintained, exhibiting analogy to those in the pregnant rat (Sommerville, 1971) (Figure 32). The similarity in the hormonal preconditioning of the rat uterus and the human for implantation imposes similar changes in the behaviour of the endometrial cells. Thus the glandular epithelium of the human uterus shows marked proliferation immediately following the ovulatory secretion of oestrogen, whereas the stromal cells remain quiescent. The continuous secretion of progesterone during the luteal phase prevents further glandular mitosis (Figure 33). In contrast, stromal oedema reaches a peak at the time of maximal oestrogen secretion. Stromal mitosis during the menstrual cycle and gestation is initiated immediately following the peak of oestrogen secretion and reaches a peak of activity during the cycle in the 12th post-ovulatory day (Noyes, 1973) (Figure 33). The human cycle differs from rat pseudopregnancy in that the stroma gives rise to decidua- lised cells, the 'predecidual cells', in the absence of a decidual stimulus. By day 10 post-ovulation, these cells form a cuff around the spiral arterioles, the following day they form isolated areas of predecidua beneath the luminal epithelium, and by day 13 they have coalesced to form a solid mass (Noyes, 1973). The predecidua, being subepithelial, have a similar location to that of the rodent SEAM stroma. The formation of these cells are contemporary with stromal mitosis. Stromal mitoses are only found in conjunction with predecidua, which implies that a link exists between stromal mitosis and the occurrence of pre- decidual cells (Noyes, 1973) (Figure 33).

The stromal (S) cells of the rat endometrium having undergone mitosis in the presence of progesterone become presumptive decidual (PD) cells, but this potential is not expressed in the formation of decidual (D) cells unless exposed to a decidual stimulus (Figure 26). In contrast, the presensitised stromal cell of the human uterus is capable of spontaneously forming 'predecidual' cells (Figure 34). In the light of the rodent model, it is conceivable that this is a quantal mitosis and that the presumptive decidua so formed do not require further induction and become differentiated functional decidual cells. These cells have passed from the S condition through a quantal mitosis and the daughter cells immediately express their cellular reprogramming as decidualized cells (D) omitting the sensitive condition as the presumptive decidual cell (PD). During myogenesis, Dienstman and Holtzer (1975) have shown that as the developing muscle cell passes through the compartments of the myogenic lineage by a series of quantal mitoses, the synthetic activity of each cell type is expressed post- mitotically, and is not contingent on the presence of an inducer. Similarly, human decidual differentiation do not appear to require the post-mitotic presence of an inducer for the expression of decidual function. The transformation of the stromal cell to a 'predecidual cell', as with all terminal differentiation is an irreversible process (Tsanev, 1975). Should a fertilised ovum not be

228

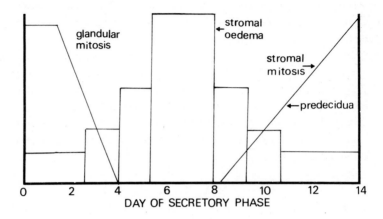

Figure 33: Glandular mitosis, stromal oedema and mitosis and predecidua formation during the secretory phase of the human menstrual cycle. Adapted from Noyes, (1973).

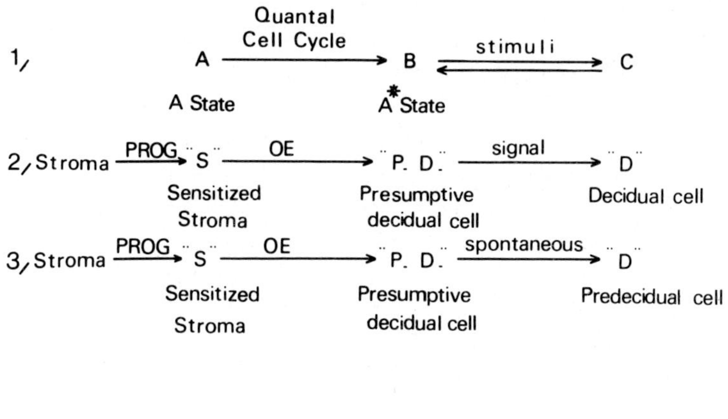

1/   A $\xrightarrow{\text{Quantal Cell Cycle}}$ B $\underset{}{\overset{\text{stimuli}}{\rightleftharpoons}}$ C

A State $\overset{*}{A}$ State

2/ Stroma $\xrightarrow{\text{PROG}}$ S $\xrightarrow{\text{OE}}$ P. D. $\xrightarrow{\text{signal}}$ D

      Sensitized   Presumptive   Decidual cell
      Stroma   decidual cell

3/ Stroma $\xrightarrow{\text{PROG}}$ S $\xrightarrow{\text{OE}}$ P. D. $\xrightarrow{\text{spontaneous}}$ D

      Sensitized   Presumptive   Predecidual cell
      Stroma   decidual cell

Figure 34: The Quantal cell cycle and differentiation. 1) The generation of daughter cells and the expression of the differentiated state. 2) Differentiation of decidual cells from stromal cells in the rat uterus. 3) Differentiation of predecidual cells from stromal cells in the human uterus.

present, the predecidua must be removed before a new population of stroma cells can be formed. This removal occurs at the time of menstruation, and a new population of stroma is generated during the proliferative place of the cycle in association with glandular regeneration (Noyes, 1973). Rat deciduomata suffer a similar fate, regression occurring even in the presence of elevated levels of progesterone (Selye et al., 1942; Atkinson, 1944). Following the regression of deciduomata, previously induced in pseudopregnant rats, there is a failure in any pregnancies initiated during the following two weeks of oestrous cycle but when sufficient stromal regeneration has occurred, the formation of a competent decidua is assured (De Feo, 1967).

The fate of the presumptive decidual cells (PD) formed in the SEAM stroma of the pseudopregnant rat is unknown. If the production of these cells after stromal mitosis represents a terminal differentiation, they may regress during the next oestrous cycle. The sensitised stromal cells formed during the oestrous cycle of animals with a long luteal phase but which have no menstrual loss, may suffer a similar fate.

The predecidua of the human uterus differs from the rodent decidua in that prolonged support of the human tissue with oestrogen and progesterone leads to the formation of extensive and persistant stromal decidualization (Eichner et al., 1951). This decidual tissue is atrophic in appearance and its formation is accompanied by glandular involution but nevertheless persists until menstruation is precipitated by progesterone withdrawal (Kistner, 1973).

Thus, although there is cirumstantial evidence that the stroma of the human uterus requires a similar regime of hormonal conditioning and quantal division if it is to decidualise, there are several points of divergence between the human and rodent systems. In the human, the expression of decidual structure and function of the reprogrammed daughter cell derived from the stroma, does not require induction by stimuli ultimately derived from blastocyst-epithelial interaction. Neither do the decidual cells derived from the human endometrium have the same preprogrammed limitation in lifespan which is characteristic of the rodent decidua.

ACKNOWLEDGEMENTS

We are indebted to Professor P.J. Heald and Dr. B. Green for stimulating discussion and continuing support in the development of this work. J.E. O'Grady is indebted to the Medical Research Council for support, being part of generous grants (G 863/1-2) to Professor Heald for the establishment of a Research Group in the Biochemistry of Reproduction. S.C.B. is indebted to the Ford Foundation for a Research Fellowship and for research facilities.

REFERENCES

ABELL, C.W. & MONAHAN, T.M. (1973) The role of 3',5'-cyclic monophosphate in the regulation of mammalian cell division. J. Cell Biol. 59, 549-558.

ALDEN, R.H. (1947) Implantation of the rat egg. 2) Alteration in osmophilic epithelial lipids of the rat uterus under normal and experimental conditions. Anat. Rec. 97, 1-13.

ALLEN, E. & DOISY, E.A. (1923) An ovarian hormone. J. Am. Med. Assoc. 81, 819-821.

ALLEN, E., SMITH, G.M. & GARDNER, W.U. (1937) Accentuation of the growth effect of theelin on genital tissues of the ovariectomized mouse by arrest of mitosis with colchicine. Amer. J. Anat. 61, 321-341.

ALLEN, W.M. (1931) Cyclical alterations of the endometrium of the rat during the normal cycle, pseudopregnancy and pregnancy. Anat. Rec. 48, 65-84.

ANDERSON, J.N., PECK, E.J. Jr., & CLARK, J.H. (1975) Estrogen-induced uterine responses and growth: Relationship to receptor estrogen binding by uterine nuclei. Endocrinology 96, 160-167.

ARIAS-STELLA, J. (1973) Gestational endometrium. In: H.J. Norris (Ed.), The Uterus, Williams and Wilkins Co., Baltimore, pp.185-212.

ATKINSON, W.B. (1944) The persistence of deciduomata in the mouse. Anat. Rec 88, 271.

BARRACLOUGH, C.A., COLLIE, R., MASA, R. & MARTINI, L. (1971) Temporal relationships which exist amongst plasma LH, ovarian secretion rates, and peripheral plasma progesten concentrations in the rat. Effects of and exogenous gonadotrophins. Endocrinology 88, 1437-1447.

BARNEA, A. & GORSKI, J. (1970) Estrogen-induced protein. Time course of synthesis. Biochemistry, N.Y. 9, 1899-1904.

BATTELLINO, L.J., SABULSKY, J. & BLANCO, A. (1971) Lactate dehydrogenase isoenzymes in rat uterus: changes during pregnancy. J. Reprod. Fert. 25, 393-399.

BEALL, J.R. (1972) Uterine lipid metabolism - A review of the literature. Comp. Biochem. Physiol. 42, 175-195.

BEALL, J.R. & WERTHESSEN, N.T. (1971) Lipid metabolism of the rat uterus after mating. J. Endocr. 51, 637-644.

BELL, S.C., REYNOLDS, S. & HEALD, P.J. (1967a) Presumptive induced protein synthesis in the rat uterus during early pregnancy. J. Endocr. 68, 34-35P.

BELL, S.C., REYNOLDS, S. & HEALD, P.J. (1976b). Uterine protein synthesis during the early stages of pregnancy in the rat. J. Reprod. Fert. (in press).

BEERMANN, W., HOLTZER, H., REINERT, J. & URSPRUNG, H. (1975) Preface. In: J. Reinert and H. Holtzer (Eds.), Cell Cycle and Cell Differentiation, Springer Verlag, Berlin - Heidelberg - New York.

BEIER, H.M. (1974) Oviducal and uterine fluids. J. Reprod. Fert. 37, 221.

BERGSTROM, S. & NILSSON, O. (1975) Embryo-endometrial relationship in the mouse during activation of the blastocyst by oestradiol. J. Reprod. Fert. 44, 117-120.

BINDON, B.M. (1969) The incorporation of ^{131}I-iodinated human serum albumin in the ovary and uterus before implantation in the mouse. J. Endocr. 45, 543-548.

BINDON, B.M. (1969) Follicle-stimulating hormone content of the pituitary gland before implantation in the mouse and rat. J. Endocr. 44, 349-356.

BLANDAU, R.J. (1961) Biology of eggs and Implantation. In: W.C. Young (Ed.) Sex and Internal Secretions, Vol.2. (3rd ed.) Williams and Wilkins, Baltimore, p.797.

BOOT, L.M., & MUHLBOCK, O. (1953) Transplantations of ova in mice. Acta Physiol. Pharm. Neerl. 3, 133.

BOSHIER, D.P. & HOLLOWAY, H. (1973) Effect of ovarian steroid hormone on histochemically demonstrable lipids in the rat uterine epithelium. J. Endocr. 56, 59-67.

BROUHA, L. (1928) Production of placentomata in rats injected with anterior hypophysial fluid. Proc. Soc. Exp. Biol. Med. 25, 488-489.

BURGER, V.H. (1966) Uber den Lipidstoffwechsel im Endometrium der Frau. Geburtsh. Frauenheilk. 26, 663-668.

CANIVENC, R. & M. LAFFARGUE (1957) Survie des blastocysts de rat en l'absence d'hormones ovariennes. Compt. Rend. 245, 1752-1754.

CHALLIS, J.R.G., HEAP, R.B. & ILLINGWORTH, D.V. (1971) Concentrations of oestrogen and progesterone in the plasma of pregnant, non-pregnant and lactating guinea pigs. J. Endocr. 51, 333-345.

CHAUDHURY, R.R. & SETHI, A. (1970) Effects of an intra-uterine contraceptive device on mitosis in the rat uterus on different days of pregnancy. J. Reprod. Fert. 22, 33-40.

CHRISTIE, G.A. (1966) Implantation of the rat embryo: glycogen and alkaline phosphatases. J. Reprod. Fertil. 12, 279-294.

CLARK, J., ANDERSON, J.N. & PECK, E.J. Jr. (1972) Receptor-estrogen complex in the nuclear fraction of rat uterine cell during the estrous cycle. Science 176, 528-530.

COCHRANE, R.L., & MEYER, R.K. (1957) Delayed implantation in the rat induced by progesterone. Proc. Soc. Exptl. Biol. Med. 96, 155-159.

CORNER, G.W. & WARREN, S.L. (1919) Influence of the ovaries upon the production of artificial deciduomata: confirmatory studies. Anat. Rec. 16, 168-169.

COWELL, T.P. (1969) Implantation and development of the mouse eggs transferred to the uteri of non-progestational mice. J. Reprod. Fert. 19, 239-245.

DAS, R.M. (1972) The effects of oestrogen on the cell cycle in epithelial and connective tissues of the mouse uterus. J. Endocr. 55, 21-30.

DAS, R.M. & MARTIN, L. (1973) Progesterone inhibition of mouse uterine epithelium proliferation. J. Endocr. 59, 205-206.

DASS, C.M.S., MOHLA, S., & PRASAD, M.R.N. (1969) Time sequence of action of oestrogen and protein synthesis in the uterus and blastocyst during delayed implantation in the rat. Endocrinology. 85, 528-536.

DEANESLY, R. (1960) Implantation and early pregnancy in ovariectomized guinea-pig. J. Reprod. Fert. 1, 242-248.

DEFOE, V.J. (1967) Decidualization. In: R.N. Wynn. (Ed.), Cellular Biology of the Uterus, Appleton-Century-Crofts, New York, pp.190-240.

DICKMAN, Z & NOYES, R.W. (1960) The fate of ova transferred into the uterus of the rat. J. Reprod. Fertility 12, 197-212.

DICKMAN, Z., SEN GUPTA, J. & DEY, S.K. (1976) Does implantation depend on 'blastocyst oestrogen'? Proc. Society for the Study of Fertility. (in press).

DIENSTMAN, S.R. & HOLTZER, H. (1975) Myogenesis: A cell lineage interpretation. In: J. Reinert and H. Holtzer (Eds.), Cell Cycle and Cell Differentiation, Springer Verlag, Berlin-Heidelberg-New York, pp.1-25.

DODSON, K.S., COUTTS, J.R.T. & MACNAUGHTON, M.C. (1975) Plasma sex steroids and gonadotrophin patterns in human menstrual cycles. Brit. J. Obstet. Gynaec. 82, 602-614.

DOYLE, L.L., GATES, A.H., & NOYES, R.W. (1963) Asynchronous transfer of mouse ova. Fertility and Sterility 14 : 215.

DUPON, C & KIM, M.H. (1973) Peripheral plasma levels of testosterone, androstenedione, and oestradiol during the rat oestrous cycle. J. Endocr. 59, 653-654.

DUPONT-MAIRESS, N. & GALAND, P. (1975) Estrogen action: Induction of the synthesis of a specific protein (IP) in the myometrium, the stroma and the luminal epithelium of the rat uterus. Endocrinology 96, 1587-1591.

EICHNER, E., GOLER, G.G., REED, J. & GORDON, M.B. (1951) The experimental production and prolonged maintenance of decidua in the non-pregnant women. Amer. J. Obst. Gynec. 61, 253-264.

ELFTMAN, H. (1963) Estrogen-induced changes in the golgi apparatus and lipid of the uterine epithelium of the rat in the normal cycle. Anat. Rec. 146, 139-143.

EL-SHERSHABY, A.M. & HINCHLIFFE, J.R. (1976) Epithelial autolysis during implantation of the mouse blastocyst: an ultrastructural study. J. Embryol. exp. Morph., in press.

ENDERS, A.C. & SCHLAFKE, S. (1967) A morphological analysis of the early implantation stages in the rat. Amer. J. Anat. 120, 185-226.

ENDERS, A.C. & SCHLAFKE, S. (1969) Cytological aspects of trophoblast-uterine interaction in early implantation. Amer. J. Anat. 125, 1-30.

FAINSTAT, T. (1963) Extracellular studies of the uterus. 1) Disappearance of the discrete collagen in endometrial stroma during various reproductive states in the rat. Amer. J. Anat. 112, 337-367.

FARQUAHAR, M.G. & PALADE, G.E. (1963) Junctional complexes in various epithelia. J. Cell Biol. 17, 375-412.

FEHERTRY, P., ROBERTSON, D.M., WAYNFORTH, H.B. & KELLIE, A.E. (1970) Changes in the concentration of high-affinity oestradiol receptors in rat uterine supernatant preparations during the oestrous cycle, pseudopregnancy, pregnancy maturation and after ovariectomy. Biochem. J. 120, 837-844.

FIEL, P.D. & BARDIN, C.W. (1975) Cytoplasmic and nuclear progesterone receptors in the guinea pig uterus. Endocrinology 97, 1398-1407.

FINN, C.A. (1966) The initiation of the decidual cell reaction, in the uterus of the aged mouse. J. Reprod. Fert. 11, 423-428.

FINN, C.A. (1971) Biology of decidual cells. In: M.W.H. Bishop (Ed.) Adv. Reprod. Physiol. Vol.5, Elek, London, pp.1-26.

FINN, C.A. & HINCHLIFFE, J.R. (1964) The reaction of the mouse uterus during implantation and deciduoma formation as demonstrated by changes in the distribution of alkaline phosphatase. J. Reprod. Fert. 8, 331-338.

FINN, C.A. & KEEN, P.M. (1962a) Studies on deciduomata formation in the rat. J. Reprod. Fert. 4, 215-216.

FINN, C.A. & LAWN, A.M. (1968) Transfer of cellular material between the uterine epithelium and trophoblast during the early stages of implantation. J. Reprod. Fert. 15, 333-336.

FINN, C.A. & MCLAREN, A. (1967) A study of the early stages of implantation in mice. J. Reprod. Fert. 13, 259-267.

FINN, C.A. & MARTIN, L. (1967) Patterns of cell division in the mouse uterus during early pregnancy. J. Endocr. 39, 593-597.

FINN, C.A. & MARTIN, L. (1966a) Hormone secretion during early pregnancy in the mouse. J. Endocr. 45, 57-65.

FINN, C.A. & MARTIN, L. (1970) The role of oestrogen secreted before oestrus in the preparation of the uterus for implantation in the mouse. J. Endocr. 47, 431-438.

FINN, C.A. & MARTIN, L. (1971) Endocrine control of the proliferation and secretion of uterine glands in the mouse. Acta Endocr. Suppl. 155, 139.

FINN, C.A. & MARTIN, L. (1972) Endocrine control of the timing of endometrial sensitivity to a decidual stimulus. Biol. Reprod. 7, 82-86.

FINN, C.A. & MARTIN, L. (1973) Endocrine control of gland proliferation in the mouse uterus. Biol. Reprod. 8, 585-588.

FINN, C.A. & MARTIN, L. (1974) The control of implantation. J. Reprod. Fert. 39, 195-206.

FINN, C.A. & POLLARD, R.M. (1973) The influence of the oestrogen secreted before oestrus on the timing of endometrial sensitivity and insensitivity during implantation. J. Endocr. 56, 619-620.

FINN, C.A., MARTIN, L., & CARTER, J. (1969) A refractory period following oestrogenic stimulation of cell division in the mouse uterus. J. Endocr. 44, 121-126.

FUXE, K. & NILSSON, O. (1963) The effect of oestrogen on the histology of the uterine epithelium of the mouse. Exp. Cell Res. 32, 109-117.

GALASSI, L. (1968) Autoradiographic study of the decidual cell reaction in the rat. Develop. Biol. 17, 75-84.

GIDLOWSKI, J.A. & MULDOON, T.G. (1974) Estrogenic regulation of cytoplasmic receptor populations in estrogen-responsive tissues of the rat. Endocrinology 95, 1621-1629.

GILLMAN, J. (1940) Fat: an index of oestrogen and progesterone activity in the human endometrium. Nature, 146, 402.

GILLMAN, J. (1941) The lipids in the human endometrium during the menstrual cycle and pregnancy and their relationship to the metabolism of oestrogen and progestagen. S. Afr. F. Med. Sci. 6, 59-81.

GLASSER, S.R. (1972) The uterine environment in implantation and decidualization. In: H.A. Balin and S.R. Glasser (Eds.), Reproductive Biology, Excerpta Medica Fdn., Amsterdam, pp.776-883.

GLASSER, S.R. & CLARK, J.H. (1975) A determinant role for progesterone in the development of uterine sensitivity to decidualization and ovoimplantation, pp.311-345. Acad. Press Inc., New York.

GLENISTER, T.W. (1965) The behaviour of trophoblast when blastocysts effect nidation in organ culture. In: The Early Conceptus, Normal and Abnormal. Ed. W.W. Park, Livingstone, Edinburgh, 24-33.

GORE-LANGTON, R.E. & SURANI, M.A.H. (1976) Uterine luminal proteins of mice. J. Reprod. Fert. 46, 271-274.

GOSWAMI, A., KAR, A.B. & CHOWDHURY, S.R. (1963) Uterine lipid metabolism in mice during the oestrous cycle: effect of ovariectomy and replacement therapy. J. Reprod. Fert. 6, 287-295.

GROBSTEIN, C. & COHEN, J. (1965) Collagenase: effect on morphogenesis of embryonic salivary epithelium in vitro. Science 150, 626-628.

GROSS, M., & GOLDWASSER, E. (1970) On the mechanism of erythropoietin-induced differentiation. VII. The relationship between stimulated deoxyribonucleic acid synthesis and ribonucleic acid synthesis. J. Biol Chem. 245, 1632-1636.

GURDON, J.B. (1975) Nuclear transplantation and the cyclic reprogramming of gene expression. In: J. Reinert and H. Holtzer (Eds.)., Cell Cycle and Cell Differentiation, Springer Verlag, Berlin-Heidelberg-New York, pp.123-131.

GURDON, J.B. & WOODLAND, H.R. (1970) On the long term control of nuclear activity during cell differentiation. Curr. Top. Develop. Biol. 5, 39-70.

HALL, K. (1969) Uterine mitosis, alkaline phosphatase and adenosine triphosphatase during development and regression of deciduomata in pseudopregnant mice. J. Endocr. 44, 91-100.

HAM, K.N., HURLEY, J.V., LOPATA, A. & RYAN, G.B. (1970) A combined isotopic and electron microscope study of the response of the rat uterus to exogenous oestradiol. J. Endocr. 46, 71-81.

HARKNESS, M.L.R., HARKNESS, R.D. & MORALEE, B.E. (1956) Loss of collagen from the uterus of the rat after ovariectomy and from the non-pregnant horn after parturition. Quart. J. Exp. Physiol. 51, 254.

HEALD, P.J. & O'GRADY, J.E. (1970) The uptake of ^{3}H-uridine into the nucleic acids of the rat uterus during early pregnancy. Biochem. J. 117, 65-71.

HEALD, P.J. & O'HARE, A. (1973) Changes in rat uterine nuclear RNA during early pregnancy. Biochim. biophys. Acta. 324, 86-92.

HEALD, P.J., O'GRADY, J.E. & MOFFAT, G.E. (1972a) The incorporation of ^{3}H-uridine into nuclear RNA in the uterus of the rat during early pregnancy. Biochim. biophys. Acta. 281, 347-352.

HEALD, P.J., GOVAN, A.D.T. & O'GRADY, J.E. (1975a) A simple method for the preparation of suspensions of luminal epithelial and stromal cells from rat uterus. J. Reprod. Fert. 42, 585-593.

HEALD, P.J., O'GRADY, J.E., O'HARE, A. & VASS, M. (1972b) Changes in uterine RNA during early pregnancy in the rat. Biochim. biophys. Acta 262, 66-74.

HEALD, P.J., O'GRADY, J.E., O'HARE, A. & VASS, M. (1975b) Nucleic acid metabolism of cells of the luminal epithelium of the rat uterus in early pregnancy. J. Reprod. Fert. 45, 124-138.

HENDRICKX, A. & KRAEMER, D.C. (1968) Preimplantation stages of baboon embryos. Anat. Record. 162, 111-120.

HINCHLIFFE, J.R. & EL-SHERSHABY, A.M. (1965) Epithelial cell death in the oil-induced decidual reaction of the pseudopregnant mouse: an ultra-structural study. J. Reprod. Fert. 45, 463-468.

HOLMES, P.V. & BERGSTROM, S. (1975) Induction of blastocyst implantation in mice by cyclic AMP. J. Reprod. Fert. 43, 329-332.

HOLTZER, H., WEINTRAUB, H., MAYNE, R., & MOCHAN, B. (1972) The cell cycle, cell lineage and cell differentiation. In: A. Moscons and A. Monroy (Eds.), Current Topics in Developmental Biology, Vol.6, Academic Press Inc. pp.229-256.

HOOKER, C.W. (1945) A criterion of luteal activity in the mouse. Anat. Rec. 93, 333-347.

HOOKER, C.W. & FORBES, T.R. (1947) A bio-assay for minute amounts of progesterone: method for biological determination of progestins. Endocrinology 41, 158.

HORI, T., IDE, M. & MIYAKE, T. (1968) Ovarian oestrogen secretion during the estrous cycle and under the influence of exogenous gonadotrophins in rats. Endocrinol. Japon. 15, 215-222.

HUMPHREY, K.W. (1968) Observations on transport of ova in the oviduct of the mouse. J. Endocrinol. 40, 267-273.

HSUEH, A.J.W., PECK, E.J. Jr., & CLARK, J.H. (1975) Progesterone antagonism of the oestrogen receptor and oestrogen-induced uterine growth. Nature 254, 337-339.

HSUEH, A.J.W., PECK, E.J. Jr., & CLARK, J.H. (1976) Control of uterine receptor levels by progesterone.   98, 438-444.

IACOBELLI, S. (1973) Induced protein synthesis and oestradiol binding to the nuclei in the rat uterus. Nature, 245, 154-155.

JEFFREY, J., COFFEY, R.J. & EISEN, A.Z. (1971) Studies on uterine collagenase in tissue culture II. Effect of steroid hormones on enzyme production. Biochim. Biophys. Acta. 252, 143-149.

JENSEN, E.V. & DESOMBRE, E.R. (1972) Mechanism of action of the female sex hormones. Annual Rev. Biochem. 41, 204-230.

JOLLIE, W.P. & BENSCOMBE, S.A. (1965) Electron microscope observations on primary decidua formation in the rat. Amer. J. Anat. 116, 217-236.

KATZENELLENBOGEN, B.S. (1975) Synthesis and inducibility of the uterine estrogen-induced protein, IP, during the rat estrous cycle: clues to the uterine estrogen sensitivity. Endocrinology 96, 289-297.

KISTNER, R.W. (1973) Endometrial alterations associated with estrogen and estrogen-progestin combinations. In: H.J. Norris (Ed.), The Uterus, Williams and Wilkins Co., Baltimore, pp.227-254.

KREHBIEL, R.H. (1937) Cytological studies of the decidual reaction in the rat during pregnancy and in the production of deciduomata. Physiol. Zool. 10, 212-238.

LAN, N.C. & KATZENELLENBOGEN, B.S. (1976) Temporal relationships between hormone receptor binding and biological responses in the uterus: Studies with short- and long-acting derivatives of estriol. Endocrinology, 98, 220-227.

LARSEN, J.F. (1961) Electron microscopy of the implantation site in the rabbit. Am. J. Anat. 109, 319-334.

LARSEN, J.F. (1975) Ultrastructural studies of the implantation process. In: E.M. Coutinho and F. Fuchs (Eds.), Physiology and genetics of Reproduction, Part B, Plenum Press, New York, pp.287-296.

LAWN, A.M. (1973) The ultrastructure of the endometrium during the sexual cycle. Adv. Reprod. Physiol. Vol.6, Elek, London, pp.61-97.

LEAVITT, W.D., TOFT, D.O., STROTT, C.A. & O'MALLEY, B.W. (1974) A specific progesterone receptor in the hamster uterus: Physiologic properties and regulation during the estrous cycle. Endocrinology, 94, 1041-1053.

LEROY, F. & GALAND, P. (1969) Radioautographic evaluation of mitotic parameters in the endometrium during the uterine sensitivity period in pseudopregnant rat. Fertil. Steril. 20, 980-992.

LEROY, F., BOGAERT, C., VAN HOECK, J. & DELCROIX, C. (1974a) Cytophotometric and autoradiographic evaluation of cell kinetics in decidual growth in rats. J. Reprod. Fert. 38, 441-449.

LEROY, R., VAN SANDE, J., SHETGEN, G. & BRASSEUR, D. (1974b). Cyclic AMP and the triggering of the decidual reaction. J. Reprod. Fert. 39, 207-211.

LOBEL, B.L., TIC, L. & SHELESNYAK, M.C. (1965a) Studies on the mechanism of nidation. XVII. Histochemical analysis of decidualization in the rat. Part 1: Framework: Oestrous cycle and pseudopregnancy.

LOBEL, B.L., TIC, L., & SHELESNYAK, M.C. (1965b) Studies on the mechanism of nidation. XVII. Histochemical analysis of decidualization in the rat. Part 2: Induction. Acta Endocr. 50, 469-485.

LOBEL, B.L., TIC, L., & SHELESNYAK, M.C. (1965c) Studies on the mechanism of nidation. XVII. Histochemical analysis of decidualization in the rat. Part 3: Formation of deciduomata. Acta Endocr. 50, 517-536.

LOCKSHIN, R.A. & BEAULATON, J. (1975) Programmed cell death. Life Sci. 15, 1549-1565.

LOEWENSTEIN, W.R. (1966) Permeability of membrane junctions. Ann. New York Acad. Sci. 137, 441-472.

LJUNGKVIST, I. (1972) Attachment reaction of rat uterine luminal epithelium. IV. The cellular changes in the attachment reaction and its hormonal regulation. Fert. Steril. 23, 847-865.

LJUNGKVIST, I. & NILSSON, O. (1974) Blastocyst-endometrial contact and pontamine blue reaction during normal implantation in the rat. J. Endocr. 60, 149-154.

MACMANUS, J.P. & WHITFIELD, J.F. (1974) Cyclic AMP, prostaglandins, and the control of cell proliferation. Prostaglandins 6, 475-487.

MAJNO, G. & LEVENTHAL, M. (1967) Pathogenesis of histamine-type vascular leakage. Lancet 2, 99-100.

MAJOR, J.S. & HEALD, P.J. (1974) The effects of ICI 46, 474 on ovum transport and implantation in the rat. J. Reprod. Fert. 36, 117-124.

MARCUS, G.J. (1970) A cellular basis for implantation failure due to massive decidualization in the rat. 3rd Meeting, Society for the Study of Reproduction, Abstract 46, Academic Press, New York.

MARCUS, G.J. (1974a) Mitosis in the rat uterus during the estrous cycle, early pregnancy and early pseudopregnancy. Biol. Reprod. 10, 447-452.

MARCUS, G.J. (1974b) Hormonal control of proliferation in the guinea-pig uterus. J. Endocr. 63, 89-97.

MARTIN, L. (1963) Interactions of oestrogen and progestins in the mouse. J. Endocr. 26, 31-39.

MARTIN, L. & FINN, C.A. (1968) Hormonal regulation of cell division in epithelial and connective tissues of the mouse uterus. J. Endocr. 41, 363-371.

MARTIN, L. & FINN, C.A. (1969) Duration of progesterone treatment required for a stromal response to oestradiol-17 in the uterus of the mouse. J. Endocr. 44, 279-280.

MARTIN, L. & FINN, C.A. (1971) Oestrogen-gestagen interactions on mitosis in target tissues, in Basic Action of Sex Steroids on Target Organs, Karger, Basel, pp.172-188.

MARTIN, L., DAS, R.M. & FINN, C.A. (1973a). The inhibition by progesterone of uterine epithelial proliferation in the mouse. J. Endocr. 57, 549-554.

MARTIN, L., EMMENS, C.W. & COX, R.I. (1961) The effects of oestrogens and anti-oestrogens on early pregnancy in mice. J. Endocr. 20, 299-306.

MARTIN, L., FINN, C.A. & CARTER, J. (1970) Effects of progesterone and oestradiol-17β in the luminal epithelium of the mouse uterus. J. Reprod. Fert. 21, 461-469.

MARTIN, L., FINN, C.A. & TRINDER, G. (1973b) Hypertrophy and pyperslasia in the mouse uterus after oestrogen treatment: an autoradiographic study. J. Endocr. 56, 133-144.

MARTIN, L., HALLOWES, R.C., FINN, C.A. & WEST, D.G. (1973c). Involvement of the uterine blood vessels in the refractory state of the uterine stroma which follows oestrogen stimulation in progesterone-treated mice. J. Endocr. 56, 309-314.

MAYER, G., NILSSON, O. & REINIUS, S. (1967) Cell membrane changes of uterine epithelium and trophoblast during blastocyst attachment in the rat. Z. Anat. Entwicklung, 126, 43-48.

MAYOL, R.F. (1975) Studies in the synthesis of estrogen-specific uterine proteins, comparison of methods of quantitative evaluation of double-isotope peaks. Mol. Cell. Endocrinol. 2, 133-147.

McCORMACK, J.T. & GREENWALD, G.S. (1974) Evidence for a preimplantation rise in oestradiol-17 levels on day 4 of pregnancy in the mouse. J. Reprod. Fert. 41, 297-301.

McLAREN, A. (1973) Blastocyst activation, in Regulation of Mammalian Reproduction. Eds. S.J. Segal, R. Crozier, P.A. Corfman and P.G. Condliffe, C.C. Thomas, Springfield, Illinois, pp.321-328.

McLAREN, A. & MICHIE, D. (1956) Studies on the transfer of fertilized mouse eggs to uterine foster-mothers. 1. Factors affecting the implantation and survival of native and transferred eggs. J. Exptl. Biol. 33, 394-416.

MEHROTRA, S.N. & FINN, C.A. (1974) Cell proliferation in the uterus of the guinea-pig. J. Reprod. Fert. 37, 405-409.

MESTER, I., MARTEL, D., PSYCHOYOS, A. & BAULIEU, E.E. (1974) Hormonal control of oestrogen receptor in uterus and receptivity for ovo-implantation in the rat. Nature, 250, 776-778.

MEYERS, K.P. (1970) Hormonal requirements for the maintenance of oestradiol-induced inhibition of uterine sensitivity in the ovariectomized rat. J. Endocr. 46, 341-346.

MILGROM, E., ATGER, M., PERROT, M. & BAULIEU, E.E. (1972) Progesterone in uterus and plasma. VI. Uterine progesterone receptors during the estrous cycle and implantation in the guinea pig. Endocrinology 90, 1071-1078.

MILGROM, E., LUU, T., ATGER, M. & BAULIEU, E.E. (1973) Mechanisms regulating the concentration and conformation of progesterone receptor(s) in the uterus. J. Biol. Chem. 248, 6366-6374.

MILLER, B.G. & EMMENS, C.W. (1969) The effects of oestradiol and progesterone on the incorporation of ^{3}H-uridine into the genital tract of the mouse. J. Endocr. 43, 427-436.

MILLONIG, G. & PORTER, K.R. (1960) Structural elements of rat liver cells involved in glycogen metabolism. In: Proceedings of the European Regional Conference on Electron Microscopy, Delft., pp.655-659.

MORIN, R.J. & CARRION, M. (1968) In vitro incorporation of acetate-^{14}C into the phospholipids of rabbit and human endometria. Lipids 3, 349-353.

NICOLLETTE, J.A. & BABLER, M. (1974) The role of protein in the estrogen-stimulated in vitro RNA synthesis of isolated rat uterine nucleoli. Arch. Biochem. Biophys. 163, 263-270.

NILSSON, O. (1958a). Ultrastructure of mouse uterine surface epithelium under different oestrogenic influences 1) Spayed animals and oestrous animals. J. Ultrastr. Res. 1, 375-396.

NILSSON, O. (1958b) Ultrastructure of mouse uterine surface epithelium under different estrogenic influences 2) Early effect of estrogen administered to spayed animals. J. Ultrastr. Res. 2, 73-95.

NILSSON, O. (1958c) Ultrastructure of mouse uterine surface epithelium under different estrogenic influences 3) Late effect of estrogen administered to spayed animals. J. Ultrastr. Res. 2, 185-199.

NILSSON, O. (1959) Ultrastructure of mouse uterine surface epithelium under different estrogenic influences. 4) Uterine secretion. J. Ultrastr. Res. 2, 331-341.

NILSSON, O. (1966a) Estrogen-induced increase of adhesiveness in uterine epithelium of mouse and rat. Exp. Cell Res. 43, 239-241.

NILSSON, O. (1966b) Structural differentiation of luminal membranes in the rat uterus during normal and experimental implantations. Z. Anat. Entwick. Gesch. 125, 152-159.

NILSSON, O. (1967) Attachment of rat and mouse blastocysts on to uterine epithelium. Int. J. Fert. 12, 5-13.

NILSSON, O. (1970) Some ultrastructural aspects of ovo-implantation, in Ovo-implantation, Human Gonadotrophins and Prolactin, Eds. P.O. Hubinot, F. Leroy, C. Robyn and P. Leleux, Marger, Basel, Munchen and New York, pp.52-72.

NILSSON, O. (1974) The morphology of blastocyst implantation. J. Reprod. Fert. 39, 187-194.

NIMROD, A., LADANY, S. & LINDNER, H.R. (1972) Perinidatory Ovarian Oestrogen Secretion in the Pregnant Rat, determined by Gas Chromatography with Electron Capture Detection. J. Endocr. 53, 249-260.

NOTIDES, A. & GORSKI, J. (1966) Estrogen-induced synthesis of a specific uterine protein. Proc. Nat. Acad. Sci. USA. 56, 230-235.

NOYES, R.N. (1973) Normal phases in the endometrium. In: H.J. Norris (Ed.), The Uterus, Williams & Wilkins Co., Baltimore, pp.110-135.

NOYES, R.W. & DICKMAN, Z. (1960) Relation of ovular age to endometrial development. J. Reprod. Fert. 1, 186-196.

O'GRADY, J.E. & HEALD, P.J. (1976) Uterine nucleic acid and phospholipid metabolism in the early stages of rat pregnancy. J. Endocr. 68, 33-34P.

O'GRADY, J.E., ARMSTRONG, E.M., MOORE, I.A.R. & VASS, M.A. (1974b). Effect of Tamoxifen (ICI 46,474) on mitosis in the uterus of the rat during the early stages of pregnancy. J. Endocr. 63, 19P.

O'GRADY, J.E., HEALD, P.J., BOYCE, E., KANE, K. & VASS, M.A. (1976) The synthesis of DNA and phospholipids by the luminal epithelium, stroma and myometrium of the rat uterus during early pregnancy. J. Reprod. Fert. (in press).

O'GRADY, J.E., MOFFAT, G.E., McMINN, L., VASS, M.A., O'HARE, A. & HEALD, P.J. (1975) Uterine chromatin template activity during the early stages of pregnancy in the rat. Biochim biophys. Acta 407, 125-132.

O'GRADY, J.E., ALAM, M., ANDERSON, F.B., COOKE, B.A., HISSEY, P., MARTIN, B. & WATSON, J. (1974a) Plasma and pituitary hormone levels during early stages of pregnancy in the rat. J. Endocr. 61, XV-XVI.

O'MALLEY, B.W. & MEANS, A.R. (1974) Female Steroid Hormones and target cell Nuclei. Science 183, 610-620.

OTTEN, J., JOHNSON, G.S. & PASTAN, I. (1972) Effect of cell density and agents which alter cell growth on cyclic adenosine 3', 5' monophosphate levels in fibroblasts, J. Biol. Chem. 247, 2082.

OWENS, I.S., VONDERHAAR, B.K. & TOPPER, Y.J. (1973) Concerning the necessary coupling of development to proliferation of mouse mammary epithelial cells. J. Biol. Chem. 248, 472-277.

OKA, T. & SCHIMKE, R.T. (1969a) Progesterone antagonism of estrogen-induced cytodifferentiation in chick oviduct. Science (Wash. D.C.) 163, 83-85.

OKA, T. & SCHIMKE, R.T. (1969b) Interaction of oestrogen and progesterone in chick oviduct development. I. Antagonistic effect of progesterone on estrogen-induced proliferation and differentiation of tubular gland cells. J. Cell Biol. 41, 816-831.

OKA, T. & SCHIMKE, R.T. (1969a) Interaction of estrogen and progesterone in chick oviduct development. II. Effects of estrogen and progesterone on tubular gland cell function. J. Cell Biol. 43, 123-137.

PAUL, J. & HUNTER, J.A. (1968) DNA synthesis is essential for increased haemoglobin synthesis in response to erythroprotein. Nature (Lond.). 219, 1362-1363.

PHELPS, C.H. & PFEIFFER, S.E. (1975) Neurogenesis and the cell cycle. In: A.Moscona and A. Monroy (Eds.), Cell Cycle and Cell Differentiation, Springer Verlag, Berlin-Heidelberg - New York, pp.63-83.

POPE, G.S. & WAYNFORTH, H.B. (1970) Secretion of oestrogen into the ovarian venous blood of pregnant rats. J. Endocr. 48, i-ii.

POLLARD, R.M. & FINN, C.A. (1972) Ultrastructure of the uterine epithelium during the hormonal induction of sensitivity and insensitivity to a decidual stimulus in the mouse. J. Endocr. 55, 293-398.

POLLARD, R.M. & FINN, C.A. (1974) Influence of the trophoblast upon differentiation of the uterine epithelium during implantation in the mouse. J. Endocr. 62, 669-674.

POLLARD, J.W., FINN, C.A. & MARTIN, L. (1976) Actinomycin D and uterine epithelial protein synthesis. J. Endocr. 69, 161-162.

PORTER, K.R. & BONNEVILLE, M.A. (1963) In: An Introduction to the fine structure of cells and tissues. Lea and Febiger, Philadelphia.

POTTS, M. (1966) The attachment phase of ovo-implantation. Amer. J. Obst. Gynec. 96 (8), 1122-1128.

POTTS, M. (1969) The Ultrastructure of egg implantation. In: Advances in Reprod. Physiol. 4, Ed. A. McLaren, Logos, London, pp.241-267.

POTTS, M. & PSYCHOYOS, A. (1967) Evolution de l'ultrastructure des relations ovo-endometriales sous l'influence de l'oestrogene chez la ratte en retard experimental de nidation. C.R. Acad. Sc. Paris, 264, 370-373.

PSYCHOYOS, A. (1960) La reaction deciduale est precedee de modifications precoces de la permeabilite capillaire de l'uterus. C.R. Seanc. Soc. Biol. 154, 1384.

PSYCHOYOS, A. (1961) Permeabilite capillaire et decidualisation uterine. C.r. Hebd. Seanc. Acad. Sci. Paris, 252, 1515.

PSYCHOYOS, A. (1963) Precisions sur l'etat de "non-receptivite" de l'uterus. Compt. Rend. 257, 1153-1156.

PSCYOYOS, A. (1965) Control de la nidation chez les mammiferes. Arch. Anat. Microscop. Morphol. Exptl. 54, 85-104.

PSYCHOYOS, A. (1966a) Etude des relations de l'oeuf et de l'endometre au cours du retard de la nidation ou des premieres phases du processus de la nidation chez la ratte. Compt. Rend. 263, 1755-1758.

PSYCHOYOS, A. (1966b) Recent research of egg-implantation. In: W. Wolstenholme and M. O'Connor, (Eds.), Ciba Foundation Study Group on Egg Implantation, Churchill, London, pp.4-28.

PSYCHOYOS, A. (1967) The hormonal interplay controlling egg implantation in the rat. Adv. Reprod. Physiol. 2, Ed. A. McLaren, Logos-Academic, London, pp.257-278.

PSYCHOYOS, A. (1969a) Hormonal factors governing decidualisation. Excerpta Med. Found. Intern. Congr. Ser. 184, 935-938.

PSYCHOYOS, A. (1969b) Hormonal requirements for egg-implantation. In: Advance in Biosciences. IV. Mechanisms Involved in Conception. Edited by G. Raspe, London: Pergamon Press, p.275-290.

PSYCHOYOS, A. (1970) Hormonal requirements for egg implantation. In: Adv. in Biosciences, 4 Ed. G. Raspe, Pergamon, Vieweg, pp.275-290.

PSYCHOYOS, A. (1973a) Hormonal control of ovo-implantation. Vitamins and Hormones 31, 201-256.

PSYCHOYOS, A. (1973b) Endocrine Control of Egg Implantation. In: Handbook of Physiology, Section 7, Endocrinology. Vol. II, Part 2 (R.O. Greep and E.B. Astwood, Eds.), pp.187-215. American Physiological Society, Washington, D.C.

PSYCHOYOS, A. & BITTON-CASMIRI, V. (1969) Captation in vitro d'un precurseur d'acide ribonucleique (ARN) (uridine-5^{3}H) par le blastocyst du rat differences entre blastocysts normaux et blastocysts en diapause. C. R. Acad. Sci., Paris, 268, 188-190.

PSYCHOYOS, A. & MANDON, P. (1971) Scanning electron microscopy of the surface of the rat uterine epithelium during delayed implantation. J. Reprod. Fert. 26, 137-138.

RAY, S.C. & MORIN, R.J. (1965) Lipid composition of the non-gravid and gravid rabbit endometrium. Proc. Soc. exp. Biol. Med. 120, 849-853.

REEL, J.R. & SHIH, Y. (1975) Oestrogen-inducible uterine progesterone receptors. Characteristics in the ovariectomized immature and adult hamster. Acta Endocr. 89, 344-354.

REID, R.J. (1971) Ph.D. Thesis, University of Strathclyde, Glasgow.

REID, R.J. & HEALD, P.J. (1970) Uptake of ^{3}H-leucine into proteins of rat uterus during early pregnancy. Biochim. biophys. Acta 204, 278-279.

REID, R.J. & HEALD, P.J. (1971) Protein metabolism of the rat uterus during the oestrous cycle, pregnancy and pseudopregnancy and as affected by an anti-implantation compound, ICI 46,474. J. Reprod. Fert. 27, 73-82.

REINIUS, S. (1967) Ultrastructure of blastocyst attachment in the mouse. Zeit. fur Zellforschung 77, 257-266.

ROSENFELD, M.G. & O'MALLEY, B.W. (1970) Steroid hormones: effects on adenyl cyclose activity and adenosine 3', 5'- monophosphate (cyclic AMP), Science 168, 253-255.

ROTHCHILD, I. & MEYER, R.K. (1942) Studies of the pretrauma factors necessary for placentoma formation in the rat. Physiol. Zool. 15, 216-223.

ROWLATT, C. (1969) Subepithelial fibrils associated with basal lamina under simple epithelia in mouse uterus: possible tropocollagen aggregates. J. Ultrastr. Res. 26, 44-51.

SACCO, A.G. & MINTZ, B. (1975) Mouse uterine antigens in the implantation period of pregnancy. Biol. Reprod. 12, 498-503.

SALDARINI, R.J. & YOCHIM, J.M. (1967) Metabolism of the uterus of the rat during early pseudopregnancy and its regulation by estrogen and progestogen. Endocrinology 80, 453-466.

SAUNDERSON, R. & HEALD, P.J. (1974) Ornithine decarboxylase activity in the uterus of the rat during early pregnancy. J. Reprod. Fert. 39, 141-143.

SCHLAFKE, S. & ENDERS, A.C. (1975) Cellular basis of interaction between trophoblast and uterus at implantation. Biol. Reprod. 12, 41-65.

SCHULTZ, R.H., BURCALOW, H.B., FAHNING, M.L., GRAHAM, E.F. & WEBER, A.F. (1969) A karyometric study of epithelial cells lining the glands of the bovine endo-metrium. J. Reprod. Fert. 19, 169-171.

SEYLE, H., BORDUAS, A. & MASSON, G. (1942) Studies concerning the hormonal control of deciduomata and metrial glands. Anat. Rec. 82, 199-209.

SHAIKH, A. (1971) Estrone and estradiol in the ovarian venous blood from rats during estrous cycle and pregnancy. Biol. Reprod. 5, 297-307.

SHAIKH, A.A. & ABRAHAM, G.E. (1969) Measurement of the estrogen-surge during pseudo-pregnancy in rats by radioimmunossay. Biol. Reprod. 1, 378-380.

SHELESNYAK, M.C. (1933a) The production of deciduomata in immature rats by pregnancy urine treatment. Amer. J. Physiol. 104, 693-699.

SHELESNYAK, M.C. (1933b) The production of deciduomata in spayed immature rats after oestrin and progestin treatment. Anat. Rec. 56, 211.

SHELESNYAK, M.C., KRAICER, P.F. & ZEILMAKER, G.H. (1963a) Studies on the mechanism of decidualization. I. the oestrogen surge of pseudo-pregnancy and progravidity and its role in the process of decidualization. Acta Endocrinol. 42, 225-232.

SHELESNYAK, M.C. & TIC, L. (1963b) Studies on the mechanism of decidualization. IV. Synthetic processes in the decidualizing uterus. Acta Endocrinol. 42, 465-472.

SHELESNYAK, M.C. & TIC, L. (1963c) Studies on the mechanism of decidualization. V. Suppression of synthetic processes of the uterus (DNA, RNA and protein) following inhibition of decidualization by an antioestrogen ethanoxytriphetol. Acta Endocrinol. 43, 462-468.

SHORT, J., TSOKADA, K., RUDERT, W.A. & LIEBERMAN, A. (1974) J. Biol. Chem. 249, 1427-1431.

SIM, K. (1974) Effect of IUD on uterine cyclic AMP and the activities of adenyl cyclase and phosphadiesterase during the oestrous cycle and early pregnancy in rats. J. Reprod. Fert. 39, 399-402.

SINGHAL, R.L., TSANG, B.K. & SUTHERLAND, D.J.B. (1976) Regulation of cyclic nucleotide and prostaglandin metabolism in sex steroid-dependent cells. In: R.L. Singhal and J.A. Thomas (Eds.), Cellular Mechanisms Modulating Gonadal Action, Vol. 2, HM + M Medical and Scientific Publishers, Aylesbury, England, pp. 325-424.

SINGHAL, R.L. & LAFRENIERE, R.T. (1972) Metabolic control mechanisms in mammalian systems. XV. Studies on the role of adenosine 3', 5' monophosphate in estrogen action on the uterus. J. Pharmacol. Exp. Ther. 180, 86-91.

SOMMERVILLE, B.W. (1971) Daily variation in plasma levels of progesterone and estradiol throughout the menstrual cycle. Am. J. Obstet. Gynecol. 111, 419-426.

SOMJEN, D., SOMJEN, G., KING, R.J.B., KAYE, A.M. & LINDNER, H.R., (1973) Nuclear binding of oestradiol-17 and induction of protein synthesis in the rat uterus during postnatal development. Biochem. J. 136, 15-33.

SMITH, J.A. & MARTIN, L. (1973) Do cells cycle? Proc. Nat. Acad. Sci. USA, 70, 1263-1267.

SMITH, J.A. & MARTIN, L. (1974) In: G.M. Padilla, Cameron, I.L., and Zimmerman, A. (Eds.), Cell Cycle Controls, Academic Press Inc., New York, pp.43-60.

SMITH, J.A., MARTIN, L., KING, L.J.B. & VERTES, M. (1970) Effects of oestradiol-17 and progesterone on total and nuclear-protein synthesis in epithelial and stromal tissues of the mouse uterus and of progesterone on the ability of these tissues to bind oestradiol-17β . Biochem. J. 119, 773-784.

SURANI, M.A.H. (1975a) Zona pellucida denudation: blastocyst proliferation and attachment in the rat. J. Embryol. exp. Morph. 33, 343-353.

SURANI, M.A.H. (1975b) Hormonal regulation of proteins in the uterine secretion of ovariectomized rats and the implications for implantation and embryonic diapause. J. Reprod. Fert. 43, 411-417.

SURANI, M.A.H. & HEALD, P.J. (1971) The metabolism of glucose by rat uterus tissue in early pregnancy. Acta Endocrinol. 66, 16-24.

SWIFT, A.D. & O'GRADY, J.E. (1976) Uterine cyclic AMP in early pregnancy in the rat. J. Endocr. 68, 35-36P.

242

SZEGO, C.M. & DAVIS, S. (1967) Adenosine 3', 5'-monophosphate in rat uterus: acute elevation by oestrogen. Proc. natn. Acad. Sci. USA. 58, 1711-1718.

TACHI, C. & TACHI, S. (1975) Cellular aspects of ovum implantation and decidualization in the rat. In: E.M. Coutinho and F. Fuchs (Eds.), Physiology and genetics of Reproduction, Part B, Plenum Press, New York, pp.263-386.

TACHI, C., TACHI, S. & LINDNER, H.R. (1970) Ultrastructural features of blastocyst attachment and trophoblastic invasion in the rat. J. Reprod. Fert. 21, 37-56.

TACHI, C., TACHI, S. & LINDNER, H.R. (1972) Modification by progesterone of oestradiol-induced cell proliferation, RNA synthesis and oestradiol distribution in the rat uterus. J. Reprod. Fert. 31, 59-76.

TOBERTY, J.A. (1976) A study of the possible role of prostaglandins in decidualization using a nonsurgical method for the instillation of fluids into the rat uterine lumen. J. Reprod. Fert. 47, 391-393.

TSANEV, R. (1975) Cell cycle and liver junction. In: J. Reinert and H. Holtzer (Eds.) Cell Cycle and Cell Differentiation, Springer Verlag, Berlin-Heidelberg-New York, pp.197-248.

VOKAER, R. (1952) Recherches histophysiologiques sur l'endometre du rat en particulier sur le conditionnement hormonal de ses proprietes athrocytaires, Arch. Biol. 63, 3-84.

VONDERHAAR, B.K. & TOPPER, Y.J. (1974) A role of the cell cycle in hormone dependent differentiation. J. Cell Biol. 63, 707-712.

WATSON, J., ANDERSON, F.B., ALAM, M., O'GRADY, J.E. & HEALD, P.J. (1975) Plasma hormones and pituitary luteinising hormone in the rat during the early stages of pregnancy and following post-coital treatment with Tamoxifen ICI 46, 474, J. Endocr. 65, 7-17.

WEBB, F.T.G. (1975) Implantation in ovariectomized mice treated with dibutyryl adenosine 3', 5' monophosphate (dibutyryl cAMP), J. Reprod. Fert. 42, 511-517.

WEBB, F.T.G. & SURANI, M.A.H. (1975) Influence of environment on blastocyst proliferation, differentiation and implantation. In: G.P. Talwar (Ed.) Regulation of Growth and Differentiated Function in Eukoryote Cells, Raven Press, New York, pp.519-522.

WEBER, G., QUEENER, S.F. & FERDINANDUS, J.A. (1972) Control of gene expression in carbohydrate pyrimidine and DNA metabolism. Adv. Enzyme Reg. 9, 63-95.

WEITLAUF, H.M. (1973) In vitro uptake and incorporation of amino acids by blastocysts from intact and ovariectomized mice. J. Exp. Zool. 183, 303-308.

WESSELLS, N.K. (1968) Problems in the analysis of determination, mitosis and differentiation. Epithelial-Mesenchymal Interactions. R. Fleischmajer and R.E. Billingham, editors. The Williams & Wilkins Co., Baltimore, Md. 132-151.

WESSELLS, N.K. & RUTTER, W.J. (1969) Phases in cell differentiation. Sci. Am. 220, 36-44.

WILLIAMS, M.F. (1948) The vascular architecture of the rat uterus as influenced by estrogen and progesterone. Am. J. Anat. 83, 247-307.

WILSON, J.D. (1963) The nature of the RNA response to estradiol administration by the uterus of the rat. Proc. Natn. Acad. Sci. USA 50, 93.

YOCHIM, J.M. (1971) Intrauterine oxygen tension and metabolism of the endometrium during the preimplantation period. In: R. Blandau (Ed.) Biology of the Blastocyst, University of Chicago Press, Chicago, pp.363-382.

YOCHIM, J.N. (1975) Development of the progestational uterus: Metabolic aspects. Biol. Reprod.

YOCHIM, J.M. & DEFEO, V.J. (1963) Hormonal control of the onset magnitude of duration of uterine sensitivity in the rat by steroid hormones of the ovary. Endocrinology 72, 317-326.

YOSHINAGA, K. (1972) Rabbit anti-serum to rat deciduoma. Biol. Reprod. 6, 51-57.

YOSHINAGA, K. (1974) Interspecific cross-reactivity of deciduoma antiserum: Interaction between mouse deciduoma and anti-serum to rat deciduoma. Biol. Reprod. 11, 50-55.

YOSHINAGA, K., HAWKINS, R.A. & STOCKER, J.F. (1969) Estrogen secretion by the rat ovary in vivo during the estrous cycle and pregnancy. Endocrinology 85, 103-112.

ZHINKIN, L.N. & SAMOSHKINA, N.A. (1967) DNA synthesis and cell proliferation during formation of deciduomata in mice. J. Embryol. Exp. Morph. 17, 593-603.

ZYBINA, E.V. & GRIAHENKO, T.A. (1972) Spectrophotometrical estimation of ploidy level in decidual cells of the endometrium of the white rat. Dokl. Akad. Nauk. SSSR, 14, 284.

CELLULAR AND MOLECULAR APPROACHES TO BLASTOCYST
UTERINE INTERACTIONS AT IMPLANTATION

M. Azim H. Surani

Physiological Laboratory
Cambridge CB2 3EG
U.K.

The metabolic response of embryos to the environment of the uterine lumen and the interactions involving the cell surfaces of trophectoderm and uterine epithelium are two important facets of the interrelations between blastocysts and uterus at implantation. Long range communication between blastocysts and uterus are mediated by humoral factors in the uterine lumen. Short range cell surface interactions occur when blastocysts are apposed at the epithelial surface at the initiation of implantation. Considerable advances have been made in the understanding of the metabolic response of cells to the changes in the extracellular environment and on the trophic action of macromolecules after their binding to the surface receptors. Cell-cell interactions are important in cooperation between cells, in recognition, specific adhesions, and control of growth and morphogenesis. The purpose of this chapter is to analyse available evidence which establishes a basis, at the cellular and molecular levels, for these two types of interactions at implantation. This may provide new concepts and approaches for studies on implantation.

Differentiation of the 1 cell egg to the blastocyst is accompanied by alterations in gene expression, appearance of distinct cell populations with specialized functions, changes in the cell surface topography and constituents, and apparent changes in embryonic response to the extracellular environment with a shift towards a behaviour recognizable as more typical of that in other mammalian cells. The uterus in response to the ovarian steroids, progesterone and oestrogen, provides a conducive environment for blastocysts by the production of luminal components. Cell surface interactions between the sensitized uterus and blastocysts commence at apposition. These events can be interrupted experimentally by ovariectomizing females and administering exogenous steroids. Most of the studies described here apply mainly to the rat and the mouse.

EVIDENCE FOR UTERINE CONTROL OF IMPLANTATION

Rat embryos undergo cleavage in the oviduct and enter the uterine lumen late on day 4 of pregnancy. At blastulation, on day 5 of pregnancy, at least two cell types, trophectoderm and inner cell mass cells, are recognized. The zona pellucida is removed and trophoblast is apposed to the uterine epithelium (Enders & Schlafke, 1975; Surani, 1975a). The uterus is sensitized by oestrogen and progesterone (Psychoyos, 1973) and imposes a rigid control over the metabolism of blastocysts and implantation. Cleavage stage embryos are relatively free of constraints from environmental and endocrine conditions whereas blastocysts exhibit a reliance on extracellular macromolecules for postimplantation development (Spindle & Pedersen, 1973; Webb & Surani, 1975).

Blastomeres divide exponentially during cleavage with a cell doubling time of about ten hours in the mouse and the rat (Figure 1; Table 1) (Graham, 1973; Surani, 1975a,b) and cell metabolism reaches a peak level after blastulation (Monesi & Molinaro, 1971). In females ovariectomized on day 3 of pregnancy, cell doubling time increases to 48 hours (Sanyal & Meyer, 1970, 1972) and eventually ceases. The metabolic rate also declines and blastomeres arrest in the G_1 phase of the cell cycle without loss of viability (Surani, 1975a; Webb & Surani, 1975). Metabolic activity in quiescent blastocysts is renewed immediately after an injection of oestradiol into these females (Prasad et al., 1968). Oestradiol may act in various ways to release blastocysts from quiescence (McLaren, 1973), either directly or through their influence on the uterus. Hormones may influence embryos by inducing changes in the luminal proteins. Although postimplantation development of blastocysts can occur in extrauterine sites (Kirby, 1969a) and in vitro (Gwatkin, 1966b; Spindle & Pedersen, 1973; Hsu et al., 1974; McLaren & Hensleigh, 1975), the rigid control over development observed in utero is absent, and embryos in extrauterine sites can develop regardless of the endocrine state of the female (Kirby, 1969a).

EMBRYONIC RESPONSE AND THE INFLUENCE OF THE ENVIRONMENT AT THE CELLULAR AND MOLECULAR LEVELS

Considerable changes in macromolecular synthesis and gene expression are detected during the differentiation of mammalian eggs (Wolf & Engel, 1972; Herbert & Graham, 1974; Church & Schultz, 1974; Epstein, 1975; Schultz & Cooper, this volume). The response of embryos to environmental conditions may also vary during differentiation although this has not been critically evaluated. The appearance of the G_1 phase of the cell cycle in mammalian embryos is one of the crucial events of early differentiation and may have a marked effect on the response of embryos to environmental conditions.

BLASTOCYST CELL PROLIFERATION
DURING DAY 5 OF PREGNANCY

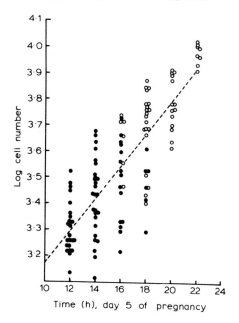

Figure 1: Blastocyst cell proliferation in the rat on day 5 of pregnancy; log mean cell number versus time (log y = 2.550 + 0.0621x). ● Zona encased blastocysts; o zona free blastocysts.

<u>Appearance of properties of the eukaryote cell in blastomeres</u>

One of the characteristic features of early mammalian embryos is that the G_1 phase of the cell cycle is very short during early cleavage. Studies on the cell cycle of mammalian embryos after labelling with ^{3}H-thymidine have suggested that the G_1 phase is not detectable until the late morula or blastocyst stage in the rat (Dalcq & Pasteels, 1955) and the mouse (Gamow & Prescott, 1970; Mukherjee, 1976). Other studies, using microdensitometry and labelling with thymidine, suggest the presence of the G_1 phase at the 8 cell stage (Barlow et al., 1972; Graham, 1973). The proportion of the total G_1 and G_2 phases of the cell cycle during early cleavage is thought to be short and only at the late blastocyst stage is a greater proportion of the cell cycle occupied by these phases (Graham, 1973). Recent studies (Mukherjee, 1976) suggest that in the mouse, the G_1 phase is not detectable until the morula stage, after which there is a lengthening of the G_1 phase to 20 min, and to 1.5 hours by the late blastocyst stage with concomitant decrease in the G_2 phase.

The G_1 phase varies in duration in mammalian cells and critical events may occur during this part of the cell cycle (Prescott, 1968; Epifanova & Terskikh,

248

TABLE 1

RECOVERY OF FREE EMBRYOS AND CELL COUNTS OF BLASTOCYSTS* DURING DAY 5 OF PREGNANCY

Time	Number animals	Number embryos recovered	Number embryos for air-dried preparation	Mean number embryos recovered per animal ± S.D.	% zona-free eggs	Mean number cells per embryo ± S.D.	Number mitotic cells per embryo	
							Mean	Range
12.00	6	72	23	12 ± 1.79	2.6	27.52 ± 4.01	1.6	0-5
14.00	4	48	27	12 ± 1.41	4.2	30.99 ± 4.54	2.5	1-7
16.00	7	58	21	8.29 ± 2.06	46.0	34.38 ± 5.32	2.0	0-5
18.00	7	54	26	7.71 ± 1.90	87.0	38.89 ± 6.05	2.3	0-6
20.00	4	25	14	6.25 ± 1.71	100.0	44.07 ± 4.46	3.3	1-6
22.00	4	17	8	4.25 ± 1.73	100.0	53.69 ± 2.58	2.3	0-4
Total	32	274	119					

* Blastocysts were flushed out on day 5 of pregnancy and the number of cells determined.

1969). These include carcinogenesis (Paul & Hickey, 1974), differentiation of mammary gland cells prior to synthesis of milk proteins (Vonderhaar & Topper, 1974) and synthesis of RNA and proteins involved in the initiation of DNA synthesis (Baserga, 1968). Since several biochemical events appear to be determined in the G_1 phase, this phase of the cell cycle is under close investigation (Naha et al., 1975).

Signals from mitogens are perceived by cells during the early G_1 or G_0 phase (G_0 is a specific part of the G_1 phase) (Holley, 1972; Mannio & Burger, 1975). Since cell surface topography undergoes changes at different phases of the cell cycle (Fox et al., 1971), presumably during the G_1 or G_0 phase surface topography and receptors are arranged for binding of extracellular trophic factors (Edelman et al., 1973; Baker, 1976). Cells stimulated at this phase show an increase in cell metabolism (Hershko et al., 1971) and undergo cell division. In the absence of extracellular trophic factors, cell metabolism declines (Hershko et al., 1971) and cells arrest in the G_1 (G_0) phase (Pardee, 1974; Baker, 1976). Cells remain in quiescence until appropriate stimulus is received at the cell surface. Apart from humoral factors, contact between cells also generates signals which affect cell metabolism and proliferation (Dulbecco, 1970).

The appearance of the G_1 phase of the cell cycle in blastocysts brings fundamental changes in the response of embryos to environmental conditions. At this stage, the response is apparently similar to that observed in other mammalian cells (Surani, 1975b,c, 1976; Webb & Surani, 1975; Gore-Langton & Surani, 1976). The reversible metabolic enhancement and the ability to enter into quiescence are two prominent features of the cellular response to environmental changes and will be considered in detail below. These responses shown by blastocysts, but not perhaps by early cleavage stage embryos, may explain the nature of uterine control over embryos.

Environmental conditions necessary for development

Critical biochemical data is needed to define how the cleavage stage embryos differ in their response to extracellular environmental changes, compared with blastocysts. As may be expected, environmental conditions for the development of the 2 cell stage to the blastocyst are not exacting, presumably to allow rapid cell division with an absence of the G_1 phase (Mukherjee, 1976). Mouse embryos, and to a lesser extent rat embryos, will develop from the 2 cell stage to the blastocyst stage in a chemically defined medium consisting of a balanced salt solution, pyruvate, lactate and/or glucose and bovine serum albumin (Folstad et al., 1969; Brinster, 1969; Whitten, 1971; Whittingham, 1971). Development to the blastocyst stage can also occur in vivo under a variety of endocrinological conditions (Dickmann, 1970), in the oviduct as tube locked

embryos (Kirby, 1969) and after transfer to foreign species (Tarkowski, 1962).
At the blastocyst stage, however, the cell doubling time of about ten hours is
not maintained in vitro in this simple medium (Bowman & McLaren, 1970). In a
medium consisting of amino acids and a high molecular weight serum fraction,
development of blastocysts past the implantation stage can be observed (Cole
& Paul, 1965; Gwatkin, 1966b; Spindle & Pedersen, 1973; Hsu et al., 1974; McLaren
& Hensleigh, 1975). Similarly, postimplantation development also occurs in
extrauterine sites (Billington et al., 1968; Kirby, 1969), perhaps because of
the presence of serum. These experiments provide some evidence suggesting an
increasing reliance by blastocysts on extracellular macromolecules for
implantation and postimplantation development. This shift in the requirements
for development past the blastocyst stage may be associated with the differentiation of the embryo and the appearance of the G_1 phase. In the uterine lumen,
serum and uterine specific proteins may provide such an environment for
implantation.

Embryonic response: Reversible metabolic enhancement

Evidence for the reversible control of blastocyst metabolism can be
obtained from studies on delay of implantation which is widespread in mammals
and is a result of seasonal adaptation or lactation (Enders, 1967; Daniel, 1971b;
Sadlier, 1972; Aitken, 1974 and this volume). Lataste (1891) first described
delay of implantation in rodents. During delay of implantation, blastocyst
cell proliferation declines (Daniel, 1968; Aitken, 1974; Surani, 1975b) and,
in rodents, the embryos enter into quiescence. In the rat and mouse, delay of
implantation can also be induced by ovariectomy during the first 3 or 4 days
of pregnancy, followed by daily treatment with progesterone (Cochrane & Meyer,
1957). This model has been extensively used to study the hormonal requirements
for implantation (Psychoyos, 1973) and the biochemical changes which occur
during embryonic diapause (McLaren, 1973; Weitlauf, 1974a; Webb & Surani, 1975).

In diapausing blastocysts, incorporation of precursors into RNA (Prasad et
al., 1968; Dass et al., 1969; Gulyas & Daniel, 1969; Psychoyos & Bitton-Casimiri,
1969; Jacobson et al., 1970) and into proteins (Weitlauf & Greenwald, 1968;
Weitlauf, 1969, 1974a; Gulyas & Daniel, 1969; Dass et al., 1969) is very low.
Carbon dioxide production is depressed (Menke & McLaren, 1970; Weitlauf, 1974a;
Torbit & Weitlauf, 1974) and the total weight and protein content changes little
in quiescent embryos (Hensleigh & Weitlauf, 1974). DNA synthesis also ceases
(Gulyas & Daniel, 1969; Sanyal & Meyer, 1970, 1972), as does mitotic activity
(Sanyal & Meyer, 1970, 1972; Surani, 1975a), and the cells apparently arrest in
the G_1 phase of the cell cycle (Sherman & Barlow, 1972; Sanyal & Meyer, 1972;
Surani, 1975a).

Removal of the litter during lactational delay, or injection of oestrogen
into ovariectomized females, restores normal metabolic activity in embryos.
Incorporation into RNA increases within an hour (Prasad et al., 1968; Dass et
al., 1969; Jacobson et al., 1970) and the incorporation of precursors into
proteins increases within three hours (Weitlauf & Greenwald, 1968; Prasad et al.,
1968; Dass et al., 1969; Sanyal & Meyer, 1970, 1972). Carbon dioxide production also increases (Weitlauf, 1974a; Torbit & Weitlauf, 1974) and giant
cell transformation of trophoblast (Dickson, 1969) is followed by implantation
(Psychoyos, 1969). Transfer of blastocysts between physiologically different
females shows that the influence of the uterine environment is reversible.
Hence, protein synthesis in normal embryos declines on transfer to ovariectomized animals given progesterone (Weitlauf, 1974a) but the embryos remain
quiescent and viable (Psychoyos, 1961; Dickmann & De Feo, 1967). Diapausing
blastocysts transferred to pseudopregnant recipients implant normally (Dickmann
& De Feo, 1967; Weitlauf & Greenwald, 1968).

Embryos transferred to extrauterine sites show some postimplantation
development (Billington et al., 1968; Kirby, 1969a) the trophoblast becomes
polyploid (Barlow & Sherman, 1972) and generally the embryos behave independently
of the hormonal state of the mother, presumably because of the presence of serum
in these sites. Normal and diapausing embryos placed in vitro in the presence
of serum also show an increase in the incorporation of precursors of RNA and
protein (Ellem & Gwatkin, 1968; Psychoyos & Bitton-Casimiri, 1969), in glucose
utilization (Menke & McLaren, 1970b) and in the ploidy of the trophoblast
nuclei (Barlow & Sherman, 1972).

The schematic diagram (Figure 2) summarizes a large number of studies on
rodent blastocysts and shows reversibility of metabolic enhancement under
different conditions. Thus, mechanistically unrelated biochemical events act
coordinately during metabolic enhancement or depression, which is the definition
of a pleiotypic response (Hershko et al., 1971). Thus blastocysts, as distinct
from early cleavage stages, exhibit a pleiotypic response in utero and in vitro.
Morulae transferred to females experiencing delay of implantation either proceed to the blastocyst stage or degenerate (Dickmann, 1970). Embryos enter
into quiescence only when they reach the blastocyst stage.

An explanation for embryonic quiescence: the importance of the G_1 phase

Two general hypotheses have been put forward to explain the embryonic
diapause. Brambell (1937) postulated that during lactation the uterus may
produce an inhibitory substance to keep blastocysts in quiescence. The
observations that diapausing embryos, when removed from the uterus, resume normal
metabolic activity in vitro (Psychoyos, 1973; Weitlauf, 1973b) and in ectopic
sites (Kirby, 1969a), have been cited as evidence for this postulated repressive

252

influence of the uterus during lactation.   Supernatants of homogenized uteri
and luminal flushings have been tested for inhibitory substances but the
findings are inconclusive (Psychoyos, 1969, 1973; Weitlauf, 1976).   In
ovariectomy-induced delay of implantation, a single dose of actinomycin D (15 µg
per animal) induces implantation.   This has been interpreted to suggest that
actinomycin D blocks a uterine inhibition to embryo implantation (Finn, 1974),
although other explanations can be given.   Diapausing mouse embryos explanted
in vitro resume normal metabolic activity (Weitlauf, 1973b);   resumption of
incorporation of amino acids into proteins is blocked by high doses of
actinomycin D, but when used at levels normally sufficient to interfere with
the transcription of RNA, the drug is not effective (Weitlauf, 1974b).

An alternative explanation for embryonic diapause is based on the response
of blastocysts to environmental changes and has arisen as a result of analytical
work on uterine fluids (see later).   After ovariectomy, there is a decline in
total luminal proteins with a substantial reduction in their synthesis and
secretion (Surani, 1975c; 1976).   During the same period, cell doubling time
in the blastocyst increases to 48 hours (Sanyal & Meyer, 1972).   With further

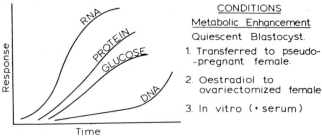

CONDITIONS
Metabolic Enhancement
Quiescent Blastocyst.
1. Transferred to pseudo-
   -pregnant female.
2. Oestradiol to
   ovariectomized female
3. In vitro (+ serum)

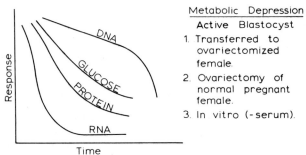

Metabolic Depression
Active Blastocyst
1. Transferred to
   ovariectomized
   female.
2. Ovariectomy of
   normal pregnant
   female.
3. In vitro (-serum).

Figure 2: Coordination of biochemical response of blastocysts under different
conditions. Initial response is shown in RNA metabolism, followed by protein,
energy metabolism and finally in DNA synthesis.

decline in protein content, embryonic mitotic activity declines and ceases
(Surani, 1975a, 1976), metabolic activity falls to basal levels (McLaren, 1973;
Weitlauf, 1974a; Webb & Surani, 1975), and cells arrest in the G_1 phase (Sherman
& Barlow, 1972).   This capacity for adaptation may enable embryos to survive
during suboptimal environmental conditions not conducive for growth or
implantation.   Similar adaptation by embryos to uterine environmental conditions may occur in other species, such as the roe deer (Aitken, 1974) and the
northern fur seal(Daniel, 1971b).   In both these species, intraluminal protein
content is very low during embryonic quiescence but increases when implantation
resumes.   In the northern fur seal, blastocyst cell doubling time increases to
50-60 days during quiescence, and at the time of resumption of metabolic
activity and implantation, a high molecular weight protein peak appears together
with an increase in total macromolecules (Daniel, 1971b).

This explanation for embryonic quiescence is contrary to the suggestion that
the embryos are maintained in quiescence by a unique uterine inhibitor(s) acting
at the transcriptional level (Psychoyos, 1973; Weitlauf, 1974b).   Although
there is a resumption of normal metabolic activity in quiescent embryos
explanted in vitro (Bitton-Casimiri et al., 1976), the stimulation may occur
because of the presence of serum and essential ions, such as calcium, in the
medium.   Apart from humoral factors, cell-cell interactions between the trophoblast and uterine epithelium may be important for contact inhibition of growth
during quiescence.   When quiescent blastocysts are removed from the lumen, the
cell-cell interactions and topoinhibition (Dulbecco, 1970) may be destroyed,
resulting in the resumption of normal metabolic activity.   The inability of
blastocysts to remain in quiescence and viable in vitro may be due to the
absence of contact inhibition.

Other eukaryote cells show an inherent capacity for arresting in the G_1
phase of the cell cycle when environmental conditions are not optimal (Hershko
et al., 1971; Pardee, 1974).   Withdrawal or lack of serum leads to translational inhibition (Hassel & Engelhardt, 1973) and brings about quiescence,
which is not, therefore, unique to embryonic cells.   In contrast, neoplastic
and virally transformed cells continue cell division even during suboptimal
environmental conditions, having lost the ability to arrest in the G_1 phase of
the cell cycle, and this ultimately results in cell death (Pardee, 1974).
Furthermore, the cyclic changes in the cell surface topography detected in normal
cells are not observed in transformed cells (Fox et al., 1971).

Most of the studies on the cellular response to environmental conditions
have been carried out comparing normal cells with virally transformed cells.
In view of the apparent differences between the response of the early cleavage
stage embryos to that shown by blastocysts, the early mammalian embryo provides
an alternative system for study.   Molecular studies on the G_1 phase, and the

likely influence of this on the behaviour and differentiation of the cell, can also be studied.

HUMORAL FACTORS INVOLVED IN THE CONTROL OF BLASTOCYST METABOLISM AND IMPLANTATION

Uterine luminal fluids, like other body fluids, consist of ions, amino acids and other metabolites, serum proteins and enzymes, and hormones such as steroids. However, uterine luminal fluids differ from other body fluids due to the presence of uterine specific proteins. These components constitute the environment of the preimplantation embryo and since some of them fluctuate with the endocrine state of the female, they may influence the metabolic state of the blastocyst. These fluctuations in the levels of metabolites should be considered as a part of the total luminal environment. The primary role of regulating blastocyst metabolism may be assigned to uterine specific proteins, since these macro-molecules distinguish uterine luminal fluids from other fluids.

Ions

Relatively few studies have been carried out on the ionic composition of the uterine lumen (see Hamner, 1971). In the luminal fluids of the rat, potassium ion may be present in amounts as much as ten times greater than in serum (Howard & De Feo, 1959; Ringler, 1961). In ovariectomized animals treated with progesterone, potassium content is low but increases following an injection of oestradiol (Clemetson et al., 1970a,b). Since the rat blastocyst, after loss of the zona pellucida on day 5 of pregnancy, behaves as a negatively charged body (Clemetson et al., 1970a,b), the authors postulated that the delay of implantation in the ovariectomized females given progesterone alone may occur because the negatively charged blastocyst is repelled by the high negative membrane potential of the endometrium. In other studies, fluctuations in the levels of potassium were not found (Setty et al., 1973). In the roe deer, calcium content in the uterine lumen is low during embryonic diapause, but the levels increase when implantation resumes (Aitken, 1974).

Amino acids

An earlier hypothesis suggested that blastocyst metabolism and implantation may be governed by fluctuating levels of essential amino acids in the lumen. Mouse blastocysts, in the absence of arginine or leucine, cease development and undergo diapause in vitro (Gwatkin, 1966a). Analyses of intraluminal extracts show no differences in the amino acids during normal implantation compared with the material obtained from animals in delay of implantation (Gwatkin, 1969). More significantly, whereas blastomeres in utero during diapause arrest pre-dominantly in the G_1 phase, those in vitro, in the absence of arginine and leucine, arrest in both the G_1 and G_2 phases of the cell cycle (Sherman & Barlow, 1972).

Steroid hormones

Steroids may enter the lumen and have a direct influence on blastocysts (see also Warner, this volume). In ovariectomy-induced delay of implantation, resumption of normal metabolic activity occurs within one hour after an injection of oestradiol (Prasad et al., 1968; Dass et al., 1969). Oestrogen was thus thought to have a direct effect on the metabolic activity of blastocysts (Prasad et al., 1968). In this respect, when 2 cell embryos are incubated in the presence of 10^{-8}M oestradiol, followed by transfer to ovariectomized females given progesterone alone, the embryos implant without requiring systemic oestrogen (Smith, 1968). The possibility that small amounts of oestradiol are carried with embryos on transfer cannot be discounted. Oestradiol enhances the incorporation of uridine into RNA (Lau et al., 1973; Harrer & Lee, 1973) as well as amino acid uptake and incorporation into proteins (Smith & Smith, 1971) when added to the medium in vitro. The suggestion that oestradiol stimulation of RNA synthesis occurs by 'gene induction' (Lau et al., 1973) may, nevertheless, be premature. Other similar studies do not show any stimulation of embryo metabolism in the presence of oestradiol (Weitlauf, 1974a) and although there is a reported increase in the volume of the embryos (Lau et al., 1973; Bowman & McLaren, 1970), oestradiol has no effect on cell proliferation (Bowman & McLaren, 1970). An influx of precursors through non-specific effects of the hormone on the cell membrane (Wilmer, 1961), as indicated by increase in the volume of the embryos (Bowman & McLaren, 1970; Lau et al., 1973), can give the same result. Furthermore, progesterone has been found to inhibit cleavage in the rabbit, but the result may be due to the accumulation of the hormone on the cell surface (Daniel & Levy, 1964) affecting membrane permeability. The target cells for steroids possess cytoplasmic receptors but no intracellular localization of oestradiol can be detected in rat embryos (Prasad et al., 1974). Also, post implantation development of embryos in extrauterine sites occurs regardless of the endocrine state of the animal (Kirby, 1969a). Steroidal effects on embryos may be non-specific.

There is evidence that embryos themselves may synthesize steroids (Dickmann & Dey, 1974; Perry et al., 1976). Uteroglobin, a progesterone-dependent protein in the rabbit uterine lumen, has been found to bind progesterone (Beato & Baier, 1975; Fowler et al., 1976; Beato, this volume). Steroidogenic enzymes, notably Δ^5-3 hydroxysteroid dehydrogenase, have been detected in some species (Dickmann & Dey, 1975; Dickmann et al., 1974) and the pig embryo is able to convert neutral steroids, and possibly progesterone, to oestrogen in vitro (Perry et al., 1976).

Proteins

Of all the luminal components, proteins of the uterine lumen have been most extensively studied. It is apparent that there are considerable differences in the types of macromolecules found in different species. There is a fundamental difference between species in the steroidal requirements for inducing implantation, some, such as the rabbit, needing progesterone alone (Nalbandov, 1971), and others, such as the rat and mouse, which require both oestrogen and progesterone (Psychoyos, 1969). In the rabbit, uterine specific proteins are produced in response to progesterone (Daniel, 1971a; Beier & Beier-Hellwig, 1973), and in the rat to oestrogen (Surani, 1975b,c, 1976). The interactive action of the two hormones in the rat are complex.

Extensive studies have been conducted on uteroglobin, a progesterone dependent protein in the rabbit (Schwick, 1965; Daniel, 1971; Beier & Beier-Hellwig, 1973; Johnson, 1974). The properties of uteroglobin are discussed by Beato (see this volume). The northern fur seal (Daniel, 1971), mink (Daniel, 1968), pig and dog (Daniel & Krishnam, 1969), and human (Shirai et al., 1974) are also claimed to contain traces of uteroglobin, although the presence of uteroglobin in the human is not confirmed (Beier & Beier-Hellwig, 1973; Shirai & Iizuka, 1974). In the marsupial, a protein of electrophoretic mobility similar to uteroglobin has been found when embryos are present in the uterus (Renfree, 1973). In the pig uterine lumen, two protein fractions are found on day 15 of the oestrous cycle. One of these is a cathode migrating glycoprotein of molecular weight 45,000, and the second fraction consists of six protein bands (Squire et al., 1972). In the cow, a quantitatively minor cathode migrating component is detected in embryo-bearing horns and is suggested to function as an antiluteolysin (Roberts & Parker, 1974a). In the baboon, no marked differences in uterine luminal proteins are detected in cycling animals (Peplow et al., 1974b). In the hamster, one or two minor evidently uterine specific proteins have been detected (Noske & Daniel, 1974).

Serum components may form a part or the bulk of luminal proteins. Some of the changes in luminal components of serum origin are dependent on the endocrine state of females. In the rabbit, injection of oestradiol suppresses uterine-specific proteins and increases serum components (Beier, 1974). During normal pregnancy in the rabbit, a large influx of serum proteins only occurs after implantation on day 8 (Kirchner, 1972). The endometrial epithelium is a barrier to the diffusion of δ-globulins and the lumen is essentially inaccessible to large molecular weight serum proteins during the early stages of pregnancy (Symons & Herbert, 1971). The proteins of the human uterine secretion include several serum components through selective secretion of plasma proteins (Beier & Beier-Hellwig, 1973), with only a pre-albumin and a post-transferrin band forming a small proportion of evidently specific proteins (Beier & Beier-

Hellwig, 1973).   There is a significantly lower concentration of α-macro-
globulin in the uterine fluid, compared with albumin, which reflects selectivity
based on molecular size (Chandra et al., 1974).

In rodents, few studies have been carried out and these suggest that the
uterine-specific proteins may form a minor proportion of the total protein.
Moving boundary electrophoresis of pooled uterine oestrous fluid reveals four
major components similar to those found in blood plasma, whereas paper electrophoresis shows a component migrating between β and δ-globulin of serum, but
albumin is not detected (Junge & Blandau, 1958).   Other studies using the same
techniques reveal six peaks which differ from serum components (Ringler, 1956).
Exogenous steroid administration evidently modified the luminal proteins in some
studies; β, δ-1 and δ-2 globulins and albumin are found in fluids collected
from ligated horns, with a fast-migrating component being detected only in
animals receiving oestrone (Ringler, 1961).   Immunoelectrophoresis and
Ouchterlony gel diffusion studies reveal a component migrating with β-globulin
that is uterine-specific (Albers et al., 1961);   four other components are
shared with rat serum.   In rats which receive oestradiol-17β , nine protein
species are identified in fluids from ligated horns by disc-gel electrophoresis;
five of the nine are evidently uterine-specific (Kunitake et al., 1965).   Total
protein content in uterine fluids of ovariectomized rats receiving oestrone
increases threefold, and 3 to 4 bands are detected in δ-globulin;   progesterone
administration together with oestrone reduces the total amount of protein, and
some bands in the δ-globulin zone are not detected (Hasegawa et al., 1973).
Presence of an IUD also increase total protein in the uterine flushings, most
likely due to inflammation (Peplow et al., 1974a,b).

Immunological studies using antisera against uterine fluid and blastocysts
from day 5 pregnant females show a single specific antibody, and the injection
of antiserum reduces the number of embryo implants in experimental animals (Ying
& Greep, 1972).   Antisera prepared after the removal of blastocysts and cellular
debris from the uterine fluid does not show any uterine specific antibody, or
impede implantation (Beck & Boots, 1973).

These studies have shown little clear indication of changes in the luminal
macromolecules in the rat.   The material that has been analysed has been pre-
dominantly oestrous fluid, or fluid obtained after uterine ligation.   Ligation
introduces difficulties,for the components of uterine fluid resemble those of
plasma with increasing length of time after ligation (Ringler, 1961).   However,
these studies do indicate that the uterine fluid may not simply be a serum
transudate, and that secretory products of the endometrium conditioned by steroids
may be present.   Analyses of static samples as well as detection of synthesis
and secretion of uterine proteins are essential.   Analyses of static samples
by themselves could be misleading if uterine specific proteins are in trace

amounts, which may be the case in the rat as judged from immunological studies (Ying & Greep, 1972; Beck & Boots, 1973; Peplow et al., 1974b). Physiological functions of macromolecules are not dependent merely on the quantity of the components but rather on the turnover rates and biological activities of the proteins.

HORMONAL REGULATION OF UTERINE LUMINAL PROTEINS IN THE RAT
Endocrinology of the oestrous cycle and early pregnancy

Quantitative changes in the levels of progesterone and oestrogen are thought to be responsible for regulating luminal proteins as well as uterine growth and cytodifferentiation prior to blastocyst implantation. The ovarian steroid levels in the rat and mouse differ considerably, both temporally and quantitatively, during the oestrous cycle and early pregnancy. Levels of oestradiol in the rat during proestrus are about 15-20 times higher than during pregnancy (Yoshinaga et al., 1969; Nimrod-Zmigrod et al., 1972; Schwartz, 1974). Progesterone secretion is minimal at the time of peak oestrogen secretion during early proestrus, and progesterone increases to peak levels later in proestrus. Both oestradiol and progesterone increase during pregnancy (Hasmimoto et al., 1968; Nimrod-Znigrod et al., 1972). A diagrammatic representation of the ovarian steroid secretions, constructed using the above references, is shown in Figure 3. The diagram gives general indications of a changing endocrine state.

The temporal and synergistic aspects of steroid secretion are critical, since progesterone/oestrone ratios of 2000-2500/1 (W/W) produce maximum sensitivity to decidualization (Yochim & De Feo, 1963; De Feo, 1967). Daily administration of hormones to ovariectomized females at these levels produces a pattern of endometrial mitosis similar to that in the pregnant uterus before implantation (Finn & Martin, 1974). An increase in oestrogen above these levels inhibits implantation and decreases the decidualization response of the uterus (Psychoyos, 1969; De Feo, 1967). The synergistic actions of the two hormones on the uterus are not clearly understood, although several studies on biochemical changes show that progesterone modifies oestrogen-mediated effects (O'Grady & Bell, this volume). Progesterone is essential for the sensitization of the uterus for implantation (Psychoyos, 1973) and presumably helps to control growth and functional differentiation of the uterus. Luminal proteins could serve as markers for the uterine response to steroids.

The formation and retention of intraluminal fluids is also under hormonal control (Ringler, 1961; Armstrong, 1968; Kennedy & Armstrong, 1972; Kennedy, 1974); high levels of oestrogen and low levels of progesterone during proestrus cause the formation of proestrous fluids. Simultaneous injections of progesterone, in amounts which do not affect uterine growth, effectively inhibit oestrogen-mediated accumulation of fluids in prepubertal rats (Armstrong, 1968).

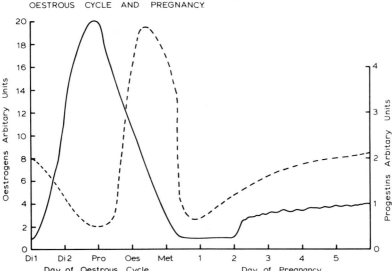

CHANGES IN OESTROGENS AND PROGESTINS LEVELS DURING OESTROUS CYCLE AND PREGNANCY

Figure 3: Schematic diagram of changes in the levels of oestrogens (————) and progestins (------) during the oestrous cycle and pregnancy in the rat. Implantation commences on day 5 of pregnancy.

The loss of luminal fluid at oestrus is probably due to the high levels of progesterone during late proestrus (Figure 3). In ovariectomized rats, 10 μg oestrone daily for four days evokes a 25-fold increase in total proteins, but the increase is only two-fold if 4 mg progesterone is also injected daily during the last two days (Hasegawa et al., 1973; Surani, unpublished). The low amount of oestrogen and increasing amount of progesterone during pregnancy prevents the accumulation of easily detectable amounts of intraluminal fluids in the rat.

Analyses of static samples

Analyses of static samples illustrate quantitative and qualitative changes in the luminal proteins at any particular stage of pregnancy or the oestrous cycle. Uterine luminal proteins can be compared with serum but only the macromolecules present in bulk can be reliably quantified.

Quantitative analyses

Quantitative estimation of the total protein shows that the greatest amount is present during the morning of proestrus when up to 2000 μg protein/animal is estimated. This is also the only time when a large quantity of uterine fluid, between 7 to 200 μl, can be aspirated from the lumen. During early pregnancy the total luminal protein content is much lower (Table 2). The protein content increases to a maximum on day 5 of pregnancy, the day of implantation. If the females are ovariectomized on day 3 and maintained on progesterone alone, there is a gradual decline in the levels of protein until the tenth day as blastocysts enter into quiescence (Surani, 1976). These changes may partly be due to the influence of the hormones on serum transudation as well as on the synthesis of uterine specific proteins.

TABLE 2

TOTAL INTRALUMINAL PROTEIN CONTENT DURING PREGNANCY AND AFTER OVARIECTOMY*

Day of pregnancy	No. of animals	µg protein/ rat	Day after ovariectomy	No. of animals	µg protein/ rat
1	9	100.05 ± 9.0			
2	16	41.25 ± 2.5			
3	12	30.06 ± 3.1			
4	15	49.70 ± 5.0	4	15	40.0 ± 5.0
5	16	72.92 ± 7.6**	5	21	39.9 ± 8.0*
6***	9	65.42 ± 3.2	6	16	40.0 ± 6.9
7***	8	60.01 ± 6.1	7	11	33.0 ± 7.1
			8	12	30.0 ± 7.2
			9	15	27.1 ± 9.0
			10	11	20.2 ± 8.3

* Uterine lumen was flushed with saline during pregnancy and after ovariectomy on day 3 of pregnancy. The total protein content was estimated by the method of Lowry et al. (1951).

** Significantly different, $P < 0.05$ (Student's t test).

*** Animals with ligated uterotubal junctions.

Qualitative analyses

Qualitative analyses of uterine luminal proteins have been carried out using disc-gel electrophoresis of protein-dodecyl sulphate complexes (Surani, 1975b,c, 1976; Gore-Langton & Surani, 1976). There appear to be no major qualitative changes in the luminal proteins during the course of pregnancy (Figure 4a). There is, however, an increase in the protein of approximate molecular weight (M_r) 70,000 and others of higher molecular weight from day 1 to day 5 of pregnancy. If the females are ovariectomized on day 3 of pregnancy and given progesterone alone, the expected quantitative increase in the proteins on day 5 is not observed (Table 2), and qualitatively, some of the protein bands do not increase (Figure 4b). With increasing length of time after ovariectomy, the proteins of $M_r \geq 70,000$ are barely detectably by day 10, when blastocysts enter into

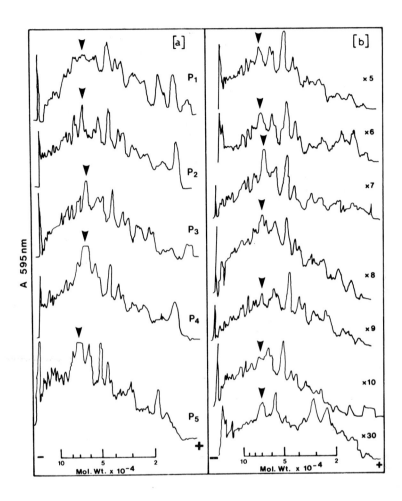

Figure 4: Rat uterine luminal proteins analysed on 7.5% acrylamide gels
(a) on days 1,2,3,4 and 5 of pregnancy (left hand series, and (b) 5,6,7,8,9,10
and 30 days after ovariectomy on day 3 of pregnancy with subsequent progesterone administration (right hand series). Compare the densitometer scan of
the pregnancy day 5 proteins (P5) with that for proteins on the fifth day after
ovariectomy (x5). Note the increase in proteins with approximate M_r 70,000
and above (arrow head) as pregnancy proceeds, and the decrease of these proteins
with days after ovariectomy.

quiescence (McLaren, 1973; Webb & Surani, 1975). When oestrogen is given
together with progesterone to ovariectomized females, all the luminal proteins
detected on day 5 of pregnancy are observed at about 13 to 20 hours after the
injection of the hormones (Figure 5). At about 18 hours after oestradiol

262

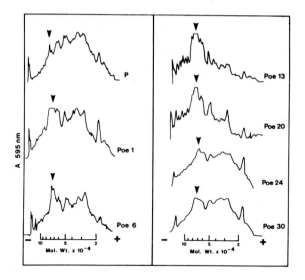

Figure 5: Densitometer scans of the protein profiles in uterine flushings taken 1,6,13,20,24 and 30 hours after injection of progesterone + oestradiol into ovariectomised females. Note similarity between profile at 13-20 hours and that on day 5 of pregnancy (Figure 4). P = progesterone injected alone.

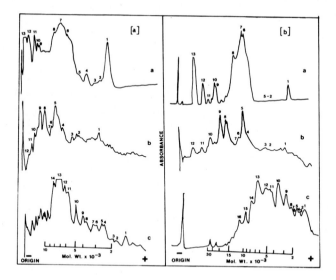

Figure 6: Densitometer scans of proteins separated on (a) 7.5% and (b) 5.0% polyacrylamide gels. a = serum; b = proestrous luminal fluid; c = luminal fluid on day 5 of pregnancy. For each sample, numbers are arbitrarily assigned to each major protein band, and mobilities and M_r for each band are given in Table 5. Note that use of the same numbers for each sample does not imply that the same proteins are being coded.

injection, embryo implantation commences, and blastocysts transferred at 24 hours after oestrogen fail to implant (Psychoyos, 1969). The changes in the luminal proteins occur regardless of the presence or absence of blastocysts.

Thus, oestrogen appears to enhance the levels of luminal proteins although both progesterone and oestrogen are required for implantation. The uterine response to oestrogen alone is of interest in determining how this hormone acts by itself. Cycling animals at proestrus have levels of oestrogen which are higher than during pregnancy when progesterone levels are very low (Figure 3). The substantial accumulation of uterine fluid and total protein during proestrous may be due to high oestrogen and low progesterone. The proteins may be derived from enhanced transudation of serum and/or through the secretory activity of the uterus. Detailed comparisons with serum show that the proestrous fluid is distinguishable by two macromolecules of approximate M_r 125,000 and 130,000 which are present in abundance (Figure 6). These two proteins are evidently absent from serum. Quantitative relationships between the major macromolecular fractions give further insight into the differences in the properties of the three fluids. Densitometer scans (Figure 6) are divided into three major fractions represented by high molecular weight proteins of $M_r > 70,000$ (HM), the major protein band of approximate M_r 70,000 (A) and the fraction with proteins of $M_r > 70,000$ (LM). The results show that the HM fraction is higher in both serum and oestrous fluids, whereas LM is higher in the material from pregnant females (Table 3). The proportion of albumin (A) is highest in serum, but the corresponding proteins of approximate M_r 70,000 are lowest in uterine extracts. The high HM in oestrous fluids is not due to an increase in serum components (see later) but due to the presence of the two proteins found in the fluid in very large amounts. The total area under each fraction is, however, subject to overall changes in profile. Ratios between the fractions further define the quantitative relationships between the fractions in these fluids (Table 4).

Estimates of mobilities and molecular weights (M_r) of proteins in 7.5% and 5% running gels are given in Table 5. Proteins of molecular weight greater than 80,000 are subject to retardation in 7.5% running gels, so that the molecular weight estimates of these proteins are much lower in 7.5% than in the 5% running gels. The mobilities and molecular weights have been estimated for comparisons in a later section with synthesized radiolabelled proteins.

Comparisons between the three biological fluids show, in general, that oestrogen appears to cause an increase in the high molecular weight protein fraction but not apparently through simple transudation of serum components. Qualitatively and quantitatively, proestrous fluid and fluid from uteri of pregnant females differ from serum. These results could be explained by selective entry and/or modification of serum proteins, or by de novo synthesis of uterine proteins or by a combination of these processes.

TABLE 3

QUANTITATIVE DISTRIBUTION OF MACROMOLECULES IN
UTERINE FLUIDS AND SERUM

Samples		Serum	Proestrous	Pregnant
Fraction* (± S.E.)	Gels Concn (%) (a) 7.5 (b) 5.0			
HM	a	32.9 ± 5.2	35.5 ± 5.9	24.5 ± 4.2
	b	45.0 ± 2.2	58.4 ± 4.1	24.4 ± 1.7
A	a	42.8 ± 6.1	20.6 ± 4.9	17.6 ± 3.8
	b	40.1 ± 3.1	13.7 ± 1.3	16.6 ± 1.1
LM	a	24.3 ± 1.9	43.8 ± 5.7	58.4 ± 3.1
	b	14.9 ± 1.0	27.9 ± 5.9	59.4 ± 2.3

* The proportion of macromolecules of M_r greater than 70,000 (HM), of approximately 70,000 (A) which is represented by albumin in serum and of less than 70,000 (LM) calculated as a % of total area of the densitometer scans of serum and uterine samples analysed in 7.5% and 5.0% gels.

TABLE 4

RATIOS BETWEEN THE MAJOR MACROMOLECULAR FRACTIONS IN
UTERINE FLUIDS AND SERUM

Samples		Serum	Proestrous	Pregnant
Ratios* (± S.E.)	Gel Concn (%) (a) 7.5 (b) 5.0			
A:HM	a	1.31 ± 0.15	0.58 ± 0.06	0.74 ± 0.08
	b	0.92 ± 0.08	0.25 ± 0.03	0.73 ± 0.05
A:LM	a	1.78 ± 0.09	0.48 ± 0.08	0.31 ± 0.05
	b	2.96 ± 0.13	0.50 ± 0.09	0.28 ± 0.02
HM+A:LM	a	3.15 ± 0.20	1.30 ± 0.15	0.72 ± 0.12
	b	5.59 ± 0.57	2.63 ± 0.50	0.69 ± 0.30
LM+A:HM	a	2.05 ± 0.17	1.82 ± 0.14	3.21 ± 0.18
	b	1.28 ± 0.23	0.72 ± 0.09	4.16 ± 0.28

* Quantitative relationships between the three major fractions HM, A and LM determined from the values of areas occupied by each of the fractions as determined in Table 3 from the densitometer scans of serum and uterine samples in 7.5% and 5.0% gels.

TABLE 5

MOBILITIES AND ESTIMATED M_r OF PROTEINS DETECTED IN UTERINE FLUIDS AND SERUM*

Sample	Band No.	Mobility in 7.5% gel	Estimated $M_r \times 10^{-3}$	Mobility in 5.0% gel	Estimated $M_r \times 10^{-3}$
Serum	1	0.72	24.5	0.93	17.0
	2	0.63	30.5	0.88	20.8
	3	0.60	32.5	0.81	25.3
	4	0.54	43.0	0.75	33.5
	5	0.50	50.5	0.71	39.0
	6	0.35	64.0	0.61	57.0
	7	0.30	72.0	0.55	73.0
	8	0.23	86.0	0.50	83.0
	9	0.21	90.0	0.35	153.8
	10	0.14	110.0	0.31	182.3
	11	0.13	120.0	0.29	198.8
	12	0.05	130.0	0.26	221.0
	13	0.03	209.0	0.20	273.3
Proestrous Fluid	1	0.67	25.5	0.90	18.5
	2	0.40	56.5		
	3	0.35	64.0		
	4	0.30	72.0	0.58	69.0
	5	0.27	76.0	0.54	74.6
	6	0.23	86.0		
	7	0.21	90.0		
	**8	0.19	96.0	0.41	125.5
	**9	0.15	105.0	0.39	130.0
	10	0.12	116.0	0.25	216.8
	11	0.10	119.0	0.19	291.3
	12	0.07	120.0	0.12	376.3
Day 5 pregnancy uterine fluid	1	0.94	14.0	0.91	15.6
	2	0.86	16.7		
	3	0.80	19.4		
	4	0.69	24.0	0.86	22.3
	5	0.65	29.0		
	6	0.59	33.7		
	7	0.56	36.6		
	8	0.52	40.6	0.83	26.0
	9	0.48	44.6	0.80	29.0
	**10	0.40	56.5	0.73	36.5
	**11	0.36	62.4	0.67	45.0
	**12	0.34	64.7	0.62	57.0
	13	0.28	74.4	0.58	69.0
	14	0.23	86.0	0.51	85.0
	**15			0.46	105.0
	**16			0.41	125.5
	17			0.22	250.0
	18			0.13	362.5

*The samples were analysed electrophorectically on 7.5% and 5.0% running gels. Their mobilities and approximate M_r were estimated. Presumptive uterine specific proteins are indicated**.

Synthesis of uterine specific proteins

Analyses of static samples has thus suggested that mere fluctuations in the quantities of serum proteins in the uterine lumen in response to steroid hormones is not an adequate explanation for profile changes. However, comparisons between serum and uterine luminal fluids based merely on electrophoretic mobilities of proteins are not satisfactory. Errors in interpretation may arise if some serum proteins are sequestered in the lumen. Quantitatively meagre but important proteins in static samples may be overlooked, especially if these are synthesized in the uterus and have a rapid turnover rate. In situ synthetic patterns must also be investigated.

Quantitative estimation of the incorporation of L- $4,5-^3H$ leucine into luminal proteins

L- $4,5-^3H$ leucine incorporation into the luminal proteins, achieved by procedures described in Table 6, is highest during proestrus and lowest during embryonic diapause. The level of incorporation increases after an injection of oestradiol to ovariectomized females (Table 6). These results may be expected since the total protein content is much higher during proestrus than

TABLE 6

INCORPORATION OF ^3H LEUCINE IN UTERINE LUMINAL FLUIDS*

Endocrine state of females (No.)	Total cpm $(x\ 10^{-2})$ ± S.E.	Incorporated cpm x 10^{-2} ± S.E.
Proestrous (4)	6980.59 ± 405.5	3101.84 ± 200.7
Day 5 of pregnancy (4)	904.25 ± 50.6	597.55 ± 51.2
Ovariectomized + progesterone (4)	46.06 ± 6.03	37.66 ± 7.2
Ovariectomized + progesterone + oestradiol (3) (12-18 hr)	702.3 ± 60.7	406.6 ± 66.6

* 5μ Ci of L- $4,5,-^3H$ Leucine (spec. act. 60 Ci/mmol), in 25μ l phosphate buffered saline (PBS) was injected into the lumen of each uterine horn. The uterine horns were flushed with 2.0 ml PBS after 6 hours and the sample centrifuged at 12,000 x g for 1 hour. 5μ l of the uterine samples were dried on four glass fibre GF/A filters. Two filters were counted to obtain total counts, and the other two treated with trichloroacetic acid to obtain the values for the incorporation of the precursor into proteins. The filters were counted in 10 ml of scintillation fluid (5.5 g Permablend/litre toluene). The values represent counts/animal.

at any other time.   The increase in incorporation of leucine may be due to
oestrogen-dependent synthesis and secretion of uterine proteins.   There are
difficulties in the estimation of the exact amount of synthesis in vivo since
the quantity of unlabelled leucine cannot be determined under different endocrine conditions.   There may also be a variable retention of the precursor in
the uterine lumen under different endocrinological conditions.   As the cells
in which the proteins are synthesized are not yet identified, in vitro labelling
of the isolated uterine cells is difficult, especially if synthesis occurs in
glandular cells rather than luminal epithelium or stromal cells since glandular
cells are deep in the stroma and difficult to isolate.

During in vivo labelling of uterine luminal proteins, the precursor might
have been distributed into serum and at least part of the radiolabelled proteins
could have been synthesized outside the uterus and entered the lumen.   Controls
show that such contribution over the time period employed in these studies is
minimal.   When the precursor is injected intraluminally or intraperitoneally,
there is an incorporation of the precursor into the serum proteins.   However,
after intraperitoneal injection of the precursor, when incorporation into serum
is higher than after intraluminal instillation, the amount of labelled proteins
in the luminal proteins is negligible (Table 7).   The radiolabelled proteins
detected after intraluminal injection of the precursor are therefore most likely
the product of uterine synthesis of proteins.   There must be considerable
dilution of the isotope after intraperitoneal injection of the isotope and
presumably the amount of precursor reaching the uterus for incorporation into
uterine proteins is very small.   The experiments do not exclude the presence of
serum proteins in the lumen.   The rate of entry of serum proteins into the
lumen and their proportion relative to other macromolecules has not yet been
estimated.

Qualitative analyses of radiolabelled proteins

Analyses of radiolabelled luminal proteins have been carried out on 6%
polyacrylamide gels using the methods described previously (Surani, 1975c).   In
the study using L-^{35}S methionine as precursor, the differences between the
luminal proteins at proestrus and pregnancy are shown (Figure 7a).   The relative
extent of incorporation into each fraction during proestrus shows incorporation
into proteins of M_r 75,000, 130,000 and 150,000.   The two proteins of
M_r 130,000 to 150,000 are also distinct in static samples of proestrous fluids
(Figure 6).   During day 5 of pregnancy, the major fractions of newly synthesized
protein are detected between M_r 30,000 to 60,000 and at about M_r 100,000
(Figure 7a).   The major protein fraction of M_r 70,000 detected in the static
samples (Figure 6a) does not appear to be labelled.   These changes can be
reproduced in ovariectomized females by altering the amounts of exogenous

TABLE 7

COMPARISON OF THE INCORPORATION OF ^{3}H LEUCINE IN SERUM AND UTERINE
FLUID AFTER INTRAUTERINE AND INTRAPERITONEAL INJECTIONS*

Endocrine state of the females (No.)	Route of isotope injection (i.u. or i.p.)	Uterine fluid (cpm x 10^{-2})		Serum cpm/5 µl sample	
		Total ± S.E.	Incorporated ± S.E.	Total ± S.E.	Incorporated ± S.E.
Proestrous (8)	i.u.	6980.6 ± 405.5	3101.8 ± 200.7	298.2 ± 32.2	198.6 ± 33.6
	i.p.	8.6 ± 0.9	6.5 ±0.6	636.2 ± 57.6	440.9 ± 44.5
Day 5 of pregnancy (8)	i.u.	904.3 ± 50.6	597.6 ± 51.2	318.7 ± 43.5	258.4 ± 48.6
	i.p.	6.7 ± 0.2	5.9 ± 0.1	469.8 ± 32.7	404.6 ± 30.9

* The intraperitoneal injection of (i.p.) 10µ Ci of the isotope was given in
50µ l PBS after injecting 25µ l PBS into the lumen of each horn. Intraluminal injection (i.u.) was as described before, and the rest of the
experimental procedure was also identical to that given in Table 6.

hormones (Figure 7b). The proestrous type of profile is obtained at 48 hours
after two injections of 1 µg oestradiol, whereas for the profile seen during
pregnancy, both progesterone and oestradiol are necessary in proportions described previously (Surani, 1975c).

The time course studies after an injection of oestradiol and progesterone
into animals experiencing delay of implantation are given in Figure 8. The time
course of changes observed in the labelled proteins are similar to those obtained
for analyses of static samples (Figure 5). Injection of 400 µg/animal cycloheximide simultaneously with the intraluminal instillation of the precursor
essentially blocks the synthesis of proteins but the effect is less severe when
the drug is administered about three hours after the injection of the precursor
(Figure 9). Further comparisons between the labelled proteins and those in the
static samples can be made with reference to Table 5.

Glycoproteins

Glycoproteins are an important constituent of the outer cell surface (Cook,
1968; Winzler, 1970; Cook & Stoddart, 1973) and are also present in serum and
other secretions. The macromolecules consist of heteropolymerized carbohydrate
sub-units covalently linked to the polypeptide backbone. The amino sugars,
N-acetylglucosamine and N-acetylgalactosamine along with mannose, form the inner

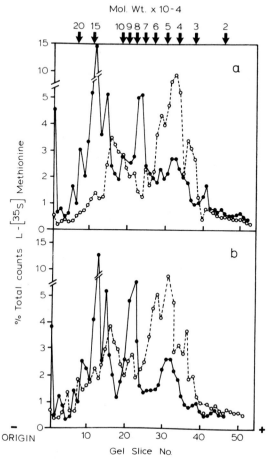

Mol. Wt. x 10⁻⁴

Figure 7: Radioactive counts, as a percentage of total counts, in fractions of a
0.16 cm polyacrylamide gel after separation of ^{35}S methionine labelled luminal
proteins on 6.0% gels.   7a shows radioactive macromolecules from pregnant
(o------o) and proestrous (●——●) females.   7b shows radioactive macromolecules from   ovariectomized females injected with progesterone (2 mg) and
oestradiol (1 μg) (o------o) or oestradiol alone (2 injections of 1 μg over
48 hours) (●——●).   Note the similarities between 7a and 7b.

residues of the glycopeptide chain.   Galactose residues are next in the chain
with fucose or sialic acid occupying the end position (Cook, 1968; Spiro, 1970;
Cook & Stoddart, 1973; Lloyd, 1975).   The structural diversity of glycoproteins
is large.   Glycosyltransferases which attach nucleotide-activated sugars in the
endoplasmic reticulum and Golgi apparatus have distinctive substrate speci-
ficities and give rise to unique glycoproteins.   The presence of sugar residues
in the macromolecules gives structural diversity and specificity and they are
important in many cellular functions (Winzler, 1970).   Glycoproteins in the

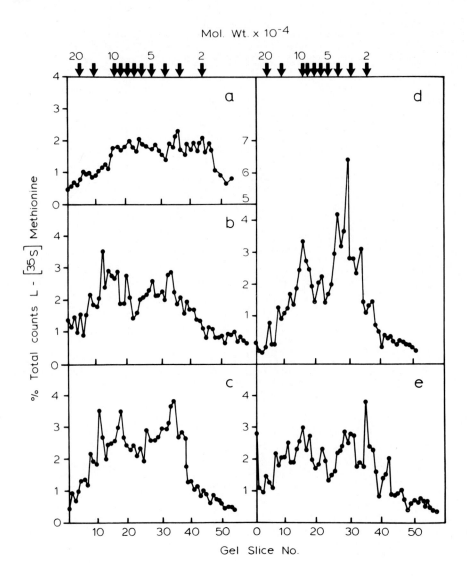

Figure 8: Time course changes in ^{35}S methionine labelled luminal proteins in ovariectomized females after (a) an injection of 2 mg progesterone alone, and at different times after injection of progesterone + 1 μg oestradiol. (b) 0-6 hr. (c) 6-12 hr (d) 12-18 hr (e) 18-24 hr. Note similarities between 8d and pregnant females (7a).

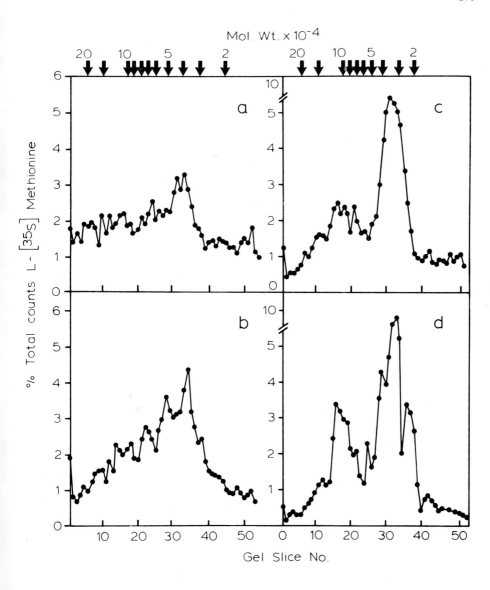

Figure 9: Effects of i.p. injection of cycloheximide (400 μg/female) into day 5 pregnant females on protein synthesis in utero over a 12 hr period. Cycloheximide was given (a) simultaneously with the radioactive precursor, (b) at 2 hr or (c) at 3 hr after the precursor. Compare with control without cycloheximide (9d).

luminal fluid could play an important role in regulating blastocyst cell metabolism if they act as trophic factors and bind to blastocyst surface receptors. They could also modify the mobility of embryonic cell surface ligands in preparation for cell surface interactions between embryos and uterus.

Fucose is a better precursor for studies on glycoprotein synthesis because it is not broken down or converted to other sugars and is only incorporated as a fucose residue into newly synthesized glycoproteins (Bennett & Leblond, 1974). Figure 10 shows that L 1-^{3}H fucose is incorporated into some luminal proteins. In proestrous luminal fluid, it is incorporated into proteins of approximate M_r 150,000 and 200,000 but it is not predominant in proteins of M_r 130,000 and 75,000. During pregnancy it is found to be associated with one major group of proteins migrating at M_r 90,000 to 100,000 whereas the lower M_r group of 30,000 to 60,000 is not labelled to the same extent. An additional group is found in the region of M_r 150,000. During delayed implantation when blastocysts are in quiescence, there is a lack of luminal proteins both quantitatively and in the incorporation of precursors (Table 6). There are no distinct macromolecules detectable at this time.

The increase in glycoproteins in the lumen during proestrus and pregnancy are presumably related to the steroid dependent changes in the endometrial enzymes involved in the synthesis of glycoproteins. Oestradiol causes an increase in the activities of the enzymes, glactosyl- and sialy-transferase (Nelson et al., 1975) and decreases the activity of UDPgalactose pyrophosphate when the hormone is given to ovariectomized rats. Progesterone acts antagonistically to oestradiol and enhances UDPgalactose pyrophosphatase activity (Jato-Rodriguez et al., 1976). The regulation of this enzyme by the ovarian steroids would determine the pool size of nucleotide sugars in the endometrium and steroids may in this way regulate synthesis of glycoproteins and glycosaminoglycans (Endo & Yosizawa, 1973).

HOW DO LUMINAL PROTEINS ACT ON BLASTOCYSTS?

Several functions for the luminal proteins have been proposed, and although at present no evidence on this question is decisive, in some instances circumstantial evidence strongly favours certain activities. Some of these functions are considered below.

Role of luminal proteins as trophic factors

A major point of controversy concerns the strict requirement for specific luminal proteins, when serum appears to be capable of supporting post-implantation development in vitro. It seems likely that one of the functions of extracellular macromolecules of either serum or uterine origin may be in regulation of blastocyst metabolism. This is evident from the analytical work

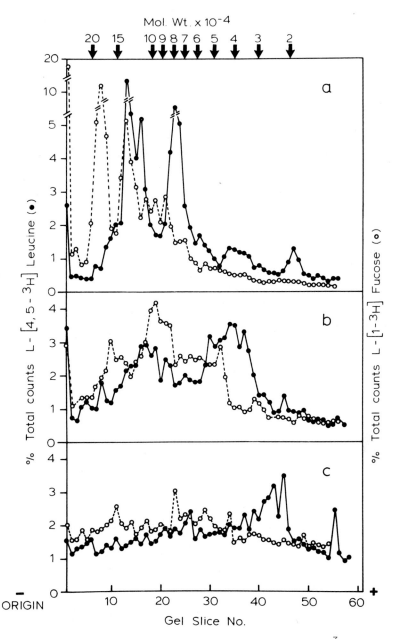

Figure 10: Radiolabelled glycoproteins after instillation of L- 1-^{3}H fucose (o------o) and protein after instillation of L- 4,5-^{3}H leucine (●———●) in luminal proteins of (a) proestrous females, (b) day 5 pregnant females, and (c) ovariectomized females during delay of implantation.

274

on the embryonic response to luminal proteins during delay of implantation and
pregnancy.   If blastocysts are to participate in the implantation process, their
functional potential will depend on metabolic enhancement.   In metabolically
active blastocysts, changes in cell surface properties and enzymes as well as
synthesis of cell specific protein will occur in preparation for cytodifferentiation and implantation.

The uterus and the embryo can be considered as an integration of two
distinct systems, the former inflexible, and the latter flexible.   Ovarian
steroids irreversibly sensitize the uterus for implantation, since its
resensitization occurs only after the original hormonal stimuli are repeated in
full.   If the two systems are to act synchronously at the time of implantation,
the blastocysts must be able to accommodate to the changing uterine state.   As
the embryos reach the blastocyst stage, they become increasingly reliant on
extracellular macromolecules of the uterus, the changes in macromolecular
secretion being mediated by steroids.   Thus, a prime feature of the response
of blastocysts is their ability to adapt to environmental changes.   This notion
is enforced by the observation that blastocysts may adapt to the existing
environment in utero, following reciprocal transfer of embryos between physiologically different females (Dickmann & De Feo, 1967; Psychoyos, 1969).
Implanting and implanted embryos revert to a state of diapause on transferring
to females experiencing delay of implantation (Psychoyos, 1961).   Since the
embryos have a demonstrably high degree of adaptability, uterine proteins or
other maternal factors would not appear to irreversibly trigger implantation in
a cascade-like manner.   How then do uterine specific proteins function during
implantation?   At the cellular level, these proteins (or serum proteins) would
be unlikely to trigger the production of embryo-specific products (through mRNA)
and commit the embryos to implant, but are much more likely to enhance embryonic
metabolism.

In Figure 11, the response of cells in a variety of systems to extracellular trophic factors is illustrated.   These factors can bind to one or more
receptors in a multivalent fashion through interaction with the cell membrane
(Cuatrecasas, 1972; Hollenberg & Cuatrecasas, 1975).   In this instance, both
serum and uterine proteins presumably evoke an identical response in blastomeres.
The consequence of these extracellular factors is to induce a coordination of
biochemical events as found at the time of cell proliferation (Hershko et al.,
1971; Alberghina, 1974).   The ultimate phenotypic response of a cell is
independent of the nature of trophic factors since the individuality of the cell
is determined by the types of macromolecules it is already primed to synthesize.
Hence, several cell types responding to the same serum factors maintain their
individuality.   Extracellular factors may not, therefore, dictate phenotypic
response of cells.

The binding of uterine or serum proteins to embryonic surface receptors may
initiate a coordinated intracellular response leading to an increase in RNA,
protein and DNA synthesis, preceded by alterations in the levels of cyclic
nucleotides.   This would be in keeping with the known action of some macromolecular mitogens (Rudland et al., 1974).   Such increases in the protein
synthesis in response to extracellular mitogens may be governed by the recruitment of ribosomes and the resulting translation of pre-existing untranslated

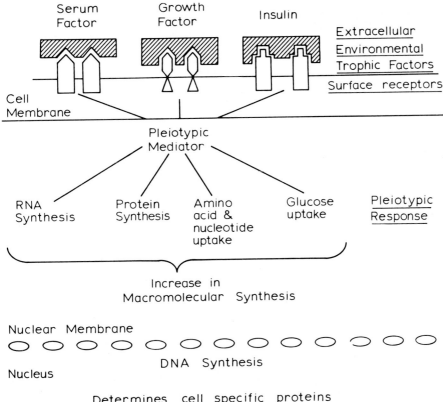

Figure 11: Cellular response after the binding of extracellular trophic factors.
Any of the factors which bind to surface receptors mediate a similar intracellular response through a pleiotypic mediator which may be cyclic nucleotides.
The intracellular biochemical events act co-ordinately (pleiotypic response)
which results in an increase in macromolecular synthesis and cell division.   The
phenotypic response of the cell is independent of the nature of trophic factors
and is determined by the cell specific protein synthesized.

mRNA, rather than by transcription (Rudland, 1974; Bandman & Gurney, 1975), because a generalized synthesis of all detectable peptides is observed after re-activation of quiescent embryos (Van Blerkom & Brockway, 1975). This scheme would explain the lability of response shown by the blastocysts to changes in the environment.

The difference between serum and uterine macromolecules may lie in the potency with which blastocyst cell metabolism is regulated. Uterine proteins may possess a high specificity for embryonic receptors and may modulate embryonic cell metabolism much more efficiently than serum. Of particular interest may be the uterine luminal glycoproteins (M_r 100,000). The presence of sugar residues in these macromolecules may provide them with a high specificity for putative surface receptors of blastomeres.

Effects on cellular interactions between ICM and trophectoderm

The metabolic regulation of embryonic cells by luminal proteins does not take into account complex interactions between the inner cell mass cells (ICM) and trophectoderm. Such interactions are presumably involved in the establishment of the longitudinal axis in the embryo and may control embryonic growth and implantation. Figure 12 is a representation of some of the likely interactions between the two-cell population in mouse blastocysts, but many details remain to be obtained. At least two areas of trophectoderm have been identified: the polar trophoblast cells overlie the ICM and the mural trophoblast surrounds the blastocoelic cavity. The ICM makes little or no cellular contribution to the trophectoderm of implanted blastocysts (Gardner et al., 1973; Rossant, 1975a,b) but its position in the blastocyst results in the regional differentiation of trophectoderm (Gardner & Papaioannou, 1975). The polar trophoblast remains diploid and proliferates under an inductive stimulus from the ICM (Gardner & Johnson, 1972), and the resulting cells contribute to the ectoplacental cone and extraembryonic ectoderm (Gardner & Papaioannou, 1975). Mural trophoblast cells and cells remote from the influence of the ICM on the other hand cease to divide (Gardner & Papaioannou, 1975) and undergo polyploidization by endoreduplication of DNA (Sherman et al., 1972). It is not known whether mural trophoblast, after blastulation, is recruited from the division of polar trophoblast. At the time of implantation, mural trophoblast adheres to the uterine epithelium and undergoes giant cell transformation (Dickson, 1969). From in vitro studies, it has been suggested that the ICM may govern the timing of the trophoblast giant cell transformation (Ansell, 1975). Cellular communications between the ICM and trophectoderm could be governed by surface to surface interactions, junctional channels, formation of chemical gradients within the embryo and through extracellular signalling molecules, such as uterine secretion proteins.

Some macromolecules are known to enter the blastocoele (Hamana & Hafez, 1970; Beier, 1974) and may influence cellular interactions between the embryonic cells, for example, by acting on both the ICM which could then regulate cell proliferation, as well as on the trophoblast directly to initiate giant cell transformation.

Blastocysts may be microsurgically dissected free of ICMs to yield both large (> 20 cells) and small (10-20) trophoblastic vesicles. On transfer to mice which are ovariectomized and given progesterone, these vesicles enter into quiescence and implant only when oestradiol is injected in the same manner as intact blastocysts (Surani & Barton, unpublished). No ICM develops (Table 8). The uterine macromolecules would not therefore appear to have an effect on blastocysts via the ICM as far as the timing of implantation is concerned. Whereas the number of cells in intact blastocysts increases from 60 to 120 before the embryos enter into quiescence, the number of cells in the vesicles remain at around 20 and only trophoblastic giant cells are detected after implantation.

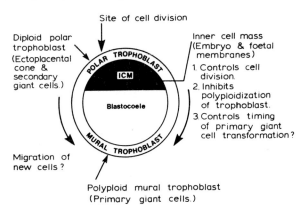

Figure 12: The organization of the preimplantation blastocysts, regional differentiation of the trophectoderm and possible interactions between the inner cell mass cells (ICM) and trophoblast cells.

This may explain why fewer trophoblastic vesicles implant compared with blastocysts, since cell proliferation is presumably dependent on the interactions between the ICM and trophectoderm.

There is an additional possibility that polar trophoblast surrounding the ICM may respond to the environmental changes and then influence the ICM. Blastocysts may be dissected free of mural trophoblast, leaving ICM fragments with just the surrounding polar trophoblast intact. These fragments fail to implant when transferred to normal pseudopregnant females. However, when transferred to recipients which have been ovariectomized and treated with exogenous progesterone to induce quiescence before implantation, about 30 to

TABLE 8

IMPLANTATION RATES OF TROPHOBLASTIC VESICLES AND BLASTOCYSTS IN INTACT AND OVARIECTOMIZED RECIPIENTS*

Recipients	Day 4 trophoblastic vesicles			Day 4 blastocysts		
	No. transferred	No. implants (%)	No. recipients	No. transferred	No. implants (%)	No. recipients
Ovariectomized	66 (small)	7 (10.6)	9	61	51 (83.6)	13
	32 (large)	18 (59.4)	6			
Intact	10	7 (70.0)	2	10	9 (90.0)	2
	Quiescent trophoblastic vesicles			Quiescent blastocysts		
Ovariectomized	13	6 (46.2)	2	4	4 (100)	1
Intact	6	4 (66.7)	1	15	9 (60.0)	2

* Intact blastocysts or microsurgically dissected trophoblastic vesicles were transferred to pseudo-pregnant mice. A group of females was ovariectomized after transfers and treated with progesterone for 6 days. Oestrogen was given for a further 3 days and implantations examined histologically.

40% of them implant upon injection of oestradiol.   Normal egg cylinders develop
by day 7.5 of pregnancy.   The fragments presumably undergo cell proliferation
and give rise to mural trophoblast before they finally enter into quiescence.
The survival of the ICM in this case may be ensured because of the blastocoele
formation which would provide the correct microenvironment for the ICM.   The
survival of the ICM could also be influenced by contact and interactions with
the polar trophoblast, itself responding to the extracellular changes.

Other specialized functions

The β-glycoprotein in the rabbit has been shown to possess proteolytic
activity (Kirchner et al., 1971).   A proteinase inhibitor is also found in the
secretion and differs from those found in serum (Beier, 1970), and a similar
component is found in the rat (Surani, unpublished).   The rat uterine fluid
contains a dipeptidase (Albers et al., 1961) and an endopeptidase (Joshi et al.,
1970).   Mouse uterine flushings show slightly increasing proteolytic activity
during pregnancy when tested against a non-specific substrate, casein (Pinsker
et al., 1974).   The detection of proteolytic enzymes in the uterine secretions
may be of interest since the zona pellucida appears to be lysed by an oestrogen-
dependent luminal component prior to implantation in the mouse (Mintz, 1971)
and the rat (Surani, 1975a).   This factor may also bring about changes in the
cell surface of the embryos for the initiation of embryo attachment (Pinsker &
Mintz, 1973).

Glycosidases have been detected in the cow luminal fluids (Roberts &
Parker, 1974b) in which β-N-acetylglucosaminidase, β-N-acetylgalactosaminidase
and α-fucosidase are particularly active during early pregnancy.   In the rabbit,
activity of α-fucosidase and several other glycosidases is elevated during
pregnancy.   These enzymes, by removing terminal sugars from glycoprotein on
the cell surface of trophoblast or endometrium, may have a role in adhesion of
embryos (Roberts & Parker, 1974b).

Numerous other enzymes have been reported to be present in the uterine
secretions.   Alkaline phosphatase and β-glucoronidase activities are found
to be much higher in rat uterine fluids, compared with plasma (Ringler, 1961).
Non-specific proteolytic enzymes, lysozyme, amylase and phosphatase have been
detected in the uterine flushings of baboons (Peplow et al., 1974b).   The
functions if any for these enzymes in implantation are not clear.

UTERINE LUMINAL PROTEINS AS MARKERS FOR HORMONAL ACTION ON UTERUS

Luminal proteins, and especially the uterine specific proteins, could
prove useful as markers for cytodifferentiation and uterine sensitization for
implantation.   In this way the molecular events associated with the uterine
preparation for implantation can be studied, as well as the synergistic and

antagonistic action of the steroids on the control of growth and functional differentiation of the uterus.

Uterine growth fluctuates with the circulating levels of steroids and the
hormones stimulate cell proliferation in preparation for implantation (O'Grady &
Bell, this volume). Both hyperplastic and hypertrophic growth is known in the
uterus, the former without an increase in mass, the latter with a weight
increase (Reynolds, 1965). Large doses of oestrogen cause a considerable
weight gain during proestrus; progesterone suppresses such an increase in
weight. There is virtually no weight increase in the uterus during the first
five days of pregnancy in the rat (Surani, unpublished). Steroids also alter
the recruitment and proliferation of uterine cells. Oestradiol causes a marked
increase in epithelial mitosis in ovariectomized females but if females are
given progesterone prior to oestradiol, stromal cells synthesize DNA but not the
epithelial cells (Galand & Leroy, 1971; Martin et al., 1973; Finn & Martin, 1974).
Progesterone on the whole antagonises the oestrogen-dependent increase in
uterine weight but is at the same time essential for implantation, possibly by
restricting growth (increase in mass) and promoting differentiation of the
recruited cells.

The changes in the growth response can be attributed to the fluctuations
in the receptor content of the uterine cells (Mester et al., 1974; Hsueh et al.,
1975). Progesterone reduces oestradiol receptors in the rat uterus (Hsueh et
al., 1975). However, maximum increase in oestradiol receptors during
pregnancy is detected on day 5 of pregnancy in the rat (Mester et al., 1974).

Oestradiol is known to cause the synthesis of specific uterine non-histone
chromosomal proteins (Cohen & Hamilton, 1975) and RNA synthesis is one of the
earliest responses to oestradiol injection (Hamilton, 1971). The increase in
genetic transcription is evident from studies on RNA polymerase activity
(Baulieu et al., 1972). In the pregnant uterus, RNA/DNA ratios increase significantly on day 5 of pregnancy (Heald, 1973). Qualitatively, mainly an
increase in ribosomal RNA is detected but hybridization experiments reveal that
some species of RNA may only be present on day 5 of pregnancy (Heald & O'Hare,
1973). There is also an increase in the soluble proteins in the endometrium
on day 5 of pregnancy (Yochim, 1975), presumably in response to oestrogen.

An oestradiol-induced protein (IP) is detected within 45 to 60 minutes after
an injection of the hormone into ovariectomized animals (Notides & Gorski, 1966;
Barnea & Gorski, 1970; Iacobelli et al., 1973). This protein may be one of the
several hormone-induced proteins (Gorski, 1973). An alternative view suggests
that the initial transcription involves a few genes which code for IP, which may
then activate synthesis of several essential components through increased
ribosomal RNA synthesis leading to a generalized increase in protein synthesis

(Baulieu et al., 1972). Experimental evidence does not entirely support this hypothesis (Katzenellenbogen & Leake, 1974).

Hormones may, therefore, induce qualitative changes in macromolecules in the uterus through the modulation of the products of existing cell types, as well as through the selection of appropriate cell types. There is a lack of markers for studies on uterine response and differentiation prior to implantation. In the chick oviduct, detection of avidin and ovalbumin has helped considerably in the studies on hormonal action in this tissue (O'Malley & Strott, 1973; O'Malley & Means, 1974; Palmiter, 1975; Tsai et al., 1975). The proteins detected in the mammalian uterine lumen may provide similar biochemical and molecular markers for uterine sensitization in preparation for implantation.

Some studies are already in progress to determine how the uterine response varies after hormonal administration to ovariectomized females, using luminal proteins as markers. The response of the uterus to administration of cyclic AMP in ovariectomized females has been studied in this way. Cyclic AMP has been shown to cause a resumption of metabolic activity (Mohla & Prasad, 1971) and implantation of quiescent blastocysts (Webb, 1975), and it could function either on the blastocyst directly or on the uterus or on both (Webb & Surani, 1975). We have shown that the progestational uterus responds to injection of cyclic AMP in the same way as to oestradiol by production of all the luminal proteins detected at the time of implantation (Surani & Webb, unpublished).

MODELS FOR STUDIES ON IMPLANTATION

Several techniques are available for studies on the binding of trophic factors to specific surface receptors and the analyses of cell surface constituents. The purpose here is only to outline some general features of in vivo and in vitro systems.

In vivo studies
Delay of implantation

Considerable use has been made of the phenomenon of delay of implantation to study the hormonal requirements for implantation of whole or dissected fragments of embryos as has been described earlier. The most useful feature of delay of implantation in rodents is that it is possible to synchronize early events of implantation and time course studies can be conducted after administration of exogenous hormones, as illustrated with studies on uterine luminal proteins and embryo metabolism.

Interspecific transfer of blastocysts

Interspecific transfer of embryos between species is a way of distinguishing between intrinsic differences in embryos and their response to the uterine

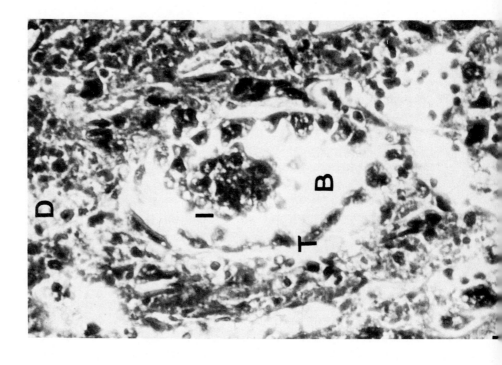

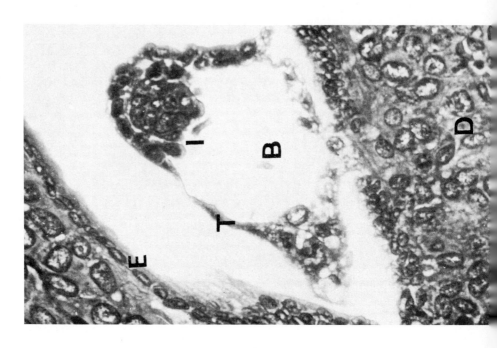

environment of other species.   Cell surface interactions between trophectoderm
and the uterine epithelium can also be assessed.

Interspecific transfer of blastocysts between the rat and mouse shows that
whilst control embryos are developing to at least the egg cylinder stage, the
interspecifically transferred blastocysts, although apparently viable and
eliciting a decidual response, fail to progress beyond the blastocyst stage
(Table 9 and Figure 13) (Surani & Barton, unpublished).   Such blastocysts,
again in contrast to within species controls, can also enter into quiescence
but fail to develop further when appropriate hormonal conditions are provided.

These findings are, on the whole, in agreement with those published
previously (Tarkowski, 1962).   The results are contrary to the suggestion that
the interspecific blastocyst can develop nearly to term if permitted to stay in
quiescence before implantation (Kirby, 1969b).   For this study and that
reported by Zeilmaker (1971), no experimental details are available for comparisons.   The restraint in development of the interspecific blastocyst is not
detected in an extrauterine site (Tarkowski, 1962).   The failure of rat blastocysts to develop in the mouse uterus may purely be due to a quantitative lack
of uterine proteins in the mouse and thus of metabolic enhancement.   The amount
of radiolabelled proteins in the mouse at the time of implantation is far less
than in the rat under identical experimental conditions (Surani, unpublished).
However, mouse blastocysts transferred to the rat uterus show a similar lack of
development; in fact the survival rate is poorer.   Alternatively, the uterine
macromolecules in the two species may not be identical.   Detailed analyses of
the mouse uterine luminal fluids are not yet complete for comparisons with that
of the rat.   Comparison of purely static samples (Surani, 1975c; Gore-Langton
& Surani, 1976) does not indicate major differences.   More subtle differences,
such as the rearrangement of terminal sugars in glycoproteins, could alter the
biological properties of these macromolecules considerably, without there being
any obvious gross analytical difference.

An alternative possibility to a general failure to stimulate metabolism
effectively may relate to the surface interactions between the trophoblast and
epithelial cell surface components.   Surface proteins and glycopeptides impart
a specificity for cell recognition prior to specific adhesions.   Clearly
surface interactions do not prevent the entry of interspecific blastocysts into
quiescence.   The failure occurs when implantation should commence.   Other
explanations for the failure of interspecific blastocysts to develop may also
be considered (Zeilmaker, 1971).

Figure 13: Photomicrographs of (a) mouse blastocyst in the rat uterus, and (b)
rat blastocyst in the mouse uterus on day 7.5 of pregnancy. Note that the
embryos have failed to progress beyond the blastocyst stage with intact blastocoel (B), mural trophoblast (T), inner cell mass (I). Decidual cells (D) and
intact uterine epithelium (E) are also present.

TABLE 9

IMPLANTATION RATES AFTER RECIPROCAL TRANSFERS OF BLASTOCYSTS BETWEEN MICE AND RATS*

Recipients	Rat Blastocysts			Mouse Blastocysts		
	No. of recipients	No. of blastocysts transferred	No. implanted (%)	No. of recipients	No. of blastocysts transferred	No. implanted (%)
Intact mice	10	55	40 (72.5)	6	24	21 (87.5)
Ovariectomized mice	6	36	26 (72.2)	4	18	18 (100)
Intact rats	-	-	-	7	54	22 (40.7)
Ovariectomized rats	-	-	-	5	41	14 (34.2)

* Rat and mouse blastocysts with and without their zonae pellucidae were transferred to intact and ovariectomized pseudopregnant recipients. Blastocysts transferred to ovariectomized recipients were permitted to enter into quiescence by daily injections of progesterone. Implantation in these females was induced by injecting oestradiol together with progesterone. Both the intact and ovariectomized females were killed on the day which was equivalent to day 7.5 of pregnancy. Implantation rate was determined by counting the decidual swellings, and the uteri were fixed for histology.

In vitro studies

An in vitro model for studies on implantation would be helpful in
establishing the role of luminal fluids and for investigations on the part played
by trophoblast cells in adhesion.   Interpretations and use of such systems
requires a judicious assessment of results.

Attempts to detect the influence of uterine fluids on blastocysts in vitro
are disappointing since the luminal fluids obtained from endocrinologically
different females are either embryotoxic or depress cell metabolism (Psychoyos,
1973; Weitlauf, 1976).   The reasons for the results are not known, but the
toxic factor is dialysable (Psychoyos, 1973) and heat resistant (Weitlauf, 1976).
The toxic substances could be breakdown products or contamination of the samples
with intracellular products having no physiological functions.

Twenty percent of the luminal fluids obtained from individual rats tested
are embryotoxic, as assessed by a bioassay of hatching of blastocysts from the
zona pellucida and trophoblast attachment and outgrowth.   Overall results,
however, show that luminal flushings from day 5 pregnant females support a high
proportion of embryos to hatch from their zonae pellucidae with subsequent outgrowths of the trophoblast.   Fluids from the ovariectomized females given
progesterone are substantially less successful in the test (Table 10).   These
results may be expected since there is a greater amount of uterine protein in
fluids from pregnant females compared with fluids from the latter females.   The
total protein in the medium is unlikely to be affected because BSA is present in
excess.   Fluids from pregnant and ovariectomized females when mixed support a
high rate of hatching of blastocysts from their zonae pellucidae and attachment.
Taken together with the analytical studies on uterine fluids, the finding implies
a compensation of macromolecules which are lacking in the fluids from ovariectomized females, with that from the pregnant females, to account for the
improved viability of blastocysts, rather than any repressive influence of
luminal fluids from the ovariectomized females.   Steroids by themselves have no
effect in this test.   In previous experiments, samples from several females were
pooled (Psychoyos, 1973; Weitlauf, 1976).   Fluid from one female, if embryotoxic, could contaminate the pool.   The improved results reported here may be
partly due to the use of flushings from individual animals.

There are several important differences in the response of blastocysts in
vitro compared with in utero.   The zona pellucida in utero appears to be removed
by complete lysis of the membrane during pregnancy (Mintz, 1971; Surani, 1975a),
which is never observed in vitro.   The other major difference is in the
survival of blastocysts in fluids from ovariectomized females given progesterone,
since they degenerated within 18 hours in culture.   Transfer of blastocysts to
the uteri of females given similar treatment results in the entry of the blasto-

TABLE 10

EFFECTS OF UTERINE LUMINAL EXTRACTS, FROM PREGNANT AND OVARIECTOMIZED
FEMALES TREATED WITH OESTRADIOL AND PROGESTERONE, ON BLASTOCYST
SURVIVAL AND ESCAPE FROM ZONA PELLUCIDA AND ATTACHMENT TO THE
PETRI DISH IN VITRO*

Medium (DMEM + 4 mg/ml BSA)	No. of blastocysts (replicates)	% hatched from zonae	% attached
Day 5 fl.	130 (10)	53.6^{bcde}	$42.7^{b*c*d*e*}$
P fl.	100 (9)	22.7^{ab}	6.8^{a*b*}
P fl. + Day 5 fl.	103 (9)	5	49.4^{e}
10% FCS	46 (4)	76.1	63.0
10% NRS	64 (5)	78.1^{d}	67.2^{d*}
POe 1 hr fl.	92 (8)	27.2	4.3
POe 14 hr fl.	114 (11)	38.9^{ac}	24.2^{a*c*}
POe 24 hr fl.	95 (9)	33.3	20.2

a	$0.01 < P < 0.05$	S	a*	$P < 0.005$	S
b	$P < 0.005$	S	b*	$P < 0.005$	S
c	$0.025 < P < 0.05$	S	c*	$P < 0.01$	S
d	$0.05 < P < 0.1$	NS	d*	$P < 0.005$	S
e	$0.05 < P < 0.9$	NS	e*	$0.05 < P < 0.1$	NS

* Flushings from each animal were tested individually. NRS and FCS were
used at 10% in DMEM + BSA medium. Flushings were obtained from pregnant
females on day 5 of pregnancy (day 5 fl.) or from ovariectomized females
treated with progesterone alone (P fl.) or from females given progesterone
and oestradiol (POe fl.) at various times after the hormone treatment.
Groups bearing common superscripts are compared using Chi-square test of
independence. S significant; NS not significant; FCS foetal calf
serum; NRS normal rat serum.

cysts into quiescence (Psychoyos, 1966; Dickmann & De Feo, 1967) in which state
they remain viable for a long time (Psychoyos, 1973). The difference could
partly be due to environmental conditions in vitro which may not be strictly
analogous to that in vivo. The ionic and amino acid content was not varied in
the medium. Some key ions, such as calcium, are likely to influence the response of blastocysts. Amino acids also affect development of blastocysts in
vitro (Gwatkin, 1966a; Spindle & Pedersen, 1973) but no variations are detected
in utero (Gwatkin, 1969). One of the major difficulties of in vitro studies is
that the tests can only be readily carried out with static samples, unlike the
continuous synthesis and secretion of uterine proteins occurring in vivo. If
these proteins are biologically important, produced in quantitatively meagre
amounts and highly labile, their activity in vitro cannot be determined meaning-
fully. Any biochemical tests on the luminal fluids must be carried out after a
brief exposure of embryos to such fluids.

A second main difference between the in vitro and the in vivo situation must be that in the former case, only environmental factors can be tested for the response of blastocysts, whereas in utero cell surface interactions between the trophoblast and the uterine epithelium could also have a marked effect on the viability, quiescence and implantation of blastocysts, and these cannot be tested in vitro. The metabolic state of blastocysts could change when they are disrupted from their contact with epithelium. The effects of serum or uterine flushings on attachment and outgrowths of trophoblast cells may also depend on cell surface interactions. Some serum macromolecules adsorb to the culture dish and providing a coating for cell adhesion (Culp & Buniel, 1976). Mouse embryos can attach to a petri dish pre-coated with collagen and show trophoblastic outgrowths, even in the absence of serum macromolecules (Jenkinson & Wilson, 1973). Changes in the substratum can influence cell behaviour, partly due to a physical effect on the cell membrane without necessarily influencing cell metabolism (Macieira-Coehlo et al., 1974). In some instances, interactions between a cell and extracellular matrix, such as collagen, could stimulate cell differentiation (Meier & Hay, 1975). The influence of humoral factors and surface interactions need to be put into perspective before their role in implantation can be assessed.

CELL SURFACE INTERACTIONS

After the denudation of the zona pellucida, the surfaces of the trophectoderm and uterine epithelium are apposed and close range cell-to-cell interactions presumably predominate over the long range communications mediated by luminal fluids. Little is known about this aspect of the implantation process.

Ultrastructural studies

Ultrastructural studies suggest a progressive increase in the contact between the trophoblast and epithelial cells (Tachi et al., 1970; Schlafke & Enders, 1975). The luminal epithelium undergoes several changes during the course of hormonal preparation for implantation. During progestation, under the influence of progesterone, there is a closure of the uterine lumen when the microvilli from the opposite sides interdigitate (Tachi et al., 1970; Ljungkvist, 1972). Oestrogen brings about changes in the morphology of endometrium (Anderson et al., 1975). Oestrogen and progesterone given together, cause even closer apposition between the epithelial cells (Potts, 1971; Nilsson, 1974). The effect of oestrogen is transitory (Nilsson, 1974). When oestrogen is given without progesterone, epithelial cells undergo mitosis (Finn & Martin, 1974) and the lumen fails to close as found during progestation.

The interdigitation between trophoblast and epithelial microvilli is similar to that observed between the epithelial cells. The contours between the epithelial cells and the trophoblast are continuous and extensive areas of

closely apposed membranes become detectable (Tachi et al., 1970; Schlafke &
Enders, 1975).   Adhesion between the trophoblast and epithelial cells is thought
to occur in all species during initial stages of implantation, even in cases
where the trophoblast does not penetrate the epithelial cells in an epithelio-
chorial placenta (Schlafke & Enders, 1975).   The progressive increase in
apposition and adhesion between blastocysts and uterus can be judged quantitatively by a decline in the rate of recovery of free blastocysts during day 4
of pregnancy in mice (Mintz, 1971) and during day 5 of pregnancy in rats
(Surani, 1975a) (Table 1).   Whether junctions are formed between trophoblast
and epithelium is not clearly established but fusion of epithelial and trophoblast cells is thought to occur in the rabbit (Schlafke & Enders, 1975), but
probably not in the rat (Tachi et al., 1970).

Cell surface components

Understanding of cell surface interactions at the molecular level requires
analysis of surface components, but little is known about cell surface components of epithelial and trophoblast cells.   Ballard and Tomkins (1969)
suggested that the action of steroids on target cells may lead to a modification
of the cell surface and hence in cell adhesiveness, as assessed by the effects
of glucocorticoids on the electrophoretic and antigenic properties of hepatoma
cells.   Similar changes in the luminal epithelium are not unlikely.   Changes
in the cell surface constituents of the embryonic surface probably occur as
differentiation proceeds.   Embryos prior to the blastocyst stage, can aggregate
even when they are from different species (Stern & Wilson, 1973; Zeilmaker, 1973),
but do not attach to substratum.   Whilst this property is lost at, or before,
blastulation (see Ducibella, this volume), trophoblast can attach to the substratum made of adsorbed serum proteins or collagen (Jenkinson & Wilson, 1973).
Enders and Schlafke (1974) have found that the mouse epithelial surface and the
surface of blastocysts can bind concanavalin A and stain with ruthenium red and
colloidal thorium, indicating a presence of acid glycoproteins.   Attempts have
been made to estimate synthesis of epithelial surface proteins and to extract
these proteins for analysis of radioactive components, from endocrinologically
distinct females (Surani, unpublished).   The surface material can be extracted
with Ca^{2+} and Mg^{2+} free Hanks medium containing 3.5 mM EDTA.   It appears that
the 'surface material' is synthesized in greatest amounts during proestrus and in
least amount during embryonic quiescence (Figure 14).   Proteins and fucose containing glycoproteins can be extracted.   The "surface components" extracted are
similar to some of the macromolecules detected in luminal fluids, but further
tests are needed to establish their origin conclusively.

The surface coat of blastocysts during quiescence has been reported not to
stain with colloidal iron Prussian blue stain, but to stain after the embryos

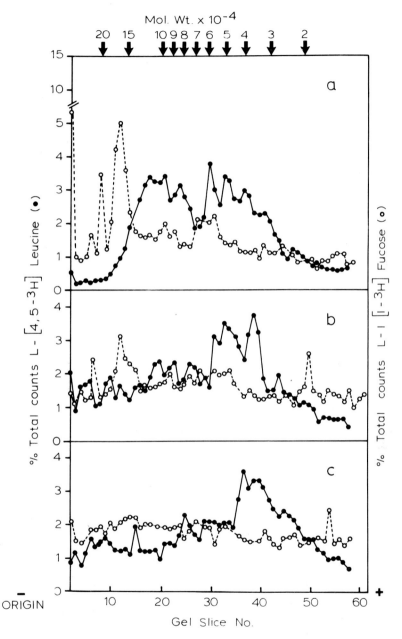

Figure 14: Epithelial cell surface proteins (●——●) and glycoproteins (o------o).
The uterine lumen was rinsed several times with saline before extraction of
surface proteins with Hanks medium plus EDTA from (a) proestrous, (b) day 5 of
pregnancy and (c) females experiencing delay of implantation.

have been stimulated to implant following an injection of oestradiol (Holmes &
Dickson, 1973).   These results have not been confirmed by others (Nilsson et
al., 1973).   There is an increase in the extractable glycopeptides as the
embryos progress from the cleavage stages to blastulation, as well as in the
types of glycopeptides present (Pinsker & Mintz, 1973).   Concanavalin A binding
sites are observable in mural trophoblast but not on polar trophoblast (Sobel &
Nebel, 1976).   Cleavage stage embryos are agglutinable, but not blastocysts,
in the presence of concanavalin A (Rowinski et al., 1976).

Cell surface interactions at the molecular level

Cell surface interaction is a dynamic process of cell communication (Cox,
1974).   When cell surface constituents interact with those of other cells,
there is probably a generation of signal(s) transmitted within the cell, similar
to those detected after the binding of trophic factors to cell membrane. The
resulting macromolecular synthesis within the cell in turn may alter cell
surface properties, and these changes would elicit further responses between
cells.   An example of surface interactions is the induction of the enzyme,
glutamine synthetase by the steroid hydrocortisone in chick retina cells. The
hormonal induction of the enzyme is stringently dependent on histotypic contact
in neuroretinal tissue (Moscona, 1972).   This continuous series of cellular
responses and interactions are involved in numerous aspects of cell communications
in recognition, specific adhesions, aggregation, morphogenesis, cell motility
and differentiation.   Such close range communication between the trophectoderm
and epithelium may be crucial during implantation.

Cell recognition and specific adhesions may be closely related and occur
before initiation of cooperation between cells.   Most of the studies on cell
recognition have been carried out on aggregation of like cells, where the cell
surface molecules important in these phenomena, are probably the carbohydrate
groupings (Chipowsky et al., 1973; Turner & Burger, 1973; Roseman, 1974; Lloyd &
Cook, 1975; Vicker, 1976).   Recognition appears to occur in two steps.   The
initial step in contact is indiscriminate allowing close association of cells
which permits their movement or consolidation of contact, where applicable, with
specific adhesions (Jones & Morrison, 1969; Van de Vyver, 1975).   Homologous
soluble aggregation factors bind specifically to cells (Lilien, 1969; Balsamo &
Lilien, 1974).   Heterologous aggregation factors in some instances cause an
irreversible change in adhesion, whilst aggregation in the presence of homologous factors is reversible (Van de Vyver, 1975).   Aggregation factors also
cause rearrangement of cell surface ligands which could influence cell-cell
interations (McDonough & Lilien, 1975).

The nature of adhesive bonds is not known but initial suggestions regarded
these as antigen-antibody type of complexes (Tyler, 1947; Weiss, 1947).

Recently it has been suggested that the specific adhesions are due to carbohydrate containing macromolecules in an enzyme-substrate type of interaction,
especially involving the glycosyltransferase enzymes (reviewed by Shur & Roth,
1975).

The contact between cells is responsible for contact inhibition of movement
and growth (Macieira-Coelho, 1967; Abercrombie, 1970; Martz & Steinberg, 1973;
Harris, 1974).   The evidence for control of cell growth on contact arises
mainly from studies comparing malignant or virally transformed cells with normal
cells (Abercrombie, 1970).   The lack of contact inhibition of growth in
malignant cells appears to be due partly to impaired cell adhesion (Harris, 1974)
or cell recognition (Dorsey & Roth, 1973).   Recent findings suggest that the
addition of a cell surface glycoprotein from normal cells to transformed
malignant cells, restores some of the properties of adhesion, morphology and
contact inhibition of movement which are found in normal cells (Yamada et al.,
1976; Pearlstein, 1976).

Cell surface interactions and implantation: an evaluation

Recognition errors between trophectoderm and uterine epithelium may be a
reason for the failure of interspecific blastocysts to implant even when the
uterus has been sensitized.   As recognition and specific adhesions appear to
be closely linked, failure of trophoblast recognition could impair specific
adhesions which is a prerequisite for giant cell transformation.   Since steroid
hormones probably alter cell surface properties of target cells (Ballard &
Tomkins, 1969), part of the uterine sensitization may involve formation or
arrangement of epithelial surface ligands for recognition of trophoblast cell
surface and initiation of cooperative interactions in implantation.
Asynchronous transfers of blastocysts to the progestational uterus has shown
that the embryo does not implant until the uterus is sensitized.   The sensitized uterus is a transitory phase so that transfers of embryos after this
period also cause a failure of implantation (Psychoyos, 1973).   The period over
which recognition of trophoblast and specific adhesions with epithelium can
occur is brief and presumably essential for coordination of implantation and
decidualization.   It may also be assumed that cell surface interactions are
important for the inhibition of blastocyst cell metabolism during delay of
implantation since blastocysts removed from such contact and placed in vitro
resume their normal metabolic activity.   It has so far not been possible to
maintain embryos in quiescence in vitro for a long period even in the absence
of extracellular macromolecules.   Viability for long periods may require
embryonic contact with uterine epithelial cells.   After administration of
oestradiol to animals experiencing delay of implantation, there are probably
rapid changes in the distribution and synthesis of cell surface proteins.   Such

activity may alter contact inhibition of growth causing resumption of metabolic
activity in blastocysts, and lead to a series of reciprocal cellular interactions
between trophectoderm and epithelium, involving their cell surface components.

CONCLUSIONS

The rat and the mouse embryos have an intrinsic potential for post-
implantation development which is realized when uterine luminal proteins or
serum macromolecules form a part of their environment. The phenotypic response
of blastocysts does not appear to be dictated by extracellular factors but
determined by the cell specific protein synthesis in embryonic cells as a result
of metabolic enhancement caused by the trophic action of macromolecules when
they bind to cell surface receptors. The uterus exerts a stringent control
over blastocyst cell metabolism when it responds to ovarian steroids and provides luminal components, partly through synthesis and secretion of specific
proteins. There is some evidence for other specialized functions of luminal
macromolecules and enzymes.

Differentiation of the one-cell embryo to the blastocyst stage is associated
with changes in macromolecular synthesis and gene expression. There are
apparent differences in the response of embryos to environmental macromolecules
during early development, so that embryonic cells by the blastocyst stage
resemble other eukaryote cells in their response to environmental changes. The
unequivocal detection of the G_1 phase at the blastocyst stage is probably fundamental for this change in embryonic response. Critical events prior to cell
proliferation and cytodifferentiation occur in this phase.

Cell surface topography and surface receptors undergo changes during the
cell cycle. Trophic factors bind to surface receptors as cells traverse
through the G_1 (G_0) phase when the distribution of receptors permits specific
binding. In the absence of these factors, cells accumulate in the G_1 (G_0)
phase and enter into quiescence. The ability of blastocysts to enter into
quiescence with cell arrest in the G_1 phase, and the reversible metabolic
enhancement in response to environmental changes are the cellular responses
which blastocysts, but not morulae, share with other mammalian cells. The
lack of constraints on cell metabolism of morulae through environmental changes
may be linked with a short or non-existent G_1 phase, and in the differences in
the cell surface topography and constituents compared with blastocysts.

Reversible metabolic enhancement of blastocysts occurs through coordination
of biochemical events, known as a pleiotypic response, as a result of changes
in environmental conditions. Embryonic quiescence during delay of implantation
appears to be related to a lack of luminal macromolecules, partly due to a
lower rate of synthesis and secretion of uterine proteins. Cell surface inter-

actions between trophectoderm and uterine epithelium could further affect embryonic metabolism and ensure their viability during quiescence.

Synthesis and secretion of uterine specific proteins and glycoproteins is enhanced by oestradiol. Progesterone suppresses the oestrogen-mediated synthesis of proteins. The interactive effects of the two hormones are complex since there are qualitative differences in synthesis and secretion of luminal macromolecules in response to oestradiol alone, compared with the synergistic action of the two hormones. Uterine specific luminal proteins can serve as useful markers for studies on cytodifferentiation and the molecular events of uterine response to the ovarian steroids.

Cell surface interactions presumably predominate as apposition of trophectoderm with luminal epithelium occurs at the initiation of implantation. At the molecular level, cell surface components may be involved in the recognition of trophoblast, specific adhesions and in intercellular communications and control of growth and metabolism. The presence of sugar residues displayed on the cell surface are likely to be involved in cell surface interactions. Epithelial cell surface components may undergo changes through synthesis and rearrangement of the surface constituents as a result of gene expression mediated by steroids. The surface changes are probably an important feature of uterine sensitization for implantation at the molecular level. The changes in the epithelial cell surface components may determine the timing of the recognition and specific adhesions of trophectoderm to epithelial cells. Attemps are now being made to analyse cell surface components of epithelium and blastocysts, but little is yet known about the molecular events of cell surface interactions between blastocysts and epithelium.

ACKNOWLEDGEMENTS

I thank Dr. R. G. Edwards for his criticisms on the manuscript, Dr. F. T. G. Webb and R. E. Gore-Langton for helpful discussions, Dr. Webb and Sheila Barton for permission to quote our unpublished results, Sheila Barton and Andrea Burling for technical assistance and Jane Hugh for typing the manuscript. The work was supported by grants from the Medical Research Council and a Ford Foundation Grant to Professor C. R. Austin.

REFERENCES

ABERCROMBIE, M., (1970) Contact inhibition in tissue culture. In Vitro 6, 128-142.

AITKEN, R.J. (1974) Delayed implantation in roe deer (Capreolus capreolus), J. Reprod. Fert. 39, 225-233.

ALBERGHINA, F.A.M. (1975) A model for the regulation of growth in mammalian cells. J. theor. Biol. 55, 533-545.

ALBERS, H.J., & NEVES E CASTRO, M. (1961) The protein components of rat uterine fluid. Fert. Steril. 12, 142.

ANDERSON, W.A., KANG, Y-H & DE SOMBRE, E.R. (1975) Estrogen and antagonist-induced changes in endometrium topography of immature and cycling rats. J. Cell. Biol. 64, 692-703.

ANSELL, J.D. (1975) The differentiation and development of mouse trophoblast. In: M. Balls and A.E. Wild (Eds.), The Early Development of Mammals, Cambridge University Press, pp.133-144.

ARMSTRONG, D.T. (1968) Hormonal control of uterine lumen fluid retention in the rat. Am. J. Physiol. 214, 764-771.

BAKER, M.E. (1976) Colchicine inhibits mitogenesis in C 1300 neuroblastoma cells that have been arrested in G_0. Nature, Lond. 262, 785-786.

BALLARD, P.L. & TOMKINS, G.M. (1969) Hormone induced modification of the cell surface. Nature, Lond. 224, 344-345.

BALSAMO, J. & LILIEN, J. (1974) Embryonic cell aggregation: Kinetics and specificity of binding of enhancing factors. Proc. natn. Acad. Sci. U.S.A. 71, 727-731.

BANDMAN, E. & GURNEY, T. (1975) Differences in the cytoplasmic distribution of newly synthesized poly (A) in serum-stimulated and resting cultures of BALB/c 3T3 cells. Expl Cell Res. 90, 159-168.

BARLOW, P., OWEN, D.A.J. & GRAHAM, C. (1972) DNA synthesis in the preimplantation mouse embryo. J. Embryol. exp. Morph. 27, 431-445.

BARNEA, A. & GORSKI, J. (1970) Estrogen-induced protein. Time course of synthesis. Biochemistry, N.Y. 9, 1899-1904.

BASERGA, R. (1968) Biochemistry of the cell cycle: A review. Cell Tiss. Kinet. 1, 167-191.

BAULIEU, E.E., WIRA, C.R., MILGROM, E. & RAYNAUD-JAMET, C. (1972) Acta endocr., Copehn. Suppl. 168, 396-419.

BEATO, M. & BAIER, R. (1975) Binding of pregesterone to the proteins of the uterine luminal fluid. Identification of uteroglobin as the binding protein. Biochim. biophys. Acta 392, 346-356.

BECK, L.R. & BOOTS, L.R. (1973) Immunologic studies on rat uterine fluid. Contraception 7, 125-131.

BEIER, H.M. (1970) Hormonal stimulation of protease inhibitor activity in endometrial secretion during early pregnancy. Acta endocr., Copenh. 63, 141-149.

BEIER, H.M. (1974) Oviducal and uterine fluids. J. Reprod. Fert. 37, 221-237.

BEIER, H.M. & BEIER-HELLWIG, K. (1973) Specific secretory protein of the female genital tract. Karolinska Symposia 6, 404-425.

BENNETT, G., LEBLOND, C.P. & HADDAD, A. (1974) Migration of glycoproteins from the Golgi apparatus to the surface of various cell types as shown by radio-autographs after labelled fucose injection into rats. J. Cell Biol. 60, 258-284.

BILLINGTON, W.D., GRAHAM, C.E. & McLAREN, A. (1969) Extra-uterine development of mouse blastocysts cultured in vitro from early cleavage stages. J. Embryol. exp. Morph. 20, 391-400.

BITTON-CASIMIRI, V., BRUN, J.L. & PSYCHOYOS, A. (1976) Uptake and incorporation of ^{3}H-uridine by normal or diapausing rat blastocysts after various periods of culture. J. Reprod. Fert. 46, 447-448.

BOWMAN, P. & McLAREN, A. (1970) Cleavage rate of mouse embryos in vivo and in vitro. J. Embryol. exp. Morph. 24, 203-207.

BRAMBELL, F.W.R. (1937) The influence of lactation on the implantation of the mammalian embryo. Am. J. Obstet. Gynec. 33, 942.

BRINSTER, R.L. (1969) In vitro cultivation of mammalian ova. Adv. Biosc. 4, 199-234.

CHANDRA, R.K., MALKANI, P.K. & BHASIN, K. (1974) Levels of immunoglobulins in the serum and uterine fluid of women using an intrauterine contraceptive device. J. Reprod. Fert. 37, 1-6.

CHIPOWSKY, S., LEE, Y.C. & ROSEMAN, S. (1973) Adhesion of cultured fibroblasts to insoluble analogues of cell-surface carbohydrates. Proc. natn. Acad. Sci. U.S.A. 70, 2309-2312.

CHURCH, R.B. & SCHULTZ, G.A. (1974) Differential gene activity in the pre-and postimplantation mammalian embryo. In: A.A. Moscona and A. Monroy (Eds.), Current Topics in Developmental Biology, Academic Press, New York, pp.179-203.

CLEMETSON, C.A.B., MALLIKARJUNESWARA, V.R., MOSHFEGHI, M.M., CARR, J.J. & WILDS, J.H. (1970a) Electrophoretic mobility of the rat blastocyst. Contraception 1, 357-360.

CLEMETSON, C.A.B., MALLIKARJUNESWARA, V.R., MOSHFEGHI, M.M., CARR, J.J. & WILDS, J.H. (1970b) The effect of estrogen and progesterone on the sodium and potassium concentration of rat uterine fluid. J. Endocr. 47, 309-319.

COCHRANE, R.L. & MEYER, R.K. (1957) Delayed nidation in the rat induced by progesterone. Proc. Soc. exp. Biol. Med. 96, 155-159.

COHEN, M.E. & HAMILTON, T.H. (1975) Effect of estradiol-17β on the synthesis of specific uterine nonhistone chromosomal proteins. Proc. natn. Acad. Sci. U.S.A. 72, 4346-4350.

COLE, R.J. & PAUL, J. (1965) Properties of cultured preimplantation mouse and rabbit embryos, and cell straints derived from them. In: G.E.W. Wolstenholme and M. O'Connor (Eds.), Preimplantation Stages of Pregnancy, Ciba Found. Symp., pp.82-112.

COOK, G.M.W. (1968) Glycoproteins in membranes, Biol. Rev. 43, 363-391.

COOK, G.M.W. & STODDART, R.W. (1973) Surface Carbohydrates of the Eukaryote Cell. Academic Press, New York.

COX, R.P. (1974) Cell Communication. John Wiley and Sons, New York.

CULP, L.A. & BUNIEL, J.F. (1976) Substrate-attached serum and cell proteins in adhesion of mouse fibroblasts. J. Cell. Physiol. 88, 89-106.

DALCQ, A.M. & PASTEELS, J. (1955) Determination photometrique de la tenour relative en DNA des nayaux dans les oeufs en segmentation du rat et de la souris. Expl Cell Res. 3, 72-97.

DANIEL, J.C., Jr. (1968) Comparison of electrophoretic patterns of uterine fluids from rabbits and mammals having delayed implantation. Comp. Biochem. Physiol. 24, 297-299.

DANIEL, J.C., Jr. (1971a) Uterine proteins and embryonic development. Adv. Biosc. 6, 191-206.

DANIEL, J.C., Jr. (1971b) Growth of the preimplantation embryo of the northern fur seal and its correlation with changes in uterine protein. Devl Biol. 26, 316-331.

DANIEL, J.C., Jr & LEVY, J.D. (1964) Action of progesterone as a cleavage inhibitor of rabbit ova in vitro. J. Reprod. Fert. 7, 323-329.

DASS, C.M.S., MOHLA, S. & PRASAD, M.R.N. (1969) Time sequence of action of estrogen on nucleic acid and protein synthesis in the uterus and blastocyst during delayed implantation in the rat. Endocrinology 85, 528-536.

DE FEO, V.J. (1967) Decidualization. In: R.M. Wynn (Ed.), Cellular Biology of the Uterus, North Holland Publishing Company, Amsterdam, pp. 191-290.

DICKMANN, Z. (1970) Effect of progesterone on the development of the rat morula. Fert. Steril. 21, 541-548.

DICKMANN, Z. & DE FEO, V.J. (1967) The rat blastocyst during normal pregnancy and during delayed implantation, including an observation on the shedding of the zona pellucida. J. Reprod. Fert. 13, 3-9.

DICKMANN, Z. & DEY, S.K. (1974) Steroidogenesis in the preimplantation rat embryo and its possible influence on morula-blastocyst transformation and implantation. J. Reprod. Fert. 37, 91-93.

DICKSON, A.D. (1969) Cytoplasmic changes during the trophoblastic giant cell transformation of blastocysts from normal and ovariectomized mice. J. Anat. 105, 371-380.

DORSEY, J.K. & ROTH, S. (1973) Adhesive specificity in normal and transformed fibroblasts. Devel.Biol. 33, 249-256.

DULBECCO, R. (1970) Topoinhibition and serum requirements of transformed and untransformed cells. Nature, Lond. 227, 802-806.

EDELMAN, G.M., YAHARA, I. & WANG, J.L. (1973) Receptor mobility and receptor-cytoplasmic interactions in lymphocytes. Proc. natn. Acad. Sci. U.S.A. 70, 1442-1446.

ELLEM, K.A.O. & GWATKIN, R.B.W. (1968) Patterns of nucleic acid synthesis in the early mouse embryos. Devel Biol. 18, 3111-3116.

ENDERS, A.C. (1967) The uterus in delayed implantation. In: R.M. Wynn (Ed.), Cellular Biology of the Uterus, Meredith Publishing Company, New York, pp. 151-190.

ENDERS, A.C. & SCHLAFKE, S. (1974) Surface coats of the mouse blastocyst and uterus during the preimplantation period. Anat. Rec. 180, 31-46.

ENDO, M. & YOSIZAWA, Z. (1973) Hormonal effect on glycoproteins and glycos-aminoglycans in rabbit uteri. Arch. Biochem. Biophys. 156, 397-403.

EPIFANOVA, O.T. & TERSKIKH, V.V. (1969) Review Article: On the resting periods in the cell life cycle. Cell Tiss. Kinet. 2, 75-93.

EPSTEIN, C.J. (1975) Gene expression and macromolecular synthesis during pre-implantation embryonic development. Biol. Reprod. 12, 82-105.

FINN, C.A. & MARTIN, L. (1974) The control of implantation. J. Reprod. Fert. 39, 195-206.

FOLSTAD, L., BENNETT, J.P. & DORFMAN, R.I. (1969) The in vitro culture of rat ova. J. Reprod. Fert. 18, 145-146.

FOWLER, R.E., JOHNSON, M.H., WALTERS, D.E. & PRATT, H.P.M. (1976) The progesterone and protein composition of rabbit uterine flushings. J. Reprod. Fert. 46, 427-430.

FOX, T.O., SHEPPARD, J.R. & BURGER, M.M. (1971) Cyclic membrane changes in animal cells: Transformed cells permanently display a surface architecture detected in normal cells only during mitosis. Proc. natn. Acad. Sci. U.S.A. 68, 244-247.

GALAND, P. & LEROY, F. (1971) Radioautographic evaluation of the effects of oestrogen on cell kinetics in target tissues. In: P.O. Hubinot, F. Leroy and P. Galand (Eds.), Basic Actions of Sex Steroids on Target Organs, Imprimeries Reunies S.A., Lausanne, pp. 160-172.

GAMOW, E.I. & PRESCOTT, D.M. (1970) The cell cycle during early embryo-genesis of the mouse. Expl Cell Res. 59, 117-123.

GARDNER, R.L. (1971) Manipulations on the blastocyst. Adv. Biosc. 6, 279-296.

GARDNER, R.L. & JOHNSON, M.H. (1972) An investigation of inner cell mass and trophoblast tissues following their isolation from the mouse blastocyst. J. Embryol. exp. Morph. 28, 279-312.

GARDNER, R.L. & PAPAIOANNOU, V.E. (1975) Differentiation in the trophectoderm and inner cell mass. In: M. Balls and A. E. Wild (Eds.), The Early Development of Mammals, Cambridge University Press, pp. 107-132.

GORE-LANGTON, R.E. & SURANI, M.A.H. (1976) Uterine luminal proteins of mice. J. Reprod. Fert. 46, 271-274.

GORSKI, J. (1973) Estrogen binding and control of gene expression in the uterus. In: R.O. Greep and E.B. Astwood (Eds.), Handbook of Physiology, Vol. 7 (ii), Waverley Press, Baltimore, Maryland, pp. 525-536.

GRAHAM, C.F. (1973) The cell cycle during mammalian development. In: M. Balls and F.S. Billeft (Eds.), The Cell Cycle in Development and Differentiation. Cambridge University Press, pp. 293-310.

GULYAS, B.J. & DANIEL, J.C., Jr. (1967) Oxygen consumption in diapausing blastocysts. J. cell. comp. Physiol. 70, 33-36.

GULYAS, B.J. & DANIEL, J.C., Jr. (1969) Incorporation of labelled nucleic acid and protein precursors by diapausing and non-diapausing blastocysts. Biol. Reprod. 1, 11-20.

GWATKIN, R.B.L. (1966a) Amino acid requirements for attachment and outgrowth of the mouse blastocysts in vitro. J. cell. comp. Physiol. 68, 335-344.

GWATKIN, R.B.L. (1966b) Defined media and development of mammalian eggs in vitro. Ann. N.Y. Acad. Sci. 139, 79-90.

GWATKIN, R.B.L. (1969) Nutritional requirements for post-blastocyst development in the mouse. Int. J. Fert. 14, 101-105.

HAMANA, K. & HAFEZ, E.S.E. (1970) Disc electrophoretic patterns of uteroglobin and serum proteins in rabbit blastocoelic fluid. J. Reprod. Fert. 21, 555-558.

HAMILTON, T.H. (1971) In: R.M.S. Smellie (Ed.), The Biochemistry of Steroid Hormone Action, Academic Press, London, pp. 49-84.

HAMNER, C.E. (1971) Composition of oviducal and uterine fluids. Adv. Biosc. 6, 143-163.

HARRER, J.A. & LEE, H.H. (1973) Differential effects of oestrogen on the uptake of nucleic acid precursors by mouse blastocysts in vitro. J. Reprod. Fert. 33, 327-330.

HARRIS, A. (1974) In: R.P. Cox (Ed.), Cell Communication, John Wiley and Sons, New York, pp. 147-185.

HASEGAWA, Y., SUGAWARA, S. & TAKEUCHI, S. (1973) Studies on the uterine fluid of rat. Effects of oestrogen and gestagen on protein content and disc electrophoretic patterns. Jap. J. Animal Reprod. 19, 73-77.

HASHIMOTO, A., HENRICKS, D.M., ANDERSON, L.L. & MELAMPY, R.M. (1968) Progesterone and pregn-4-en-20-ol-3-one in ovarian venous blood during various reproductive states in rats. Endocrinology 82, 333-341.

HASSEL, J.A. & ENGELHARDT, D.W. (1973) Translational inhibition in extracts from serum-deprived animal cells. Biochim. biophys. Acta 325, 545-553.

HEALD, P.J. (1973) Uterine metabolism in early pregnancy in the rat. Biochem. Soc. Trans. 1, 7-11.

HEALD, P.J. & O'HARE, A. (1973) Changes in rat uterine nuclear RNA during early pregnancy. Biochim. biophys. Acta 324, 86-92.

HENSLEIGH, H.C. & WEITLAUF, H.M. (1974) Effect of delayed implantation on dry weight and lipid content of mouse blastocysts. Biol. Reprod. 10, 315-320.

HERBERT, M.C. & GRAHAM, C.F. (1974) Cell determination and biochemical differentiation of the early mammalian embryo. In: A.A. Muscona and A. Monroy (Eds.), Current Topics in Developmental Biology, Academic Press, New York, pp. 152-179.

HERSHKO, A., MAMMONT, P., SHIELDS, R. & TOMKINS, G.M. (1971) Pleiotypic response. Nature, Lond. 232, 206-211.

HOLLENBERG, E.J. & CUATRECASAS, P. (1975) Insulin and epidermal growth factor. Human fibroblast receptors related to deoxyribonucleic acid synthesis and amino acid uptake. J. biol. Chem. 250, 3845-3853.

HOLLEY, R.W. (1972) A unifying hypothesis concerning the nature of malignant growth. Proc. natn. Acad. Sci. U.S.A. 69, 2840-2841.

HOLMES, P.V. & DICKSON, A.D. (1973) Estrogen-induced surface coat and enzyme changes in the implanting mouse blastocyst. J. Embryol. exp. Morph. 29, 639-645.

HOWARD, E. & DE FEO, V.J. (1959) Potassium and sodium content of uterine seminal vesicle secretions. Am. J. Physiol. 196, 65.

HSU, Y.C., BASKAR, J., STEVENS, L.C. & RASH, J.E. (1974) Development in vitro of mouse embryos from the two-cell egg stage to the early somite stage. J. Embryol. exp. Morph. 31, 235-245.

HSEUH, A.J., PECK, E.J. & CLARK, J.H. (1975) Progesterone antagonism of the oestrogen receptors and oestrogen-induced uterine growth. Nature, Lond. 254, 337-339.

IACOBELLI, S., PAPARATTI, L. & BOMPIANI, A. (1973) Oestrogen-induced protein (IP) of rat uterus. Isolation and preliminary characterization. FEBS, Letters, 32, 199-203.

IGARASHI, Y. & YAOI, Y. (1975) Growth-enhancing protein obtained from cell of cultured fibroblasts. Nature, Lond. 254, 248-250.

JACOBSON, M.A., SANYAL, M.K. & MEYER, R.K. (1970) Effect of estrone on RNA synthesis in preimplantation blastocysts of gonadotrophin-treated immature rats. Endocrinology 86, 982-987.

JATO-RODRIGUEZ, J.J., NELSON, J.D. & MOOKERJEA, S. (1976) Effect of estradiol and progesterone on UDPgalactose pyrophosphatase activity in the endometrium of ovariectomized rats. Biochim. biophys. Acta 428, 639-646.

JENKINSON, E.J. & WILSON, I.B. (1973) In vitro studies on the control of trophoblast outgrowth in the mouse. J. Embryol. exp. Morph. 30, 21-30.

JOHNSON, M.H. (1974) Studies using antibodies to the macromolecular secretions of the early pregnant uterus. In: A. Centaro and N. Carretti (Eds.), Immunology in Obstetrics and Gynaecology, Excerpta Medica, Amsterdam, pp. 124-133.

JONES, B.M. & MORRISON, G.A. (1969) A molecular basis for indiscriminate and selective cell adhesion. J. Cell. Sci. 4, 799-813.

JOSHI, M.S., YARON, A. & LINDNER, H.R. (1970) An endopeptidase in the uterine secretion of the proestrous rat and its relation to a sperm decapitating factor. Biochem. biophys. Res. Commun. 38, 52-57.

JUNGE, J.M. & BLANDAU, R.J. (1958) Studies on the electrophoretic properties of the cornual fluids of rats in heat. Fert. Steril. 9, 353.

KATZENELLENBOGEN, B.S. & LEAKE, R.E. (1974) Distribution of the oestrogen-induced protein and of total protein between endometrial and myometrial fractions of the immature and mature rat uterus. J. Endocr. 63, 439-449.

KENNEDY, T.G. (1974) Effect of relaxin on oestrogen-induced uterine luminal fluid accumulation in the ovariectomized rat. J. Endocr. 261, 347-353.

KIRBY, D.R.S. (1969a) The extra-uterine mouse egg as an experimental model. Adv. Biosc. 4, 255-273.

KIRBY, D.R.S. (1969b) In: G.E.W. Wolstenholme and M. O'Connor (Eds.), Foetal Autonomy, Ciba Found. Symp., J. and A. Churchill Ltd. London, p. 57.

KIRCHNER, C., HIRSCHHAUSER, C. & KIONKE, M. (1971) Protease activity in rabbit uterine secretions 24 hours before implantation. J. Reprod. Fert. 27, 259-260.

KULANGARA, A.C. (1972) Volume and protein concentration of rabbit uterine fluid. J. Reprod. Fert. 28, 419-425.

KUNITAKE, G.M., NAKAMURA, R.M., WELLS, B.G. & MOYER, D.L. (1965) Disc electrophoresis and disc gel Ouchterlony analyses of rat uterine fluid. Fert. Steril. 16, 120-124.

LATASTE, F. (1891) Des variations de duree de la gestation chez les mammiferes et des circonstances qui determinent ces variations. Theorie de la gestation retardee. C. r. Seanc. Soc. Biol. 43, 21-31.

LAU, N.I.F., DAVIS, B.K. & CHANG, M.C. (1973) Stimulation in vitro of ^{3}H-uridine and RNA synthesis in mouse blastocysts by 17β -estradiol. Proc. Soc. exp. Biol. Med. 144, 333-336.

LILIEN, J.E. (1969) Towards a molecular explanation for specific cell adhesion. In: A.A. Moscona and A. Monroy (Eds.), Current Topics in Developmental Biology, Vol. 4, Academic Press, New York, pp. 169-195.

LJUNGKVIST, I. (1972) Attachment reaction of rat uterine luminal epithelium. IV. The cellular changes in the attachment reaction and its hormonal regulation. Fert. Steril. 23, 847-865.

LLOYD, C. (1975) Sialic acid and the social behaviour of cells. Biol. Rev. 50, 325-350.

LLOYD, C.W. & COOK, G.M.W. (1975) A membrane glycoprotein-containing fraction which promotes cell aggregation. Biochim. biophys. Res. Commun. 67, 696-700.

MACIEIRA-COELHO, A. (1967) Dissociation between inhibition of movement and inhibition of division in RSV transformed human fibroblast. Expl Cell Res. 47, 193-200.

MACIEIRA-COELHO, A., BERUMEN, L. & AVRAMEAS, S. (1974) Properties of protein polymer as substratum for cell growth in vitro. J. Cell Physiol. 83, 379-388.

MANNINO, R.J. & BURGER, M.M. (1975) Growth inhibition of animal cells by succinylated concanavalin A. Nature, Lond. 256, 19-22.

MARTIN, L., DASS, R.M. & FINN, C.A. (1973) The inhibition by progesterone of uterine epithelial proliferation in the mouse. J. Endocr. 57, 549-554.

MARTZ, E. & STEINBERG, M.S. (1973) Contact inhibition of what? An analytical review. J. Cell Physiol. 81, 25-38.

McDONOUGH, J. & LILIEN, J.E. (1975) Inhibition of mobility of cell-surface receptors by factors which mediate specific cell-cell interactions. Nature, Lond. 256, 416-417.

McLAREN, A. (1973) Blastocyst activation. In: S.J. Segal, R. Crozier, P.A. Corfman and P.G. Condliffe (Eds.), The Regulation of Mammalian Reproduction, Charles C. Thomas, Springfield, Illinois, pp. 321-328.

McLAREN, A. & HENSLEIGH, H.C. (1975) Culture of mammalian embryos over the implantation period. In: M. Balls and A.E. Wild (Eds.), The Early Development of Mammals, Cambridge University Press, pp. 45-60.

MEIER, S. & HAY, E.D. (1975) Stimulation of corneal differentiation by inter-action between cell surface and extracellular matrix. I. Morphometric analysis of transfilter "induction". J. Cell. Biol. 66, 275-292.

MENKE, T.M. & McLAREN, A. (1970) Carbon dioxide production by mouse blastocysts during lactation delay of implantation or after ovariectomy. J. Endocr. 47, 287-294

MESTER, I., MARTEL, D., PSYCHOYOS, A. & BAULIEU, E.E. (1974) Hormonal control of oestrogen receptor in uterus and receptivity for implantation in the rat. Nature, Lond. 250, 776-778.

MINTZ, B. (1971) Control of embryo implantation and survival. Adv. Biosc. 6, 317-342.

MOHLA, S. & PRASAD, M.R.N. (1971) Stimulation of RNA synthesis in the blastocyst and uterus of the rat by adenosine 3'5'-monophosphate (cyclic AMP). J. Reprod. Fert. 23, 327-329.

MONESI, V. & MOLINARO, M. (1971) Macromolecular synthesis and effect of metabolic inhibitors during preimplantation development in the mouse. Adv. Biosc. 6, 101-120.

MOSCONA, A.A. (1972) Induction of glutamine synthetase in embryonic neural retina: A model for the regulation of specific gene expression in embryonic cells. In: A. Monroy (Ed.), Symposium on Biochemistry of Cell Differentiation, 7th Meeting Federation of European Biochemical Societies, Varna, Bulgaria, Vol. 24, Academic Press, London, pp. 1-23.

MUKHERJEE, A.B. (1976) Cell cycle analysis and X-chromosome inactivation in the developing mouse. Proc. natn. Acad. Sci. U.S.A. 73, 1608-1611.

MULLER, W.E.G., MULLER, I. & ZAHN, R.K. (1976) Species specific aggregation factor in sponges. V. Influence on programmed syntheses. Biochim. Biophys. Acta 418, 217-225.

NAHA, P.M., MEYER, A.L. & HEWITT, K. (1975) Mapping of the G_1 phase of a mammalian cell cycle. Nature, Lond. 258, 49-51.

NALBANDOV, A.V. (1971) Endocrine control of implantation. In: R.J. Blandau (Ed.), The Biology of the Blastocyst, University of Chicago Press, Chicago, pp. 383-392.

NELSON, J.D., JATO-RODRIGUEZ, J.J. & MOOKERJEA, S. (1975) Effect of ovarian hormones on glycosyltransferase activities in the endometrium of ovariectomized rats. Arch. Biochem. Biophys. 169, 181-191.

NICOLSON, G.L. & WINKELHAKE, J.L. (1975) Organ specificity of blood-borne tumour metastasis determined by cell adhesion? Nature, Lond. 255, 230-232.

NILSSON, O. (1974) The morphology of blastocyst implantation. J. Reprod. Fert. 39, 187-194.

NILSSON, O., LINDQVIST, I. & RONQUIST, G. (1973) Decreased surface charge of mouse blastocyst at implantation. Expl Cell Res. 83, 431-433.

NIMROD-ZMIGROD, A., LADANY, S. & LINDNER, H.R. (1972) Perinidatory ovarian oestrogen secretion in the pregnant rat, determined by gas chromatography with electron capture detection. J. Endocr. 53, 249-260.

NOSKE, I.G. & DANIEL, J.C., Jr. (1974) Changes in uterine and oviducal fluid proteins during early pregnancy in the golden hamster. J. Reprod. Fert. 38, 173-176.

NOTIDES, A. & GORSKI, J. (1966) Estrogen-induced synthesis of a specific uterine protein. Proc. Natn. Acad, Sci. U.S.A. 56, 230-235.

O'MALLEY, B.W. & STROTT, C.A. (1973) The mechanism of action of progesterone. In: R.O. Greep and E.B. Astwood (Eds.), Handbook of Physiology, Vol. 7 (ii), Waverley Press, Baltimore, Maryland, pp. 591-603.

O'MALLEY, B.W. & MEANS, A.R. (1974) Female steroid hormones and target cell nuclei. Science, N.Y. 183, 610-620.

PALMITER, C. (1975) Quantitation of parameters that determine the rate of ovalbumin synthesis. Cell 4, 189-196.

PARDEE, A. (1974) A restriction point for control of normal animal cell proliferation. Proc. Natn. Acad. Sci. U.S.A. 71, 1286-1290.

PAUL, J. & HICKEY, I. (1974) The cell. Molecular pathology of the cancer cell. J. Clin. Path. 27, Suppl. 7, 4-10.

PEARLSTEIN, E. (1976) Plasma membrane glycoprotein which mediates adhesion of fibroblasts to collagen. Nature, Lond. 262, 497-499.

PEPLOW, V., BREED, W.G. & ECKSTEIN, P. (1974a) Immunochemical composition and gel filtration profiles of uterine flushings from rats with and without IUDs. Contraception 9, 161-173.

PEPLOW, V., BREED, W.G. & ECKSTEIN, P. (1974b) Studies on uterine flushings in the baboon. Immunochemical composition in animals with and without intrauterine contraceptive devices. Am. J. Obstet. Gynec. 120, 117-123.

PERRY, J.S., HEAP, R.B., BURTON, R.D. & GADSBY, J.E. (1976) Endocrinology of the early embryo and its role in the establishment of pregnancy. J. Reprod. Fert. Suppl. 25 (in press).

PINSKER, M.C. & MINTZ, B. (1973) Changes in cell surface glycoproteins of mouse embryos before implantation. Proc. Natn. Acad. Sci. U.S.A. 70, 1645-1648.

PINSKER, M.C., SACCO, A.G. & MINTZ, B. (1974) Implantation-associated proteinase in mouse uterine fluid. Devl Biol. 38, 285-290.

POLLARD, R.M. & FINN, C.A. (1974) Influence of the trophoblast upon differentiation of the uterine epithelium during implantation in the mouse. J. Endocr. 62, 669-674.

POTTS, D.M. (1971) The ultrastructure of implantation in the mouse. J. Anat. 103, 77-90.

PRASAD, M.R.N., DASS, C.M.S. & MOHLA, S. (1968) Action of oestrogen on the blastocyst and uterus in delayed implantation - an autoradiographic study. J. Reprod. Fert. 16, 97-103.

PRASAD, M.R.N., SAR, M. & STUMPF, W.E. (1974) Autoradiographic studies on (3H) oestradiol lacalization in the blastocyst and uterus of rats during delayed implantation. J. Reprod. Fert. 36, 75-81.

PRESCOTT, D.M. (1968) Regulation of cell reproduction. Cancer Res. 28, 1815-1820.

PSYCHOYOS, A. (1961) Nouvelles recherches sur l'ovoimplantation. C. r. hebd. Seanc. Acad. Sci., Paris 252, 2306-2311.

PSYCHOYOS, A. (1969) Hormonal requirements for egg implantation. Adv. Biosc. 4, 275-290.

PSYCHOYOS, A. (1973) Hormonal control of ovoimplantation. Vitamin. Horm. 31, 201-256.

PSYCHOYOS, A. & BITTON-CASIMIRI, V. (1969) Caption in vitro d'un precurseur d'acide ribonucleique (ARN) (uridine-5-3H) par le blastocyste du rat: difference entre blastocysts normaux et blastocyste en diapause. C. r. hebd. Seanc. Acad. Sci., Paris 268, 188-192.

RAJALAKSHMI, M., SANKARAN, M.S. & PRASAD, M.R.N. (1972) Changes in uterine sialic acid and glycogen during early pregnancy in the rat. Biol. Reprod. 6, 204-209.

REID, R.J. & HEALD, P.J. (1971) Protein metabolism of the rat uterus during the oestrous cycle, pregnancy and pseudopregnancy, and as affected by an anti-implantation compound, ICI 46, 474. J. Reprod. Fert. 27, 73-82.

RENFREE, M.B. (1973) Influence of the embryo on marsupial uterus. Nature, Lond. 240, 477-480.

REYNOLDS, S.R.M. (1965) Physiology of the Uterus, Hafner Publishing Company, New York.

RINGLER, I. (1956) Proteins composition of rat uterine luminal fluid. Fedn. Proc. Fedn Am. Socs exp. Biol. 15, 152-153.

RINGLER, I. (1961) The composition of rat uterine luminal fluid. Endocrinology 68, 281-291.

ROBERTS, G.P. & PARKER, J.M. (1974a) Macromolecular components of the luminal fluid from the bovine uterus. J. Reprod. Fert. 40, 291-303.

ROBERTS, G.P. & PARKER, J.M. (1974b) An investigation of enzymes and hormone-binding proteins in the luminal fluid of the bovine uterus. J. Reprod. Fert. 40, 305-313.

ROSEMAN, S. (1974) The biosynthesis of complex carbohydrates and their potential role in intercellular adhesion. In: A. Moscona (Ed.), The Cell Surface in Development, John Wiley and Sons, New York, pp. 255-272.

ROSSANT, J. (1975a) Investigation of the determinative stage of the mouse inner cell mass. I. Aggregation of isolated inner cell masses with morulae. J. Embryol. exp. Morph.33, 979-990.

ROSSANT, J. (1975b) Investigation of the determinative state of the mouse inner cell mass. II. The fate of isolated inner cell masses transferred to the oviduct. J. Embryol. exp. Morph. 33, 991-1001.

ROTH, S., McGUIRE, E.J. & ROSEMAN, S. (1971) Evidence for cell surface glycosyltransferases. Their potential role in cellular recognition. J. Cell Biol. 51, 536-547.

ROWINSKI, J., SOLTER, D. & KOPROWSKI, H. (1976) Change of concanavalin A induced agglutinability during preimplantation mouse development. Expl Cell Res. 100, 404-408.

RUDLAND, P.S. (1974) Control of translation in cultured cells, continued synthesis and accumulation of messenger RNA in non-dividing cultures. Proc. natn. Acad. Sci. U.S.A. 71, 750.

RUDLAND, P.S., SEIFERT, W. & GOSPODAROWICZ, D. (1974) Growth control in cultured mouse fibroblasts. Induction of the pleiotypic and mitogenic responses by a purified growth factor. Proc. Natn. Acad. Sci. U.S.A. 71, 2600-2604.

SADLEIR, R.M.F.S. (1972) Cycles and seasons. In: C.R. Austin and R.V. Short (Eds.), Germ Cells and Fertilization, Cambridge University Press, pp. 85-103.

SANYAL, M.K. & MEYER, R.K. (1970) Effect of estrone on DNA synthesis in pre-implantation blastocysts of gonadotrophin-treated immature rats. Endocrinology 976-981.

SANYAL, M.K. & MEYER, R.K. (1972) Deoxyribonucleic acid synthesis in vitro in normal and delayed nidation preimplantation blastocysts of adult rats. J. Reprod. Fert. 29, 439-442.

SCHLAFKE, S. & ENDERS, A.C. (1975) Cellular basis of interaction between trophoblast and uterus at implantation. Biol. Reprod. 12, 41-65.

SCHWARTZ, N.B. (1974) The role of FSH and LH and of their antibodies on follicle growth and on ovulation. Biol. Reprod. 10, 236-272.

SCHWICK, H.G. (1965) Chemisch-entwicklungsphysiologische beziehungen von uterus zu blastocyste des Kaninchens (Oryctolagus cuniculus). Wilhelm Roux Arch. EntwMech. 156, 283-343.

SETTY, B.S., SINGH, M.M., CHOWDBURY, S.R. & KAR, A.B. (1973) The role of electrolytes of the endometrium and uterine fluid during delayed implantation in rats. J. Endocr. 59, 461-464.

SHERMAN, M.T. (1975) The role of cell-cell interaction during early mouse embryogenesis. In: A.E. Wild and M. Balls (Eds.), The Early Development of Mammals, Cambridge University Press, pp. 145-166.

SHERMAN, M.I. & BARLOW, P.W. (1972) Deoxyribonucleic acid content in delayed mouse blastocysts. J. Reprod. Fert. 29, 123-126.

SHERMAN, M.I., McLAREN, A. & WALKER, P.M.B. (1972) Studies on the mechanism of DNA accumulation in giant cells of mouse trophoblast: Satellite DNA content. Nature, Lond. 238, 175-176.

SHIRAI, E. & IIZUKA, R. (1974) Serum protein components in human endometrium. Contraception 10, 311-317,

SHUR, B.D. & ROTH, S. (1975) Cell surface glycosyltransferases. Biochim. Biophys. Acta 415, 473-512.

SINGER, S.J. & NICOLSON, G.L. (1972) The fluid mosaic model of the structure of cell membranes. Science 175, 720-731.

SMITH, D.M. (1968) The effect on implantation of treating cultured mouse blastocysts with oestrogen in vitro and uptake of ^{3}H oestradiol by blastocysts. J. Endocr. 41, 17-29.

SMITH, D.M. & SMITH, A.S. (1971) Uptake and incorporation of amino acids by cultured mouse embryos: Oestrogen stimulation. Biol. Reprod. 4, 66-73.

SMITH, J.A., MARTIN, L., KING, R.J.B. & VERTES, M. (1970) Effects of oestradiol-17β and progesterone on total and nuclear protein synthesis in epithelial and stromal tissues of the mouse uterus, and of progesterone on the ability of these tissues to bind oestradiol-17β. Biochem. J. 119, 773-782.

SOBEL, J.S. & NEBEL, L. (1976) Concanavalin A agglutinability of developing mouse trophoblast. J. Reprod. Fert. 47, 399-402.

SPINDLE, A.I. & PEDERSEN, R. (1973) Hatching, attachment and outgrowth of mouse blastocysts in vitro: fixed nitrogen requirements. J. exp. Zool. 186, 305-318.

SPIRO, R. (1970) Glycoproteins. Ann. Dev. Biochem. 39, 599-638.

SQUIRE, G.O., BAZER, F.W. & MURRAY, F.A., Jr. (1972) Electrophoretic patterns of porcine uterine protein secretions during the estrous cycle. Biol. Reprod. 7, 321-325.

STERN, M.S. (1973) Chimaeras obtained by aggregation of mouse eggs with rat eggs. Nature, Lond. 243, 472-473.

SURANI, M.A.H. (1975a) Zona pellucida denudation, blastocyst proliferation and attachment in the rat. J. Embryol. exp. Morph. 33, 343-353.

SURANI, M.A.H. (1975b) Modulation of implanting rat blastocysts to macro-molecular secretions of the uterus. Dissertation, University of Cambridge.

SURANI, M.A.H. (1975c) Hormonal regulation of proteins in the uterine secretion of ovariectomized rats and the implications for implantation and embryonic diapause. J. Reprod. Fert. 43, 411-417.

SURANI, M.A.H. (1976) Uterine luminal proteins at the time of implantation in rats. J. Reprod. Fert. 48, 141-145.

SYMONS, D.B.A. & HERBERT, J. (1971) Incidence of immunoglobulins in the secretions and of IgG-globulin in the tissues of the female tract. J. Reprod. Fert. 24, 55-62.

TACHI, S., TACHI, C. & LINDNER, H.R. (1970) Ultrastructural features of blasto-cyst attachment and trophoblastic invasion in the rat. J. Reprod. Fert. 21, 37-56.

TARKOWSKI, A.K. (1962) Inter-specific transfers of eggs between rat and mouse. J. Embryol. exp. Morph. 10, 476-495.

TORBIT, C.A. & WEITLAUF, H.M. (1974) The effect of oestrogen and progesterone on CO_2 production by 'delayed implanting' mouse embryos. J. Reprod. Fert. 39, 379-382.

TSAI, S.Y., TSAI, M.J., SCHWARTZ, R., KALIMI, M., CLARK, J.H., & O'MALLEY, B.W. (1975) Effect of estrogen on gene expression in chick oviduct: Nuclear receptor levels and initiation of transcription. Proc. Natn. Acad. Sci., U.S.A. 72, 4228-4232.

TURNER, R.S. & BURGER, M.M. (1973) Involvement of a carbohydrate group in the uterine site for surface guided reassociation of animal cells. Nature, Lond. 244, 509-510.

TYLER, A. (1947) An auto-antibody concept of cell structure, growth and differentiation. Growth (Symposium) 10, 6-7.

VAN BLERKOM, J. & BROCKWAY, G.O. (1975) Patterns of protein synthesis in pre-implantation mouse embryos. II. During release from facultative delayed implantation. Devl. Biol. 46, 446-451.

VAN DE VYVER, G. (1975) Phenomenon of cellular recognition in sponges. In: A.A. Moscona and A. Monroy (Eds.), Current Topics in Developmental Biology, Vol. 10, Academic Press, New York, pp. 123-140.

VICKER, M.G. (1976) BHK 21 fibroblast aggregation inhibited by glycopeptides from the cell surface. J. Cell Sci. 21, 161-173.

VONDERHAAR, B.K. & TOPPER, Y.J. (1974) A role of the cell cycle in hormone-dependent differentiation. J. Cell Biol. 63, 707-712.

WEBB, F.T.G. (1975) Implantation in ovariectomized mice treated with dibutyryl adenosine 3',5'-monophosphate (dibutyryl cyclic AMP). J. Reprod. Fert. 42, 511-517.

WEBB, F.T.G. & SURANI, M.A.H. (1975) Influence of environment on blastocyst proliferation, differentiation and implantation. In: G.P. Talwar (Ed.), Symposium on Regulation of Growth and Differentiated Function in Eukaryote Cells, Raven Press, New York, pp. 519-522.

WEISS, P. (1947) The problem of specificity in growth and development. Yale J. Biol. Med. 19, 235-278.

WEITLAUF, H.M. (1969) Temporal changes in protein synthesis by mouse blastocysts transferred to ovariectomized recipients. J. exp. Zool. 171, 481-486.

WEITLAUF, H.M. (1974a) Metabolic changes in the blastocysts of mice and rats during delayed implantation. J. Reprod. Fert. 39, 213-224.

WEITLAUF, H.M. (1974b) Effect of actinomycin D on protein synthesis by delayed implanting mouse embryos in vitro. J. exp. Zool. 189, 197-202.

WEITLAUF, H.M. (1976) Effect of uterine flushings on RNA synthesis by 'implanting' and 'delayed implanting' mouse blastocysts in vitro. Biol. Reprod. 14, 566-571.

WHITTEN, W.K. (1971) Nutrient requirements for the culture of preimplantation embryos in vitro. Adv. Biosc. 6, 129-140.

WHITTINGHAM, D.G. (1971) Culture of mouse ova. J. Reprod. Fert. Suppl. 14, 7-21.

WILMER, E.N. (1961) Steroids and cell surface. Biol. Rev. 36, 368-398.

WINZLER, R.J. (1970) Carbohydrates in cell surfaces. Int. Rev. Cytol. 29, 77-114.

WOLF, V. & ENGEL, W. (1972) Gene activation during early development of mammals. Humangenetik 15, 99-118.

YAMADA, K.M., YAMADA, S. & PASTAN, J. (1976) Cell surface protein partially restores morphology, adhesiveness, and contact inhibition of movement to transformed fibroblasts. Proc. Natn. Acad. Sci. U.S.A. 73, 1217-1221.

YING, S.Y. & GREEP, R.O. (1972) Inhibition of implantation with antibody against uterine fluid and blastocysts in the rat. Contraception 6, 93.

YOCHIM, J.M. & DE FEO, V.J. (1963) Hormonal control of the onset, magnitude and duration of uterine sensitivity in the rat by steroid hormones of the ovary. Endocrinology 72, 317-326.

YOSHINAGA, K., HAWKINS, R.A. & STOCKER, J.F. (1969) Estrogen secretion by the rat ovary in vivo during estrous cycle and pregnancy. Endocrinology 85, 103-112.

ZEILMAKER, G.H. (1971) Blastocyst proliferation in rats and mice; effects of immunization and irradiation. J. Reprod. Fert. 27, 495-496.

ZEILMAKER, G.H. (1973) Fusion of rat and mouse morulae and formation of chimaeric blastocysts. Nature, Lond. 242, 115-116.

EMBRYONIC DIAPAUSE

R. J. Aitken

Centre d'Etudes Biologiques des Animaux Sauvages
Foret de Chize
and
Faculty of Medicine
University of Bordeaux

The terms embryonic diapause or delayed implantation describe a prolonged and sometimes variable period of gestation during which blastocysts are retained within the uterine cavity unattached, undifferentiated and metabolically inactive. This chapter will begin with a description of the general nature, distribution and possible function of this phenomenon and then go on to discuss the roles played by the blastocyst, uterus, ovary, pituitary gland and environment in achieving this remarkable control over embryonic development.

Embryonic diapause was first discovered in the roe deer (Capreolus capreolus) by 18th century German anatomists anxious to explain the five-month hiatus between mating, which takes place in July and August, and the first overt signs of pregnancy in late December (Short & Hay, 1966). Several ingenious ideas were put forward to explain this discrepancy. The existence of a second unobtrusive rut in November for the 'serious business of conceiving young' was proposed, the July rut apparently being 'for pleasure' only (Pfeil, 1807). Alternatively, it was suggested that the July rut was, in fact, the true one but that fertilisation was delayed until December (Anon, 1825). An even more bizarre explanation came from Pockels (1836) who furnished evidence to support his claim that the egg was fertilised within the follicle in July but not released into the fallopian tubes until late December. Zeigler (1843) obtained direct evidence of ovulation in August and on December 16, 1841, became the first man to find a dormant blastocyst within the roe deer uterus. Unfortunately Zeigler (1843) did not appreciate the significance of his findings and proposed that the delay of embryonic development was due to the slow passage of the fertilised ovum through the fallopian tube. The true nature of delayed implantation was eventually discovered by Bischoff in 1854. Ovulation and fertilization of the ovum occur in July or August and are followed by the rapid passage of the egg through the fallopian tube and into the uterus, which it enters as a morula surrounded by the zona pellucida. The fertilised egg rapidly develops into a blastocyst, loses its zona pellucida and then enters a 4-5 month period of dormancy which is eventually terminated in late December by the rapid elongation of the trophoblast and placental attachment.   Since Bischoff's original discovery in the roe deer, embryonic diapause has been confirmed in at least eight other mammalian orders including the Marsupalia, Chiroptera, Edentata, Rodentia, Carnivora, Insectivora and Pinnipedia (Sadlier, 1969). Despite this wide distribution, it is remarkable how little the incidence of embryonic diapause is correlated with the taxonomic relationships of a given species.   Within the Mustelidae, for example, delayed implantation is exhibited by the American otter (Lutra canadensis), mink (Mustela vison), stoat (Mustela erminea) and long-tailed weasel (Mustela frenata) but it is not shown by the European otter (Lutra lutra) or the European weasel (Mustela putorius) (Gould, 1965).   In another mustelid, the spotted skunk (Spilogale putorius), delayed implantation is confined to those subspecies occupying the western parts of America (S.p. leucoparia, S.P. latifrons, S.p. gracilis, S.p. phenax), the eastern forms (S.p. interrupta, S.p. ambarvalis, S.p. putorius) having a normal type of gestation (Mead, 1968).   Occasionally, the phenomenon is only exhibited by one species in an entire order as in the case of the roe deer and the Ungulata (Short & Hay, 1966).

Two forms of delayed implantation have been distinguished, the facultative or lactational and the obligatory or seasonal.   In the facultative form, the blastocysts derived from a post partum mating enter a short and variable of diapause while the mother is lactating heavily.   Lactational delayed implantation was first discovered in the rat by Lataste (1891) and has since been recorded in the mouse and a number of wild rodents, such as the bank vole (Clethrionomys glareolus) (Brambell & Rowlands, 1936), the Mongolian gerbil (Meriones unguiculatus) (Norris & Adams, 1971), the Indian gerbil (Tatera indica) (Bland, 1969) and deer mice (Peromyscus sp.) (Svihla, 1932) as well as in a majority of macropod marsupials (Sharman, 1954, 1970; Tyndale-Biscoe, 1963a).   In the obligatory or seasonal form of delayed implantation, the diapause phase is always terminated at a certain time of the year, the phenomenon presumably being under environmental control.   In addition, the delay period is usually of a fixed duration for a given species and may extend for as long as five (Bischoff, 1854) or even ten months (Canivenc & Bonnin-Laffargue, 1963) as in the roe deer and badger (Meles meles) respectively.   The mink (Enders & Enders, 1963) and possibly the roe deer (Stieve, 1950) are the only species with an obligatory form of delayed implantation exhibiting a variable period of diapause.   In both species the time of implantation is fixed but the duration of the delay phase is modified by wide fluctuations in the time of mating.

The fact that embryonic diapause has evolved independently in several groups of mammals is evidence of the powerful selective advantage associated with this

modified form of early pregnancy.   The lactational delay exhibited by certain
rodents, for example, clearly enhances fecundity by ensuring the rapid replacement of lost young.   Similarly the retention of a dormant blastocyst during
lactation in the macropod marsupials ensures the rapid replacement of young lost
at times of drought (Ealey, 1963) or predation (Marshall, 1967).   Such
advantages may be secondary however, and it has been suggested that embryonic
diapause evolved primarily as a consequence of either the extended gestation
period (Tyndale-Biscoe, 1968) or the post-partum oestrus (Sharman, 1970)
encountered in a majority of the Macropodidae.

One of the major advantages conferred by a long seasonal delay of implantation may be that it allows mating and parturition to occur at optimal times of
the year.   In the roe deer, for instance, delayed implantation effectively
extends the length of pregnancy from an expected five to about ten months.   As
a result, mating can occur in August when the deer are in peak condition while
parturition can take place in May or June when the young stand the best chance of
survival.   In aquatic mammals, like the southern elephant seal (Mirounga leonina)
seasonal delayed implantation serves to extend the length of gestation to such
an extent that parturition and mating closely follow one another.   As a result
the amount of time spent by these animals out of their normal environment is
reduced to a minimum (Sadlier, 1969).   Another effect of a seasonal delay phase,
although possibly not a primary function, is that it enables pregnant females to
endure the stresses of winter without being subjected to the nutritional drain
of carrying a developing embryo (Sadlier, 1969).   In several species, however,
the advantage conferred by a period of delayed implantation remains an enigma.
Hence in the equatorial fruit bat (Eidolon helvum) mating is followed by three
months of delayed implantation and birth of the young in October or November,
just before a major peak in rainfall (Mutere, 1965).   Nevertheless, there
appears to be no obvious reason why mating could not have occurred just before
parturition in October, without the intervening delay phase (Sadlier, 1969).
In the mink, implantation invariably occurs in early April and is normally preceded by a short period of diapause, the duration of which depends upon the time
of mating.   However, since normal births follow matings in late March without a
significant delay of implantation, the function of the delay phase is obscure.
It is conceivable that in the wild, delayed implantation is an adaptation to
dispersal since it permits a marked extension of the breeding season without a
change in the date of parturition, which is presumably fixed in early May to
ensure maximum survival of the young.

THE BLASTOCYST

The diapausing blastocyst

In rats, mice and macropod marsupials, lactational delayed implantation is associated with the gradual cessation of cell division as the blastocysts enter a state of metabolic inactivity (Baevsky, 1963; Clark, 1966; Clark & Poole, 1967; McLaren, 1968; Surani, this volume). The biochemical nature of this diapause state has been investigated in the rat and mouse and the findings have been reviewed by McLaren (1973); DNA synthesis is at a standstill (McLaren, 1968; Gulyas & Daniel, 1969); RNA and protein synthesis are minimal (Weitlauf & Greenwald, 1965; Gulyas & Daniel, 1969; Weitlauf, 1973) and carbon dioxide production is low (Menke & McLaren, 1970a). However, despite this apparent lack of metabolic activity the size (Baevsky, 1963; Schlafke & Enders, 1963; Dickson, 1966; Yasukawa & Meyer, 1966) weight, lipid and protein content (Hensleigh, 1971; Weitlauf, 1973) of the blastocysts increase gradually during the delay period. If delayed implantation is experimentally induced in rats and mice during early pregnancy, by ovariectomy and subsequent progesterone administration, the blastocysts enter a state of diapause which appears to be identical with that experienced during lactation. Hence DNA, RNA and protein synthesis are minimal (Prasad et al., 1968; Dass et al., 1969; Psychoyos & Bitton-Casimiri, 1969; Sanyal & Meyer, 1972) and carbon dioxide production is suppressed. Mouse blastocysts can also be induced to enter a state of diapause in vitro by the omission of certain vital amino acids or serum from the culture medium (Gwatkin, 1966a, b). However, the type of diapause induced under these conditions may not be identical with that observed during lactation, for although cell division is blocked (Bowman & McLaren, 1970) DNA synthesis is inhibited at both the G_1 and G_2 phases of mitosis, not just the G_1 state as in vivo (Sherman & Barlow, 1972). In addition, this form of diapause is not necessarily associated with a decline in carbon dioxide production (Menke & McLaren, 1970b) as in lactational delay.

The blastocysts of those species exhibiting a seasonal delay of implantation are also characterised by a general lack of metabolic activity during diapause, although there is considerable interspecies variation with respect to such parameters as the retention or loss of the zona pellucida or the degree of blastocyst expansion and cell division observed during the delay phase. The diapausing blastocyst is surrounded by a zona pellucida in the macropod marsupials, Ursidae, Mustelidae and Pinnipedia (Gould, 1965; Wimsatt, 1963; Enders, 1971), the fur seal zona pellucida being characterized by the presence of an additional sub-zonal layer which appears to be deposited by the blastocyst itself. In contrast, both the roe deer (Keibel, 1902; Short & Hay, 1966) and armadillo (Dasypus movemcinctus) (Enders, 1962) shed their zonae at the beginning of the

delay phase.   It is interesting to note that shedding of the zona pellucida
during delayed implantation only occurs in those species possessing active
corpora lutea at this time as in the mouse (Finn & Martin, 1974), rat (Yoshinaga,
1974), armadillo (Labhsetwar, 1967) and roe deer (Short & Hay, 1966; Aitken,
1974a).   Obligatory delayed implantation is also normally associated with a
complete cessation of cell division in the blastocyst (Gould, 1965; Enders, 1962).
Notable exceptions to this rule are the roe deer (Keibel, 1902), European
badger (Neal & Harrison, 1958) and spotted skunk (Mead, 1968), all of which show
some increase in blastocyst cell number during delay.   The roe deer blastocyst
is absolutely unique in not showing significant expansion (Figure 1) and cell

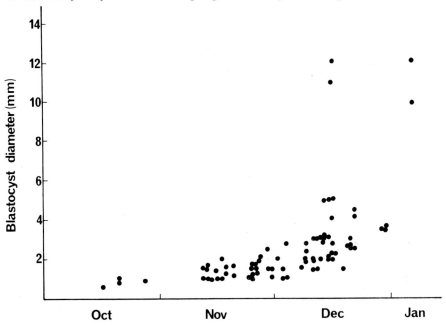

Figure 1: Growth of the roe deer blastocyst during diapause.

division (Table 1) during diapause, but also in differentiating a layer of
extra-embryonic endoderm at this time (Keibel, 1902).   Many other unique
features of the diapausing roe deer blastocyst have been revealed by scanning
and transmission electron microscope studies (Aitken et al., 1973; Aitken, 1974a,b,
1975).   The most prominent feature of the roe deer blastocyst at this time is
the presence of a large number of granular inclusions in the cytoplasm of the
trophectoderm cells (Figure 2).   These inclusions are most numerous during the
early stages of diapause and show a gradual reduction in number and electron
density as the time of implantation approaches.   These changes are consistent
with the view that the granules represent an endogenous energy reserve which

312

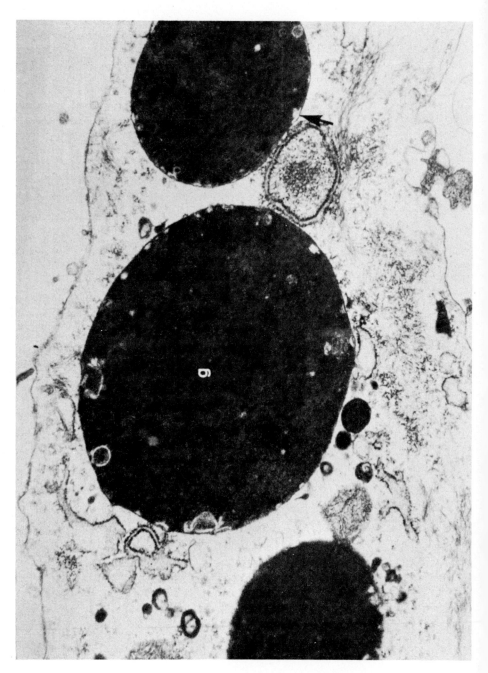

Figure 2: Electron dense granules in the trophoblast of the roe deer blastocyst during diapause: g = granules which are clearly membrane bound and perforated by clear vacuoles (arrowed). Magnification x 34,680.

is gradually utilised by the blastocyst during delay.   The blastocysts of many
other species exhibiting a seasonal delay of implantation are also characterised
by the accumulation of putative energy reserves (Enders, 1971);   in the mink and
fur seal, for example, the stored material is lipid.   The other distinctive
feature of the roe deer blastocyst during diapause, the lack of cytoplasmic
organelles, suggests a very low level of metabolic activity at this time (Aitken,
1975).

TABLE 1

CHANGES IN THE RATE OF CELL DIVISION DURING
THE LATE STAGES OF DIAPAUSE*

Animal No.	Date shot	No. of trophoblast cells examined	No. mitotic figures	Mitotic Index (No. of dividing cells/100 cells)
71/26	Dec 15	699	1	0.14
70/35	Dec 17	829	1	0.12
70/36	Dec 21	1976	12	0.61
71/34	Dec 22	2420	18	0.74
72/22	Jan 4	885	5	0.56
71/52	Jan 7	2266	33	1.46
72/30	Jan 11	4988	61	1.20
71/68	Jan 18	2572	43	1.67

* The correlation between mitotic index and date of blastocyst recovery is
   highly significant ($p < 0.001$; $r = 0.899$) and the relationship is given
   by the formula: $y = 0.167 + 0.04x$ (where x = date and Dec 15 = 1).

Scanning electron microscopy of the diapausing roe deer blastocyst revealed
the presence of an elaborate network of branched microvilli on the outer surface.
A sharp contrast is provided by the numerous short, fat, simple microvilli
observed on the surface of the mouse blastocyst during delayed implantation
(Bergstrom, 1972).   Furthermore, the various types of imprint left on the
surface of the mouse blastocyst as a result of the intimate contact between the
uterine epithelium and the trophoblast during delayed implantation (Bergstrom,
1972) were not observed in the roe deer.   This suggests that, unlike the mouse,

the roe deer blastocyst lacks any form of attachment to the uterine epithelium during diapause and lies completely free within the uterine lumen.   This conclusion is also supported by the ease with which roe deer blastocysts can be flushed from the uterine lumen during diapause and by the high incidence of transcornual migration observed in this species (Aitken, 1974a).

The activated blastocyst

The termination of embryonic diapause is associated, in all species, with a general increase in the level of metabolic activity exhibited by the blastocysts followed, secondarily, by the differentiation of the inner cell mass and extra-embryonic membranes.   The nature of these secondary changes naturally shows considerable inter-species variation, involving, for example, giant-cell transformation of the abembryonic trophectoderm in the mouse and the massive elongation of the chorion in the roe deer.

The biochemical nature of blastocyst reactivation has been most thoroughly studied in the rat and mouse.   In these species the termination of diapause is associated with increases in a number of general parameters of cellular activity, such as cell division (Bowman & McLaren, 1970), carbon dioxide production (McLaren & Menke, 1971) and RNA, DNA and protein synthesis (Weitlauf & Greenwald, 1965, 1968; Prasad et al., 1968; Dass et al., 1969).

Biochemical studies have also been carried out on the nature of blastocyst reactivation in one species of macropod marsupial, Macropus eugenii.   In this species, the first sign of blastocyst activity, after a prolonged seasonal diapause, is an increase in nucleolus associated RNA polymerase activity (Moore, 1976).   In all macropods the first morphological indications that diapause has terminated are a marked increase in blastocyst diameter and the resumption of cell division (Tyndale-Biscoe, 1963a; Clark, 1966, 1968; Sharman & Berger, 1969; Smith & Sharman, 1969).   Implantation does not take place until some time after the resumption of blastocyst activity, however, and in some species, such as Potorous tridactylus, implantation may not occur at all (Hughes, 1974).

The termination of obligatory delayed implantation in the Mustelidae is marked by a pronounced increase in the over-all size of the blastocyst and the resumption of mitotic activity in the inner cell mass and trophoblast.   The extra embryonic endoderm differentiates at this time giving rise to the bilaminar omphalopleur, the zonae are lost and the blastocysts become evenly spaced along the uterine horns (Gould, 1965).   Implantation eventually takes place about three or four days after the reactivation of the blastocysts (Enders, 1962).

In the roe deer, the termination of embryonic diapause is associated with the sudden rapid elongation of the blastocyst in conjunction with a marked rise in the rate of cell division and the appearance of numerous cytoplasmic organelles in the trophoblast cells (Keible, 1902; Short & Hay, 1966; Aitken et al., 1973; Aitken, 1975).

The role played by the blastocyst in embryonic diapause

The classic embryo transfer experiments performed by Dickman and De Feo (1967) in the rat indicate that the blastocyst is subservient to the uterus and plays no active part in either the initiation or termination of delayed implantation.   These workers found that when 'active' blastocysts were trans-ferred to recipients in a state of delayed implantation, the blastocysts responded to their new environment by entering a state of diapause.   Con-versely, when 'dormant' blastocysts, recovered from donors on day 10 of delayed implantation, were transferred to recipients on the fifth day of pseudopregnancy, the blastocysts were reactivated by the uterine environment and implanted. Hence the eventual outcome of these transfer experiments invariably depended on the state of the uterus.   Critical experiments of this nature have not been performed in species exhibiting the obligatory form of diapause.   However, since premature implantation can be induced by such procedures at ovariectomy in the armadillo (Talmage et al., 1954), ambient light and temperature changes in the badger (Canivenc et al., 1971) or late matings in the mink (Enders & Enders, 1963), the blastocysts cannot be programmed to enter a fixed period of diapause before activating.   In the absence of evidence to the contrary, it may be assumed that the blastocysts of such species are under the same stringent uterine control as observed in the rat and mouse.

Thus histological and biochemical studies in a variety of species indicate that the diapausing blastocyst is both metabolically and developmentally inactive during the delay phase.   The termination of diapause is induced by the uterus and is characterized by a general increase in cellular and mitotic activity in the blastocyst followed by the initiation of embryonic differen-tiation and development.

THE UTERUS

Ultrastructural changes in the endometrium

In order to determine whether the suppression or induction of blastocyst activity is associated with concomitant changes in the endometrium, detailed histological and ultrastructural studies of the uterine mucosa have been performed in a variety of species exhibiting delayed implantation.

In the roe deer (Aitken et al., 1973; Aitken, 1974a,b,c, 1975), such studies have revealed marked changes in the endometrial glands during the initial restraint and subsequent stimulation of embryonic growth.   During diapause, the cells of the tightly coiled glandular fundi are dominated by the presence of abundant, clear, supranuclear vesicles evidently derived from the

316

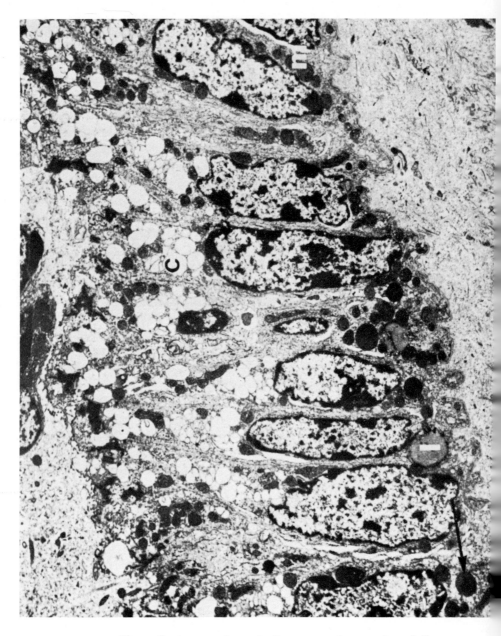

Figure 3: Clear vesicles in the supranuclear region of the endometrial gland cells during embryonic diapause in the roe deer: c = clear vesicles, l = basal lipid deposits, m = mitochondria, some of which appear swollen (arrow). Magnification x 5,500.

Golgi apparatus. The cells also contain numerous apical mitochondria, abundant free ribosomes and occasional basal lipid deposits (Figure 3). The granular endoplasmic reticulum is not well developed, however, and the apical cell membranes are largely bereft of microvilli. The lumina of the glandular ducts and fundi are generally clear at this stage but occasionally contain small amounts of an electron dense granular material and cell debris. Scanning electron microscopy of the endometrial surface during diapause revealed the presence of numerous uniform simple microvilli occasionally interrupted by small duct openings rarely exceeding 2μ in diameter. The onset of rapid embryonic elongation is associated with a decline in the height of the glandular epithelium as the clear vesicles are suddenly released into the glandular and thence the uterine lumen. Simultaneously the lumina of the endometrial glands and ducts become distended with clear vesicles, an electron dense granular material and much cellular debris. The release of the clear vesicles is also associated with a striking increase in the number and diameter of endometrial duct openings and the appearance of copious amounts of mucoid material both in the ducts and on large areas of the uterine surface (Figure 4). At this time, large numbers of apical protrusions may be seen projecting from the luminal epithelium suggesting the release of an apocrine secretion into the uterine lumen (Figure 5).

At a slightly later stage of embryonic development, during the early stages of placental attachment, the granular endoplasmic reticulum of the endometrial duct cells suddenly hypertrophies (Figures 6 and 7) and the microvilli on the luminal surface of each cell increase dramatically in number and height. This phase is also associated with the presence of numerous membrane-bound electron dense granules in the vicinity of the apical cell membranes where they appear to discharge their contents into the lumen (Figures 6 and 7).

These ultrastructural findings suggest that the termination of embryonic diapause in the roe deer is associated with the release of three, possibly four, types of endometrial secretion into the uterine lumen. In addition to the release of clear vesicles from the ductal and glandular epithelia, the endometrium appears to produce an apocrine secretion in the luminal epithelium and possibly a holocrine secretion in the glands. During the more advanced stages of embryonic growth a fourth type of secretion is synthesized in the ductal

318

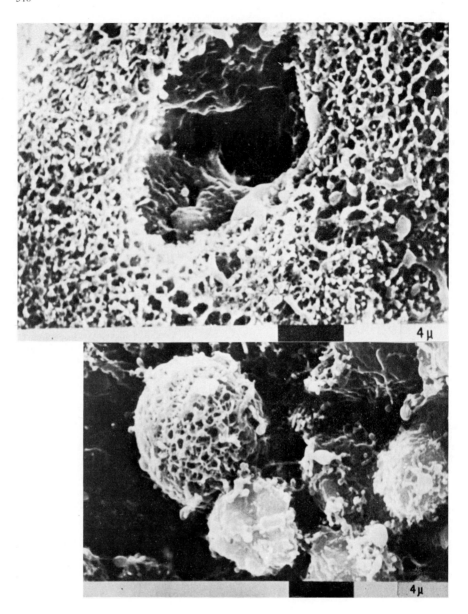

Figure 4: (upper) Scanning electron microscopy of a dilated endometrial duct opening during the rapid elongation of the roe deer blastocyst. Magnification x 5,500.

Figure 5: (lower) Scanning electron microscopy of apical protrusions on the surface of the roe deer uterus during the phase of rapid embryonic growth. Magnification x 5,000.

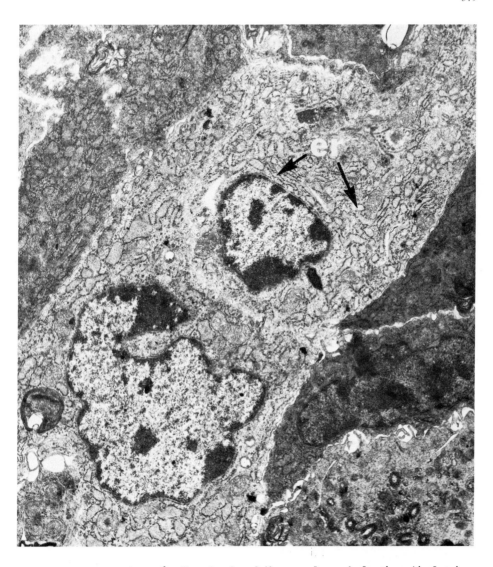

Figure 6: Hypertrophy of the granular endoplasmic reticulum in
the endometrial duct cells of the roe deer uterus during the phase of rapid
embryonic growth: er = endoplasmic reticulum. Magnification x 9,000.

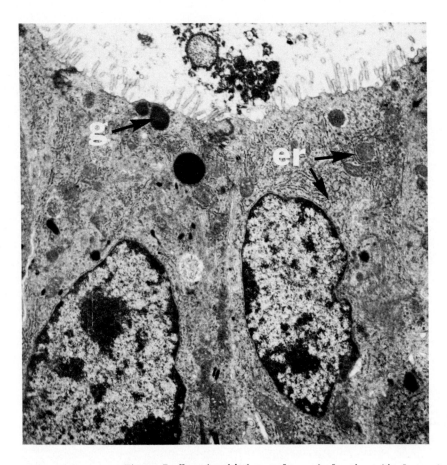

Figure 7: Hypertrophied granular endoplasmic reticulum and apical electron dense granules in the endometrial duct cells of the roe deer uterus during the phase of rapid embryonic growth: er = endoplasmic reticulum, g = apical granules. Magnification x 9,500.

epithelium by the granular endoplasmic reticulum.   The function of these
secretions is presumably to activate and maintain the rapid elongation of the
roe deer conceptus at the end of diapause.

The elegant studies of Nilsson (1970, 1974) on the ultrastructure of the
mouse endometrium during delayed implantation suggest that the termination of
diapause in this species is also associated with the release of secretory
material into the uterine lumen.   During delayed implantation, the luminal
epithelium of the mouse endometrium contains large numbers of clear apical
vesicles (similar to those observed in the roe deer), a well developed Golgi
apparatus surrounded by dense granules, many basal lipid deposits and a moderate
amount of granular endoplasmic reticulum.   Following the administration of an
implantation-inducing dose of oestradiol an electron dense secretion is released
into the uterine lumen within 8 hours.   The release of this secretion is
associated with the discharge of the clear apical vesicles from the luminal
epithelium, and an increase in the quantity of ribosomes and rough endoplasmic
reticulum visible in the cytoplasm of these cells.

Ultrastructural studies on the mink uterus (Enders et al., 1963) similarly
suggest that the termination of embryonic diapause is associated with a significant increase in glandular secretory activity.   During delayed implantation,
the endometrial gland cells contain moderate amounts of saccular, granular
endoplasmic reticulum and a large number of apical electron dense granules.   The
resumption of blastocyst activity is associated with the sudden release of these
granules into the uterine lumen and a marked dilation of the granular endoplasmic
reticulum.

Delayed implantation in the armadillo is also associated with the accumulation of numerous apical, electron dense granules in the epithelial cells of the
endometrium (Enders, 1967) which appear very similar to those observed in the
mink endometrium during diapause.   Their fate at the time of implantation,
however, is unknown.

Biochemical changes in the endometrium

A number of biochemical studies have been performed in order to determine
whether the ultrastructural changes observed in the endometrium of such species
as the roe deer, mouse and mink are, in fact, associated with the release of
secretory material into the uterine cavity.   In one such investigation,

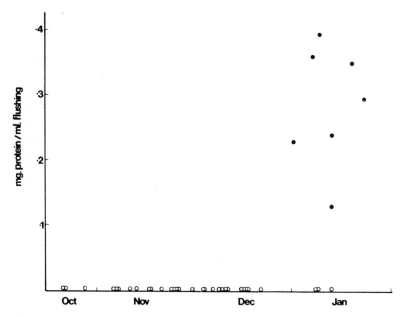

Figure 8: Changes in the protein content of roe deer uterine flushings during embryonic diapause and rapid elongation: o = blastocyst in the uterus, ● = embryo in the uterus.

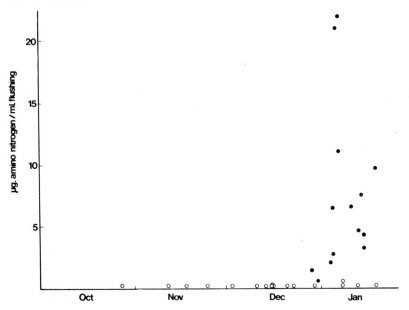

Figure 9: Changes in the α-amino nitrogen content of roe deer flushings during embryonic diapause and rapid elongation: o = blastocyst in the uterus, ● = embryo in the uterus.

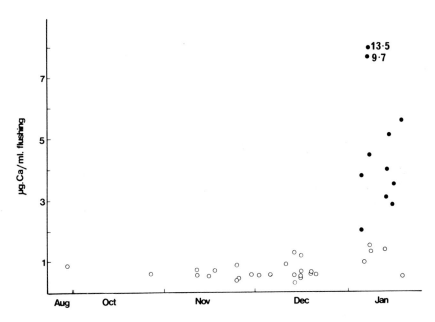

Figure 10: Changes in the calcium content of roe deer uterine flushings during embryonic diapause and rapid elongation: o = blastocyst in the uterus, ● = embryo in the uterus.

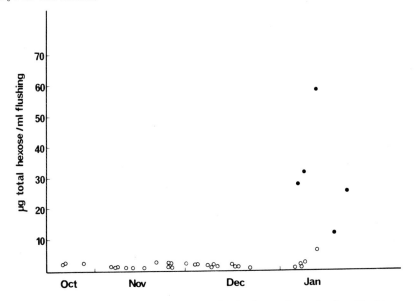

Figure 11: Changes in the total hexose content of roe deer uterine flushings during embryonic diapause and rapid elongation: O = blastocyst in the uterus, ● = embryo in the uterus.

flushings of the roe deer uterine lumen were quantitatively analysed for protein, total hexose, α-amino nitrogen and calcium (Aitken, 1974a,b,c, 1976). During embryonic diapause only trace amounts of each substance could be detected in the flushings indicating the virtual absence of secretory material within the uterine lumen at this time. During the phase of rapid embryonic growth, however, when ultrastructural signs of increased secretory activity were apparent, a dramatic increase in the concentration of each factor was observed (Figures 8 - 11).

The qualitative analysis of these flushings also revealed many interesting features. The increase in total hexose concentration at the end of diapause was associated with the presence of a free ketose subsequently identified as fructose by thin-layer chromatography (Aitken, 1976). The possible role of this sugar in the stimulation and support of rapid embryonic growth is suggested by the fact that fructose has now been identified as a component of the progestational uterine secretions of the pig (Haynes & Lamming, 1967), cow (Suga & Masaki, 1973) and human (Douglas et al., 1970). Fructose may indeed be as ubiquitous a component of the genital tract secretions of the female as it is of the male (Mann, 1946; Aitken, 1976). Fractionation of the proteins present in the uterine flushings by polyacrylamide gel electrophoresis revealed the presence of numerous serum and non-serum bands during the resumption of rapid embryonic growth. The major non-serum proteins consisted of three pre-albumins, a post albumin (Ra 0.60; Ra albumin = 1.00) and a post transferrin (Ra 0.32) (Aitken, 1974a,b). During diapause only the most dominant of these proteins could be detected (albumin and the post albumin at Ra 0.60) suggesting that blastocyst activation was associated with a quantitative rather than a qualitative change in the nature of the uterine secretions.

Similar results were obtained in a biochemical study of the protein content of mouse uterine flushings collected during diapause and at the time of implantation (R. Aitken, unpublished). These studies revealed the presence of low levels of protein on days 2 and 3 of normal pregnancy (4.57 ± 1.48 µg/mouse and 5.72 ± 2.21 µg/mouse respectively) followed by a highly significant increase to 19.96 ± 3.87 µg/mouse on day 4, the day of implantation (Figure 12). During lactational (3.83 ± 0.79 µg/mouse) and experimental (6.76 ± 1.35 µg/mouse) delayed implantation, the concentration of protein was significantly below that observed on day 4 ($p < 0.001$), but similar to that detected in ovariectomized non-pregnant mice (3.34 ± 1.99 µg/mouse). The protein peak observed on day 4 of pregnancy was not a consequence of implantation since a similar change occurred on day 4 of pseudopregnancy (Figure 13). It must be emphasised, however, that the protein peak observed during pseudopregnancy (14.98 ± 2.91 µg/mouse) was significantly lower ($p < 0.05$) than that detected during pregnancy, although the same species of proteins were present in both situations (Figures 14 and 15). The presence of active blastocysts within the uterine lumen, therefore, appeared

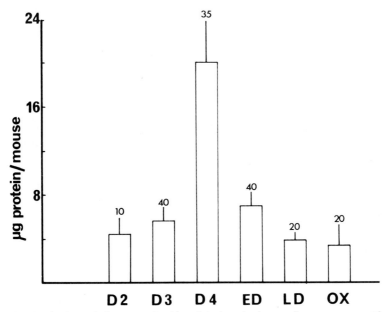

Figure 12: Luminal protein concentration in mice during early pregnancy and delayed implantation. D2, D3 and D4 = day of pregnancy; ED = experimental delay; LD = lactational delay; OX = ovariectomized animals; vertical bars = standard deviation; figures over bars = number of mice, 5 mice contributing to each sample.

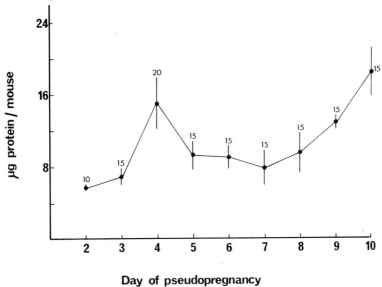

Day of pseudopregnancy

Figure 13: Changes in luminal protein concentration during pseudopregnancy. Peaks are observed on day 4 (when implantation normally takes place in pregnant animals) and days 8-10 (when the animals return to oestrus). Vertical bars = standard deviation; figures over bars = number of mice, 5 mice contributing to each sample.

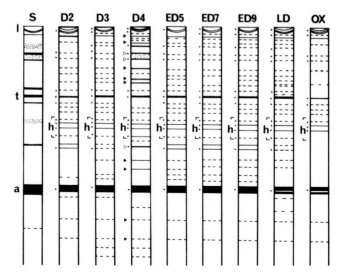

Figure 14:   Polyacrylamide gel electrophoresis of mouse uterine flushings during pregnancy, experimental and lactational delayed implantation, and following ovariectomy. S = mouse plasma; D2, D3 and D4 = day of pregnancy; open arrows = major serum proteins at Ra 0.16, 0.20 and 0.74; closed arrows = major non-serum proteins at Ra 0.06, 0.10, 0.25, 0.32, 0.34, 0.82, 0.89, 1.19 and 1.31; ED5, ED7 and ED9 = day of experimental delayed implantation; LD = lactational delayed implantation; OX = ovariectomized mice not receiving hormone treatment; l = major serum lipoprotein band; t = transferrin; h = haemoglobin bands; a = albumin.
Proteins identified by closed circles are presumed to be of serum origin.

to influence the magnitude of the uterine secretory response at the time of implantation, a point returned to later in this section.

Qualitative analysis of the proteins present in the uterine flushings on day 4 of pregnancy revealed the presence of 36 protein bands, of which 14 were of serum, and 22 of non-serum, origin.   The major non-serum proteins were observed at the origin of the gels and in the slow αglobulin (Ra 0.06, 0.10, 0.25, 0.32 and 0.34), fast α globulin (Ra 0.82 and 0.89) and prealbumin (Ra 1.19 and 1.31) regions (Figure 14).   During experimental and lactational delayed implantation, serum transferrin and albumin were the most dominant proteins present.   Nevertheless, traces of another 30 - 35 proteins could be detected in the delay flushings and a majority of these corresponded to proteins observed in larger quantities on day 4 (Figure 14).   This finding suggested, as the roe deer studies had, that blastocyst reactivation was associated with a quantitative rather than a qualitative change in uterine secretory activity.   This conclusion was also supported by the results of studies designed to determine the influence of uterine secretions obtained during diapause or at the time of implantation, on the behaviour of mouse blastocysts in vitro.   In the first series of

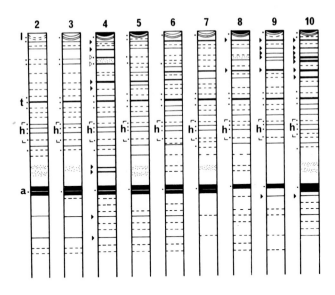

Figure 15: Polyacrylamide gel electrophoresis of mouse uterine flushings during pseudopregnancy. 2-10 = day of pseudopregnancy; on day 4, open arrows = major serum slow α globulins (Ra 0.16 and 0.20), closed arrows = major non-serum bands at Ra 0.06, 0.10, 0.32, 0.34, 0.85, 0.89, 1.19 and 1.32; day 8, closed arrows = major non-serum bands at Ra 0.06 and 0.25; day 9, closed arrows = major non-serum bands at Ra 0.06, 0.11, 0.15, 0.18, 0.25 and 1.06; day 10, closed arrows = major non-serum α globulins at Ra 0.06, 0.11, 0.15, 0.18, 0.20, 0.25, 0.32 and 1.06.
l = major serum lipoprotein band; t = serum transferrin; h = haemoglobin bands; a = albumin.
Proteins indicated by closed circles are presumed to be of serum origin.

experiments, mouse blastocysts recovered on day 3 of normal pregnancy were cultured in a modified Brinsters medium, each millilitre of which had been flushed through the uterine lumina of three mice in rapid succession. The results of this mouse to mouse comparison pointed to the presence of an embryo-trophic stimulus in the uterus on the day of implantation which was absent during diapause (Figure 16). Since significantly more protein can be recovered from the uterine lumen at the time of implantation (Figure 12), this embryo-trophic effect could have been explained simply in terms of the amount of protein added to the culture medium in each group. The experiment was therefore repeated, only this time the concentration of protein in the incubation media was kept constant (Table 2). Under these conditions, more than 90% of the blastocysts hatched from their zonae regardless of the source of uterine protein. These results again suggested that the diapausing uterus is characterised by a quantitative lack of embryotrophic material in the luminal fluids, rather than the presence of factors that directly inhibit blastocyst activity.   Another inter- pretation of the results is possible however.   The uterine flushings used in the last experiment had been extensively dialysed, prior to lyophilization, in order to remove all traces of the flushing medium, phosphate buffered saline.   Recent experimental evidence obtained by Psychoyos (1973) in the rat, and Weitlauf (1976) in the mouse, suggests that the uterine fluids of these species may con- tain a dialysable factor that actively inhibits ^{3}H uridine incorporation into RNA by preimplantation blastocysts.   It is therefore possible that the apparent embryotrophic equivalence exhibited by the uterine flushings obtained

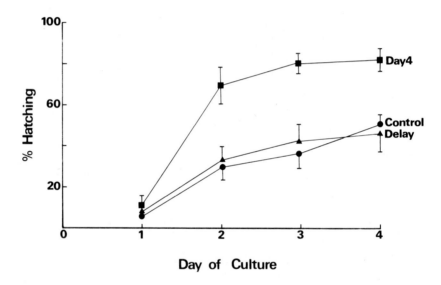

Figure 16:   Percentage of blastocysts hatching from their zonae when cultured in incubation media supplemented with uterine fluids obtained either during experimental delayed implantation or on day 4 of normal pregnancy. Each ml. of incubation medium was flushed through the uterine lumina of 3 mice;   control medium was not supplemented with uterine flushings;   vertical bars = standard error.

from 'dormant' and 'active' mouse uteri in the above experiment was influenced by the removal of an inhibitory substance during dialysis.   The existence and physiological significance of the proposed inhibitor have not yet been definitively established however (Weitlauf, 1976).   Both Psychoyos (1973) and Weitlauf (1976) found that uterine flushings obtained from 'active' uteri were as inhibitory as those from 'dormant' uteri, a result which is not consistent with a physiological role for this factor in vivo.   It is conceivable, however, that implantation is

TABLE 2

HATCHING OF BLASTOCYSTS IN PRESENCE OF 55 μg OF BSA,
DAY 4 UTERINE PROTEIN AND DELAY UTERINE PROTEIN

Protein supplement	% of blastocysts hatched on Day 3 of culture
55 μg BSA	50.67 ± 10.10
55 μg Day 4 uterine protein	91.15 ± 4.44
55 μg Delay uterine protein	94.50 ± 5.50

associated with a temporary decline in the luminal concentration of this substance since Psychoyos (1973) observed a transient but significant reduction in
inhibitory activity one hour after the administration of oestradiol-17β to
ovariectomized progesterone-treated rats. In contrast, Weitlauf (1976) did not
detect any significant change in the inhibitory properties exhibited by mouse
uterine flushings from 1 to 24 hours post-oestradiol administration. The
biochemical identification of the proposed inhibitor has also been the subject
of some speculation. Both Weitlauf (1976) and Psychoyos (1973) agree that the
material is dialysable, but disagree on its heat stability; the rat inhibitor
is destroyed by heating to 80°C for ten minutes while the mouse inhibitor remains
stable after heating to 90°C for the same period of time. Psychoyos (1973)
has put forward the interesting suggestion that the inhibitor may be lipoidal in
nature since alcohol:ether or methanol extracts of the uterine flushings retain
the inhibitory properties of the original material. Psychoyos also points to
the fact that the luminal epithelium of both the mouse (Nilsson, 1970) and rat
(Warren & Enders, 1964) endometrium contains prominent lipid deposits during
delayed implantation as does the diapausing trophoblast (Potts & Psychoyos, 1967).
Obligatory delayed implantation in the roe deer (Aitken et al., 1973; Aitken,
1974a,b; 1975), black bear (Ursus americanus) (Wimsatt, 1963) and armadillo
(Enders, 1961) is also associated with the marked accumulation of lipid reserves
in the luminal epithelium, although mustelids such as the mink (Enders & Enders,
1963), long-tailed weasel (Wright, 1963) and stoat (Enders, 1967) appear to store
glycogen instead. One possibility which has not yet been excluded is that the
uterine material responsible for inhibiting the uptake of ^{3}H uridine by rat or
mouse blastocysts in vitro is cold uridine.

In order to determine whether the oestrogen induced activation of diapausing
mouse and rat blastocysts involves the induction of uterine secretory activity,
as observed during normal pregnancy, uterine flushings have been qualitatively and
quantitatively analysed for protein at selected time intérvals following
oestradiol administration (Surani, 1975 and this volume; R. Aitken, unpublished

330

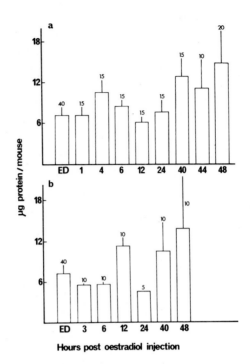

Figure 17: Oestradiol administration to pregnant mice during diapause.
(a) changes in luminal protein concentration following the administration of
1 μg oestradiol-17β . (b) changes in luminal protein concentration following
the administration of 50 μg oestradiol 17β . Vertical bars = standard
deviations; figures over bars = number of mice, 5 mice contributing to each
sample; ED = experimental delayed implantation. Note the biphasic secretory
response induced by oestradiol in diapausing pregnant mice. The major effect
of the lower oestrogen dose is to delay the initiation of the first secretory
phase by several hours.

observations).   In the pregnant mouse, the uterine response to oestradiol was
biphasic with significant increases in luminal protein concentration being
observed within 12 hours and again at 40 - 48 hours (Figure 17).   Qualitatively,
the first part of this biphasic response was found to involve the release of
both serum and non-serum proteins into the uterine lumen, a majority of proteins
migrating in the high molecular weight slow α globulin region of the gels
(Figure 19).   The presence of two major serum slow α globulines (Ra 0.16 and
0.20) in these uterine flushings may have been the result of a transient increase
in capillary permeability since a general increase in endometrial vascularity and
oedema was observed at this time.   Surani (1975) also detected the passage of
high molecular weight proteins into the rat uterine lumen, 13 - 20 hours after
an injection of oestradiol-17β and Bergstom and Nilsson (1974) have presented
ultrastructural evidence confirming the release of a uterine secretion within

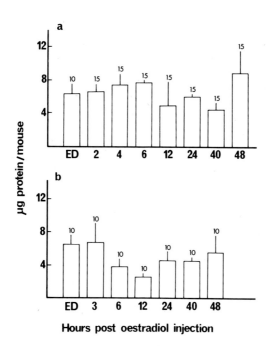

Figure 18: Oestradiol administration to pseudopregnant mice during diapause.
(a) changes in luminal protein concentration following the administration of
1 μg oestradiol-17β . (b) changes in luminal protein concentration following
the administration of 50 μg oestradiol-17β . Vertical bars = standard deviation;
figures over bars = number of mice, 5 mice contributing to each sample; ED =
experimental delayed implantation. Note that oestradiol administration to pseudopregnant mice during diapause does not induce a significant biphasic change in
uterine protein concentration.

eight hours of oestrogen administration in the mouse. Studies on the time course
of blastocyst activation by oestradiol strongly suggest that this early phase of
uterine secretory activity is responsible for the termination of diapause in the
mouse. Ultrastructural signs of blastocyst activation are apparent within eight
hours of an injection of oestradiol (Nilsson, 1974), carbon dioxide production
and protein synthesis by the blastocyst are clearly elevated within twelve hours
(Weitlauf, 1971; Torbit & Weitlauf, 1974) and a pontamine blue response is
present in the uterine stroma within 17 hours (McLaren, 1970).

The second phase of the uterine response to oestradiol occurred 40 - 48 hours
after hormone administration and involved a marked increase in the concentration
of certain non-serum proteins. The most dominant of these was a major post
transferrin (Figures 19 and 20) (Ra 0.32) which closely resembled a protein
observed 18 - 24 hours after the administration of oestradiol to rats (Surani,
1975) and in the uterine flushings of humans (Wolf & Mastroianni, 1975;

332

H. Maathuis & R. Aitken, unpublished observations), baboons (Peplow et al., 1973) and roe deer (Aitken, 1974a,b). The biological properties of this protein are, as yet, unknown.

In addition to the roe deer, rat and mouse, biochemical evidence for an increase in uterine secretory activity at the end of embryonic diapause has been obtained in the fur seal and one species of marsupial, Macropus eugenii. In both species, low levels of protein were observed in the uterine flushings throughout diapause followed by a dramatic increase during the resumption of rapid embryonic growth (Daniel, 1971; Renfree, 1973).

The influence of the blastocyst on uterine secretory activity

Earlier in this chapter, embryo transfer experiments were cited (Dickman & De Feo, 1967), the results of which indicated that the rat blastocyst does not activate spontaneously at the end of diapause but only resumes normal embryonic development when permitted to do so by the uterus. Once activated, however, such blastocysts appear to be capable of inducing profound changes in endometrial function, particularly with respect to the magnitude of the secretory response. For example, the administration of oestradiol to ovariectomized pregnant and pseudopregnant mice maintained on progesterone, results in a biphasic uterine secretory response which is qualitatively identical in both groups of animals (Figures 19 and 20). Quantitatively, however, significantly more protein is secreted by the pregnant animals (Figures 17 and 18), presumably as a consequence of the presence of active blastocysts (R. Aitken, unpublished observations). Similarly, the quantitative changes taking place in the rat uterus following the termination of diapause by oestradiol are significantly increased by the presence of active blastocysts in the uterine lumen (Singh et al., 1973). Daniel (1972)

Figure 19: (opposite, upper) Polyacrylamide gel electrophoresis of mouse uterine flushings following the administration of 1 μg oestradiol-17β . OX = ovariectomized animals; ED = experimental delayed implantation; S = serum; 1, 4, 6, 12 and 40 = hours after oestradiol administration. Note the pronounced increase in the number and intensity of bands 4-6 hr post-injection, the general decline in protein concentration at 12 hr and the subsequent increase in non-serum proteins at 40 hr. At S, 4 and 6 hr, open arrows = serum slow α globulins (Ra 0.16 and Ra 0.20); at 40 hr, closed arrows = major non-serum slow α globulin at Ra 0.32.

Figure 20: (opposite, lower) Polyacrylamide gel electrophoresis of mouse uterine flushings following the administration of 1 μg oestradiol-17β to pseudopregnant mice in diapause. S = serum; ED = experimental delayed implantation; 2, 4, 6, 12, 24 and 40 = hours after oestradiol administration. Note the pronounced increase in the number and intensity of protein bands 4-6 hr post-injection, the general decline in protein concentration at 12 hr and the subsequent increase in non-serum proteins at 24 and 40 hrs. This pattern of protein secretion is identical to that observed in pregnant mice following the administration of oestradiol (Figure 19). At S, 4 and 6 hr, open arrows = serum slow α globulins (Ra 0.16 and 0.20), at 40 hr, closed arrow = major non-serum slow α globulin at Ra 0.32.

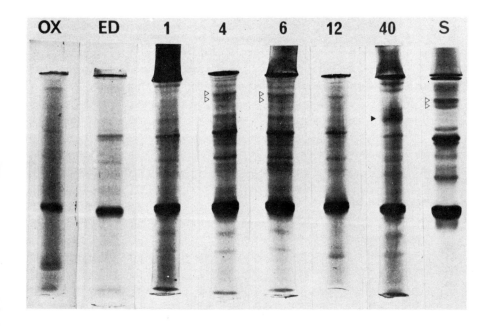

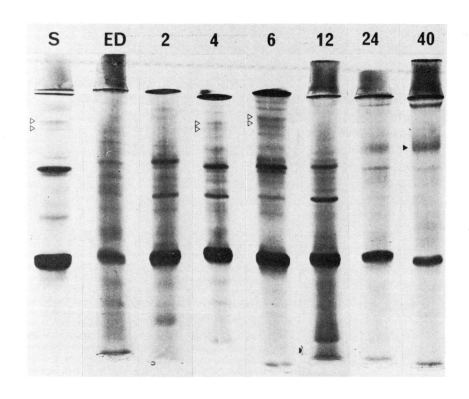

also found that the passage of proteins into the uterine lumen of the pregnant rabbit was localised in the vicinity of, and possibly induced by, the implanting blastocysts. While Renfree (1972) observed that the presence of a developing embryo in the uterus of Macropus eugenii resulted in a marked local stimulation of endometrial secretory activity. The embryonic factors responsible for stimulating the uterus are unknown but likely candidates are specific proteins (Jones et al., 1976) or oestrogenic steroids (Perry et al., 1973; Dey & Dickmann, 1974a,b) synthesized by the trophoblast. The facilitated passage of serum proteins into the mouse uterine lumen 6-12 hours after oestradiol injection could certainly be explained in terms of a local effect of embryonic oestrogens on capillary permeability (Ham et al., 1970). Embryonic oestrogens could also be responsible for enhancing uterine secretory activity. This concept is supported by observations on the pig blastocyst which is a well established source of oestrone and oestradiol-17β (Perry et al., 1973). Uterine secretory activity, comparable with that observed during the luteal phase of the oestrous cycle, can be induced in ovariectomized pigs by the simple administration of progesterone (Knight et al., 1973). If oestrogen is administered in addition to progesterone, however, the amount of protein secreted by the uterus is more than doubled. It is therefore possible that the elongating pig conceptus promotes its own survival by stimulating a hypersecretory state in the endometrium through the synthesis and release of oestrogenic hormones. The post-implantation (elongation) hypertrophy of the granular endoplasmic reticulum in the endometrial gland cells of the pig (Dempsey et al., 1955), sheep (Hoyes, 1972), mink (Enders et al., 1963) and roe deer (Aitken, 1975) appears to be induced by oestrogen (Hoyes, 1972; Aitken, 1975) and it is conceivable that the oestrogenic stimulus for this secretory activity is also derived from the blastocyst.

Surface properties of the endometrium

Throughout this review, emphasis has been placed on the relationship between the secretory state of the endometrium and the suppression or induction of blastocyst activity. In some species, however, the mere presence of embryotrophic secretions in the uterine lumen is not sufficient to bring about implantation, other conditions must also be satisfied. In the mouse, for example, the pro-oestrus stage of the oestrous cycle is associated with the accumulation of a considerable volume of protein-rich fluid in the uterine lumen (Figure 13). When mouse blastocysts are cultured in the presence of this fluid in vitro giant call transformation and outgrowth are induced (R. Aitken, unpublished observations) indicating that this secretory material is highly embryotrophic. However, when mouse blastocysts are transferred to pro-oestrous uteri they do not implant. Conversely, implantation can also be induced in the mouse in the virtual absence of uterine secretions. Such is the case when actinomycin D is administered during embryonic diapause. The intraperitoneal injection of 15 μg actinomycin D results in the gradual implantation of a small percentage of diapausing blastocysts (Finn, 1974; Finn & Downie, 1975; R. Aitken, unpublished observations), while luminal protein concentration is simultaneously depressed to extremely low levels (R. Aitken, unpublished observations) (Figure 21). A possible explanation for these observations is that normal implantation in the mouse is associated not only with an increase in uterine secretory activity, but also with a change in the surface properties of the endometrium permitting the close attachment of the blastocyst to the luminal epithelium (termed the 'attachment reaction' by Nilsson, 1970). With respect to the mechanism of action of actinomycin D, it is possible that this compound, by inhibiting the synthesis of a specific protein in the luminal epithelium

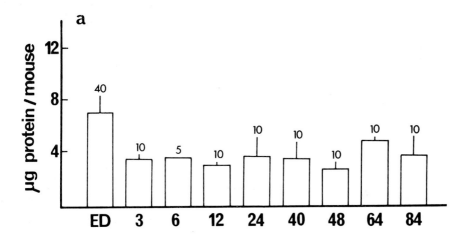

Figure 21: Changes in the protein content of mouse uterine flushings following the administration of 15 μg actinomycin D during diapause. ED = experimental delay; 3, 6, 12, 24, 40, 48, 64 and 84 = hours post drug administration; vertical bars = standard deviation; figures over bars = number of mice, 5 mice contributing to each sample. The administration of actinomycin D is followed by a highly significant fall (p < 0.001) in luminal protein concentration.

(Pollard et al., 1976) renders this cell layer susceptable to blastocyst attachment. In the absence of a simultaneous increase in uterine secretory activity, however, only a small number of blastocysts can take advantage of this surface change. Several other explanations for the implantation-inducing properties of actinomycin D are possible and the validity of the above argument will become apparent when a study is made of epithelial protein synthesis during the attachment phase of normal pregnancy. The work of Finn and Downie (1975) already suggests that the attachment response observed in actinomycin D treated animals differs from that induced by oestradiol-17β . In the presence of oestradiol, blastocyst attachment is accompanied by the close adhesion of opposing luminal surfaces elsewhere in the uterus (termed the 'second stage of closure' by Pollard & Finn, 1972). In actinomycin D treated animals, the second stage of lumen closure does not occur. The differentiation of the luminal epithelium for attachment may, therefore, involve changes other than the inhibited synthesis of certain specific proteins. In this respect, the electrochemical charge on the surface of the endometrium or the biochemical composition of the glycocalyx may also be important (Nilsson, 1974; Enders & Schlafke, 1974).

In addition to the mouse, an attachment reaction has also been observed during implantation in the rat, guinea pig and hamster (Nilsson, 1970). In all these species, the successful termination of diapause may involve changes in the surface properties of the luminal epithelium, as well as a general increase in endometrial secretory activity. In species, such as the mink (Nilsson, 1970) and roe deer where the attachment reaction is not observed, uterine secretory activity is presumably the most important factor in determining the time of implantation.

A combination of ultrastructural and biochemical techniques have thus been used to study endometrial function during diapause in a variety of species. In each of the animals so far studied in detail (roe deer, mink, armadillo, fur seal, mouse, rat and wallaby), the evidence points to a low level of uterine secretory activity during diapause which is dramatically increased during blastocyst reactivation and implantation. The magnitude and importance of the secretory changes observed at the end of diapause appear to depend upon the degree of embryonic growth preceding implantation in any given species. In the roe deer, for example, the termination of diapause is associated with the rapid elongation of the blastocyst from 5mm to 40cm before placental attachment takes place. Throughout this intense period of growth, the conceptus is entirely dependent upon embryotrophic material released into the uterine lumen from the endometrial glands and capillaries. It is not surprising, therefore, that the secretory changes observed at the end of diapause in this species are quite dramatic. Similarly, the termination of diapause in the macropod marsupial, Macropus eugenii, is associated with considerable preimplantation blastocyst expansion, from 0.25 to 15mm in diameter (Renfree & Tyndale-Biscoe, 1973) and the simultaneous initiation of intense endometrial secretory activity. In species such as the mouse and rat which do not exhibit a large degree of blastocyst expansion prior to attachment, the uterine secretory changes observed at the end of diapause are less marked.

In certain species the termination of diapause may involve changes in the surface properties of the endometrium as well as an increase in uterine secretory activity. This group of animals, which includes the rat and mouse, exhibit a characteristic 'attachment reaction' at the time of implantation as a result of which the surfaces of the trophoblast and endometrium become closely apposed.

THE OVARIES

In those mammalian species not exhibiting embryonic diapause, blastocyst implantation invariably occurs when the corpora lutea are maximally active and the endometrium has entered a progestational secretory stage. In these species, the ovarian steroid primarily responsible for the induction of implantation and the stimulation of uterine secretory activity is progesterone. In support of this concept, experimental evidence has been obtained in the guinea pig, hamster, rabbit, pig, (Nalbandov, 1971), sheep (Cumming et al., 1974) and ferret (Murphy & Mead, 1976), none of which exhibit embryonic diapause, indicating that implantation will take place in the presence of progesterone alone. In addition, studies in the rabbit (Urzua et al., 1970; Arthur & Daniel, 1972) and pig (Knight et al., 1973) indicate that the stimulation of uterine secretory activity in these animals is also under the control of progesterone.

The relationship between progesterone secretion and the induction of implantation does not, however, exist in at least one group of animals exhibiting embryonic diapause. This group, which includes the roe deer, rat, mouse and armadillo, possesses fully active corpora lutea during the delay phase. In these species, oestrogen as well as progesterone is required for the production of a fully secretory endometrium and implantation.

The weight, histological appearance and progesterone content (Short & Hay, 1966; Aitken, 1974a,b) of roe deer corpora lutea during diapause indicates that they are fully active at this time. In addition, direct measurements of plasma progesterone levels in roe deer, using a competitive protein binding assay, have revealed a mean concentration of 2.60 ± 0.66 ng/ml during delayed implantation. Following the resumption of rapid embryonic growth, the mean increased to 3.90 ± 1.85 ng/ml, a value which was not significantly different from the diapause level ($p > 0.05$) (Figure 22). When total unconjugated oestrogens were measured in these plasma samples, however, a significant ($p < 0.05$) increase in concentration was observed during embryonic elongation (Aitken, 1974a,b) (Figure 23). This circumstantial evidence suggests that in the roe deer an acceleration in the rate of ovarian oestrogen synthesis at the end of diapause may be responsible for the induction of a secretory state in the endometrium and, as a consequence, embryonic growth. However, histological examination of thirty-three serially sectioned roe deer ovaries failed to reveal any consistent change in either the number or size of ovarian follicles coincident with the termination of diapause (Aitken, 1974a). It is therefore possible that the elevated levels of plasma oestrogen observed at this time were derived from the elongating embryo itself, rather than the ovaries.

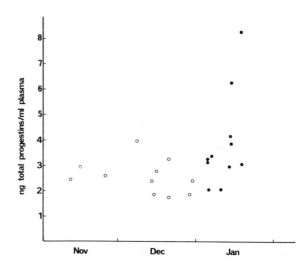

Figure 22: Changes in the plasma concentration of total progestins in roe deer during embryonic diapause and the phase of rapid embryonic growth: o = blastocyst in the uterus, ● = embryo in the uterus.

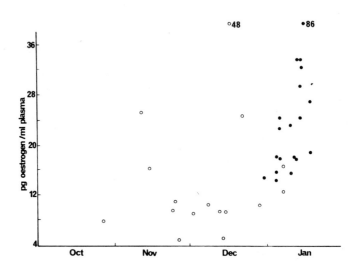

Figure 23: Changes in the plasma concentration of total unconjugated oestrogens in the roe deer during embryonic diapause and the phase of rapid embryonic growth: o = blastocyst in the uterus, ● = embryo in the uterus.

More direct experimental evidence for the participation of ovarian oestrogen in the stimulation of embryonic activity has been obtained in the rat and mouse.   In these species, ovariectomy during the first few days of pregnancy, followed by daily progesterone administration, leads to an artificially induced state of diapause which can be readily terminated by the administration of a minute dose of oestradiol-17β or oestrone (Yoshinaga, 1961; Yoshinaga & Adams, 1966; Nalbandov, 1971; Psychoyos, 1973).   The local implantation of blastocysts can even be induced by the direct application of a small quantity of oestradiol to the adipose tissue adjacent to the uterus (Yoshinaga, 1961).   Not only does the administration of exogenous oestrogen induce implantation in the rat and mouse, but it also stimulates the release of endometrial secretions into the uterine lumen (R. Aitken, unpublished observations; Surani, 1975).   Lactational embryonic diapause in the rat, and presumably the mouse, is also associated with high plasma levels of progesterone and low levels of oestrogen (K. Yoshinaga, unpublished observations; Tomogane et al., 1969; Yoshinaga, 1974).   Progesterone levels during lactational delay in the rat appear to be even higher than those observed during the early stages of normal pregnancy.   Hence, Yoshinaga (1974) found that ovarian progesterone secretion in rats suckling eight pups was 8.83 ± 1.59 $\mu g/15$ min., compared with a value of 4.56 ± 0.39 $\mu g/15$ min. for animals suckling two pups.   Since a minimum of 6-8 suckling young must be present before implantation is delayed in this species, these results clearly suggest that embryonic diapause is associated with intense luteal activity.   This, in turn, is known to be a consequence of the elevated prolactin levels observed in lactating rats (Yoshinaga, 1974; Ford & Yoshinaga, 1975; Tomogane et al., 1975). The elevated progesterone levels detected during delayed implantation may even play a causative role in the suppression of blastocyst activity since Trams et al. (1973) have found that progesterone inhibits oestrogen induced nucleic acid and protein synthesis by the rat uterus.   This concept is further supported by the fact that high levels of progestin (medroxyprogesterone acetate) administered on day 1 of normal pregnancy will induce delayed implantation in intact rats (Dickman, 1973).   The combined effect of low oestrogen and high progesterone levels during lactation in the rat must, therefore, act to ensure the thorough suppression of nucleic acid and protein synthesis in the uterus throughout diapause (Prasad et al., 1968).   If this balance is disturbed by the administration of exogenous oestrogen (Krehbiel, 1941; Weichert, 1942; McLaren, 1968; Psychoyos, 1973), synthetic activity in the uterus is resumed and the blastocysts are activated (Prasad et al., 1968; Psychoyos, 1973).

The armadillo also possesses an active corpus luteum throughout embryonic diapause (Talmage et al., 1954; Labhsetwar, 1967).   However, whether or not this species possesses an oestrogen requirement for implantation, as observed in rats and mice, is not known.   The armadillo appears to be unique in that bilateral

ovariectomy during the delay phase induces implantation within 30-34 days.
This rather enigmatic result suggests that the armadillo ovary in some way
suppresses embryonic development during diapause.   No information has yet been
obtained as to the nature of this inhibitory influence however.

All those members of the Carnivora and Marsupalia exhibiting delayed
implantation appear to possess inactive corpora lutea during diapause.   Histological evidence for luteal inactivity during delay has been obtained in bears
(Wimsatt, 1963), mustelids such as the badger (Neal & Harrison, 1958; Canivenc,
1960), mink (Hansson, 1947), stoat (Deanesly, 1935, 1943), long-tailed weasel
(Wright, 1963) American river otter (Lutra canadensis) (Hamilton & Eadie, 1964),
fisher (Martes pennanti) (Wright, 1963), wolverine (Gulo gulo) (Wright & Rausch,
1955) and American martin (Martes americana) (Wright, 1942).   Progesterone
determinations on the corpora lutea of the badger (Canivenc et al., 1966) and on
the peripheral plasma of the spotted skunk (Mead & Eik-Nes, 1969a) and mink
(Møller, 1973) during diapause have also provided biochemical evidence for luteal
inactivity at this time.   In all these species, full luteal function is resumed
shortly before implantation.   Reactivation involves a marked increase in the
size and vascularity of the corpora lutea, hypertrophy of the luteal cells and
the stimulation of progesterone secretion in the badger (Canivenc & Bonnin-
Laffargue, 1963; Canivenc et al., 1966), mink (Hansson, 1947; Enders, 1952; Møller,
1973) and spotted skunk (Mead & Eik-Nes, 1969a).   A detailed ultrastructural
study of the corpora lutea of the skunk during embryonic diapause has been carried
out by Sinha and Mead (1975).   These authors found that the corpora lutea of
diapause were composed of large numbers of small undifferentiated granulosa cells
interspersed with occasional large, differentiated, lutein cells containing large
quantities of smooth endoplasmic reticulum.   As the time of implantation
approached, the proportion of large lutein cells in the corpora lutea increased,
as did the concentration of progesterone in the peripheral plasma.   Animals in
the post-implantation phase were characterized by the presence of well
vascularised corpora lutea, composed entirely of hypertrophied luteal cells, and
plasma progesterone levels in excess of 20 ng/ml. These studies suggest that in
the mustelidae the termination of diapause is clearly associated with the
secretion of progesterone by the activated corpora lutea.   Whether there is also
an oestrogen requirement for the induction of implantation in these species is
not clear.   Mead and Eik Nes (1969b) have obtained some evidence to suggest that
plasma oestrogen levels are elevated at the time of blastocyst activation in the
skunk and Daniel (1974) detected a marked increase in plasma oestradiol-17β
concentration at the end of embryonic diapause in the fur seal.   It is therefore
possible that the oestrogen requirement for implantation extends to all species
exhibiting delayed implantation, not merely those with active corpora lutea during
the delay phase, such as the rat and mouse.   It must be emphasised, however, that

all attempts to precipitate premature implantation im mustelids by the injection
of progesterone and/or oestrogen during delayed implantation have failed (Hansson,
1947; Hammond, 1951; Enders, 1952; Cochrane & Shackelford, 1962; Holcomb, 1967,
for the mink; Canivenc & Laffargue, 1958, for the badger; Wright, 1963, for the
long-tailed weasel; Shelden, 1973, for the stoat).

In macropod marsupials, the corpora lutea of diapause are also
physiologically inactive as judged by their size, poor vascularity and lack of
mitotic activity (Sharman, 1963; Tyndale-Biscoe, 1963a; Sharman & Berger, 1969;
Tyndale-Biscoe et al., 1974). Follicular growth and development are also
suppressed at this time and in a majority of marsupials this appears to be due
to a direct inhibitory influence exerted by the quiescent corpora lutea (Tyndale-
Biscoe et al., 1974). The termination of diapause is associated with a marked
increase in luteal activity; the corpora lutea increase in size, become more
vascular and exhibit steroid dehydrogenase activity, while the luteal cells
divide and hypertrophy (Tyndale-Biscoe, 1963a; Sharman & Berger, 1969). Attempts
to induce the premature implantation of dormant marsupial blastocysts by the
exogenous administration of steroids has met with considerably more success than
that obtained with the Mustelidae. Progesterone administration to lactating
Megalei rufa (Clark, 1968), intact or ovariectomized Macropus eugenii (Berger &
Sharman, 1969) or ovariectomized Setonix brachyurus (Tyndale-Biscoe, 1963b)
successfully induced blastocyst activation, although in the latter two species,
pregnancy was not maintained. Oestradiol injections have also been found to
terminate embryonic diapause in intact and ovariectomized Macropus eugenii (Smith
& Sharman, 1969) and intact Megaleia rufa (Clark, 1968), although the efficacy
of this treatment was low in terms of both the percentage of embryos developing
and the normality of their development. It has been suggested (Tyndale-Biscoe
et al., 1974) that the ability of oestradiol to terminate diapause is achieved
indirectly via a luteotrophic effect on the corpora lutea, although this could
not be the mechanism of action in ovariectomized Macropus eugenii.

Another intriguing question which has only been partially resolved concerns
the extent to which the viability of diapausing blastocysts is dependent on
ovarian support. In macropod marsupials, blastocyst survival does not appear
to depend upon the presence of ovarian steroids since apparently normal blastocysts have been recovered as long as one and four months after ovariectomy
during diapause in Setonix bachyurus (Tyndale-Biscoe, 1963b) and Macropus
eugenii (Tyndale-Biscoe et al., 1974) respectively. Similarly, rat and mouse
blastocysts remain viable for long periods in ovariectomized hosts, although
survivability is improved by the administration of progesterone (Mayer, 1963;
Dickman, 1968; Weitlauf & Greenwald, 1968; Buchanan, 1969; McLaren, 1971;
Psychoyos, 1973). Within the Mustelidae there appears to be some variation in
the ability of blastocysts to survive ovariectomy during diapause. In general,

the blastocysts appear to survive this treatment extremely well, remaining intact up to eight or ten weeks in the spotted skunk (Mead, 1975) and European badger (Neal & Harrison, 1958) respectively.   In the short-tailed weasel (stoat), ovariectomy during diapause leads to blastocyst expansion and the resumption of mitosis in the inner cell mass and trophoblast, a result which is reminiscent of the implantation-inducing effect of bilateral ovariectomy in the armadillo (Shelden, 1972).   The stimulation of mitosis in the stoat blastocyst is not followed by normal embryonic development, however, and within four weeks the blastocysts degenerate.

In conclusion of this section then, the delay phase of species, such as the roe deer, rat, mouse and armadillo is associated with the presence of fully active secretory corpora lutea in the ovaries.   In the roe deer, rat and mouse, there is a substantial body of evidence relating the termination of diapause to an elevation of plasma oestrogen levels which, in the rat and mouse, at least, is of ovarian origin.

All the other mammalian groups exhibiting diapause, including the mustelids, seals, bears and macropod marsupials, possess inactive corpora lutea in their ovaries during the delay phase.   The termination of diapause in these species is preceded by marked increases in luteal size, vascularity and progesterone pro-duction.   In the fur seal and skunk, there is also some evidence for an increase in oestrogen secretion by the ovaries just prior to blastocyst reactivation and implantation.   Despite this apparent requirement for progesterone and possibly oestrogen in order to terminate diapause, the exogenous administration of these steroids, alone or in combination, during delay, has failed to induce premature implantation in any Eutherian mammal so far investigated.   In the Metatherian macropod marsupials, however, the exogenous administration of either progesterone or oestrogen during delay is effective in activating the blastocysts.

The blastocysts of all animals exhibiting embryonic diapause exhibit a certain degree of autonomy during the delay phase in that they will survive for several weeks following ovariectomy.   The single fascinating exception to this rule is the armadillo in which bilateral ovariectomy during diapause results in blastocyst activation and implantation.

THE PITUITARY

The luteal inactivity associated with embryonic diapause in mustelids, bears, bats and seals, appears to be associated with a deficiency in pituitary hormone secretion.   In the mink (Baevsky, 1964; Murphy, 1972), badger (Herland & Canivenc, 1960) and long-winged bat (Peyre & Herlant, 1963), for example, histochemical studies have revealed a marked paucity of gonadotrophin secreting cells in the pituitary gland during the delay phase and a subsequent increase in number at the time of blastocyst reactivation.   Similarly, in the spotted skunk (Foresman &

Mead, 1974), actual measurements of plasma LH concentration revealed a low level of hormone secretion mid-way through embryonic diapause. As the time of implantation approached, however, the plasma concentration of LH gradually increased, reaching a mean of 7.9 ng/ml during the resumption of blastocyst activity. This progressive rise in plasma LH levels did not appear to be associated with changes in either the pituitary responsiveness to LHRF (Foresman & Mead, 1973) or the metabolic clearance rate for LH. This suggested that an increase in the rate of LH secretion was involved although serial bleeds on single animals did not reveal any significant differences in either the frequency or the amplitude of LH peaks during the transition from diapause to implantation. There can be no doubt, however, that the pituitary gland does play an important role in the induction of blastocyst activity in this species, since diapause is indefinitely prolonged when skunks are hypophysectomized during the delay period (Mead, 1975). Although it is evident from these studies that the reactivation of the corpora lutea at the end of diapause involves a change in pituitary activity, it is not clear whether the corpora lutea of diapause are dependent on a low level of gonadotrophin support or are completely autonomous. Hypophysectomy of the spotted skunk during delayed implantation led to a slight decline in plasma progesterone levels without influencing the histological appearance of the corpora lutea.

In at least one species of macropod marsupial (M. eugenii), the termination of diapause is associated with the removal of an inhibitory factor secreted by the pituitary gland rather than a quantitative increase in gonadotrophin secretion (Hearn, 1973). In this species, a lack of pituitary involvement in the termination of diapause is suggested by the fact that plasma gonadotrophin levels remain unchanged throughout embryonic dormancy and subsequent implantation. When the pituitary gland is removed during diapause, however, the quiescent corpora lutea are activated and the blastocysts implant. Although this result indicates that the pituitary gland is responsible for suppressing luteal development during the delay phase, nothing is yet known of the nature of the inhibitory substance (Tyndale-Biscoe et al., 1974).

Of those species possessing active corpora lutea during the delay phase, the rat appears to be the best studied. Lactational delayed implantation in this species is associated with the inhibition of both FSH and LH secretion by the pituitary gland and the presence of extremely high levels of prolactin (Amenomori et al., 1970; Ford & Melampy, 1973; Simpson et al., 1973; Lu et al., 1976). The high plasma levels of progesterone observed during delayed implantation in the rat appear to be a consequence of the intense secretion of prolactin at this time. Accordingly, the suppression of prolactin production with such drugs as ergocornine mesylate (Ford & Yoshinaga, 1975) and ergocornine maleate (Tomogane et al., 1975) or the administration of an anti-prolactin antiserum

(Yoshinaga, 1974) leads to a concomitant fall in progesterone levels. The low levels of oestrogen observed during diapause are, in turn, thought to be a consequence of the inhibited secretion of FSH and LH observed at this time (Ford & Melampy, 1973).

Prolactin secretion during lactational delay is a direct function of the number of suckling young (Yoshinaga, 1974), the suckling stimulus serving to inhibit the release of a prolactin-inhibiting factor from the hypothalamus. The inhibition of FSH and LH secretion during lactation is also directly related to litter size (Ford & Melampy, 1973). In the case of FSH secretion, the inhibitory effect exerted by the suckling stimulus is, at least partly, mediated by the enhanced secretion of prolactin. Hence the administration of an agent (ergocornine methanesulphonate) to specifically block prolactin secretion in lactating rats results in an increase in the plasma concentration of FSH (Lu et al., 1976). The administration of a similar compound (2-Br α-ergocryptine) to women during the puerperium also results in a marked stimulation of FSH secretion by the pituitary (Seki et al., 1974). It is possible, however, that prolactin does not inhibit FSH secretion directly but does so via the stimulation of ovarian progesterone secretion. This would explain why ovariectomy during diapause leads to a highly significant increase in plasma FSH levels (Ford & Melampy, 1973). A direct effect of suckling on pituitary FSH secretion must also exist, however, since increasing the size of the litter suckled by ovariectomized lactating rats leads to a progressive reduction in plasma FSH levels (Ford & Melampy, 1973). A direct inhibitory effect of suckling on pituitary LH secretion is also suggested by the fact that intact and ovariectomized rats suckling large litters of six or twelve young exhibit a significantly lower plasma concentration of LH than rats suckling three or no pups (Ford & Melampy, 1973). In rats suckling sufficient young to induce delayed implantation, i.e. more than six, prolactin plays no part in the suppression of LH secretion since the administration of ergocornine methanesulphonate to such animals has no effect on plasma LH levels (Lu et al., 1976). In the same way, progesterone cannot be responsible for suppressing LH secretion in rats suckling large litters of six or twelve pups since ovariectomizing such animals during lactation does not elevate plasma LH levels (Ford & Melampy, 1973). When litter size is small, however, ovariectomy does stimulate a marked increase in plasma LH levels (Ford & Melampy, 1973) suggesting that under these conditions, progesterone may be responsible for suppressing pituitary LH release. The general picture that emerges from these studies is that the low level of FSH and LH secretion observed during lactation in rats is due to the combined suppressive effects of (a) the suckling stimulus and (b) prolactin secretion, the latter possibly acting through the stimulation of ovarian progesterone production. At litter sizes sufficient to induce delayed implantation, FSH secretion appears to be inhibited by the high levels of

prolactin (or progesterone), while LH secretion is suppressed as a direct result
of the suckling stimulus.   When litter size is three or less, prolactin (or
progesterone) appears to be primarily responsible for inhibiting the secretion
of both gonadotrophins.   The ability of prolactin to suppress FSH and LH in the
presence of a sub optimal suckling stimulus is indicated by the fact that rats
suckling two young and given exogenous prolactin enter a state of delayed
implantation, whereas rats suckling two young per se exhibit implantation at the
normal time.   Prolactin injections will not induce delayed implantation when
administered to non suckling rats, however (Maneckjee & Moudgal, 1975).

In rats nursing large litters and experiencing embryonic diapause,
implantation can be rapidly induced by removing the suckling young (Maneckjee &
Moudgal, 1975).   Since plasma FSH and LH levels are elevated within twelve hours
of litter removal (Ford & Melampy, 1973), these gonadotrophins are presumed to
play an important part in blastocyst reactivation at the end of diapause,
presumably through an effect on ovarian oestrogen synthesis.   The fact that
normal implantation can be blocked in rats by the administration of an antiserum
to ovine LH, also suggests that this gonadotrophin is involved in the induction
of implantation (Maneckjee & Moudgal, 1975).   Although a gonadotrophin-induced
rise in ovarian oestrogen production is an extremely plausible explanation for
the hormonal events leading to the termination of embryonic diapause in the rat,
it is not entirely consistent with the following observations.   In rats suckling
eight young, nidation is delayed so that implantation sites are only visible in
the uterus on day 12, not day 8, as during normal pregnancy (Maneckjee & Moudgal,
1975).   When prolactin, FSH and LH levels were measured in such animals, however,
no significant changes were observed between days 11 and 21 of pregnancy that
could account for the termination of diapause and the initiation of implantation
(Lu et al., 1976).   LH and FSH were consistently low throughout this period
while prolactin levels were high.   In addition, Maneckjee and Moudgal (1975)
were unable to bring about a premature end to lactational delayed implantation in
the rat by the administration of ovine LH, ovine FSH or a combination of these
hormones.   LH administration was also ineffective in terminating embryonic
diapause in rats suckling two young and given exogenous prolactin treatment.
These observations can, of course, be criticised on the grounds that, (a) single
point determinations of gonadotrophin levels do not take into account the
episodic release pattern of these hormones and (b) the biological potency of
ovine LH and FSH in the rat, is not certain.   Nevertheless, they may serve as a
warning against an oversimplified explanation of pituitary involvement in delayed
implantation.

Little data is available on the patterns of gonadotrophin secretion during
delayed implantation in the mouse.   Bindon (1970) observed high levels of FSH
in the mouse pituitary during lactational delay which fell within two hours of

litter removal, suggesting the sudden release of this hormone.   Similarly, the induction of implantation in the armadillo is associated with a sudden fall in the pituitary concentration of LH (Labhsetwar & Enders, 1969), once again suggesting that gonadotrophin release is involved in the termination of diapause. In the roe deer, however, a preliminary study (Aitken, 1974a) of pituitary function during delayed implantation did not reveal any marked changes in either the weight of the gland or the plasma levels of LH (Figures 24 and 25).

A general review of the data relating to pituitary function in Eutherian mammals during delayed implantation suggests a marked inhibition of gonado-trophin secretion during this phase.   This does not appear to be true of the roe deer, however, or Macropus eugenii, the only Metatherian species so far studied. In the latter, the pituitary gland appears to actively suppress luteal activity throughout diapause since hypophysectomy during this phase leads to the acti-vation of the corpus luteum and the resumption of embryonic growth.

ENVIRONMENTAL FACTORS AND THE PINEAL GLAND

A striking feature of seasonal delayed implantation is the way in which embryonic development is invariably resumed at a precise time of the year, despite very long and sometimes variable periods of diapause.   The role of environmental stimuli, particularly photoperiod, in synchronizing the termination of diapause has been demonstrated in several species.   In a majority of the Mustelidae, for example, blastocyst reactivation occurs in the spring when daylength is increasing.   The importance of increased daylight is indicated by the fact that premature implantation can be induced in animals given supplementary illumination during the delay phase.   This effect has been observed in the pine marten (Pearson & Enders, 1944), long-tailed weasel (Wright, 1963), mink (Pearson & Enders, 1944; Holcomb et al., 1962) and sable (Martes zibellina) (Belyaev et al., 1951).   Similarly, in the skunk (Mead, 1971) an artificial increase in photo-period during diapause induced premature activation of the corpora lutea and sub-sequent implantation;   an effect which was not observed in blinded animals.   The badger is unusual in that delayed implantation is terminated in December when daylength is decreasing.   In this case, premature implantation was induced by artificially decreasing daylength and ambient temperature during diapause (Canivenc et al., 1971).

Delayed implantation in the roe deer is usually terminated in late December or early January when daylength is increasing.   There is some doubt concerning the role of environmental stimuli in controlling embryonic diapause in this species however.   Hence, Lincoln and Guiness (1972) were unable to influence the gestation period of a pregnant roe doe by manipulating daylength during the early stages of delayed implantation.   In addition, the post mortem examination of 161 does from 1970 and 1973 revealed the presence of elongating embryos in three

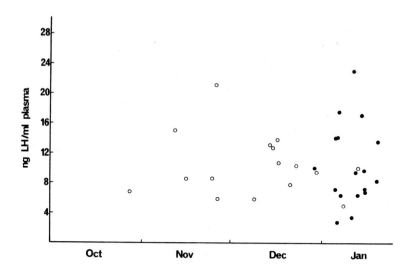

Figure 24: Changes in the plasma concentration of immunoreactive LH in roe deer during embryonic diapause and the phase of rapid embryonic growth: o = blastocyst in the uterus, ● = embryo in the uterus. This assay was performed using an equine anti-ovine LH antiserum which had been absorbed with TSH. Labelled antiserum was prepared by iodinating ovine pituitary LH (LER 1056/C$_2$) with ^{125}I, while NIH-LH-S15 was used in the preparation of the standard curves. Parallelism was observed between serial dilutions of roe deer plasma, anterior pituitary homogenate and the standard curve.

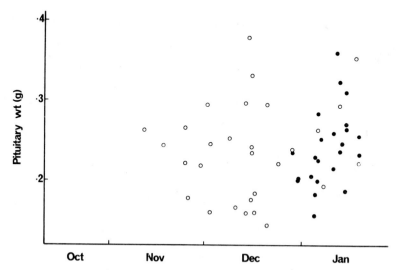

Figure 25: Changes in the pituitary weight of roe deer during embryonic diapause and the phase of rapid embryonic growth: o = blastocyst in the uterus, ● = embryo in the uterus.

animals before the winter solstice (December 22), when daylength was still decreasing.

No attempts have been made to curtail the length of seasonal embryonic diapause in the macropod marsupials by manipulating environmental stimuli. However, the fact that blastocyst activation in Macropus eugenii occurs within 24 hours of the longest day (Moore, 1976) suggests that photoperiod may also be an important factor in terminating the seasonal diapause exhibited by this species.

The photoperiodic control of reproductive function appears to be mediated by the pineal gland (Wurtman et al., 1968). The physiological function of the rat pineal gland has been found to depend upon the synthesis of a hormone, melatonin, through the activity of a unique, tissue specific, enzyme, hydroxyindole-O-methyltransferase (HIOMT). The synthesis of melatonin is stimulated by darkness and its action is inhibitory; administration of the hormone being followed by a decrease in the incidence of oestrus, reduced ovarian weight and the inhibition of LH secretion (Wurtman et al., 1963a; Wurtman et al., 1963b; Fraschini et al., 1968). Changes in photoperiod are evidently communicated to the pineal gland via the eyes and sympathetic nervous system since diurnal fluctuations in pineal weight and HIOMT activity (Axelrod et al., 1965) and the responses to constant light or darkness are abolished by blinding or superior cervical ganglionectomy (Wurtman et al., 1964). In view of the demonstrated importance of photoperiod in controlling the duration of seasonal delayed implantation in certain species, several workers have attempted to determine the role of the pineal gland in mediating this control. In the spotted skunk, Mead (1972) found that although photoperiod and blinding could exert profound effects on the duration of diapause, pinealectomy had no influence. Similarly, a morphological and histological study of the roe deer pineal (Aitken, 1974a) did not reveal any significant changes in pineal weight or pinealocyte cell size during the delay period. In the macropod marsupial, Macropus eugenii, however, preliminary evidence has been obtained to suggest that HIOMT activity in the pineal gland is lowest at the time of blastocyst activation (Kennaway & Seamark, 1976).

Thus, in a majority of species exhibiting the seasonal form of delayed implantation, environmental factors, particularly light and temperature, appear to play a major role in the induction of blastocyst activity. The role of the pineal gland in mediating changes in photoperiod is still open to question, positive results having been obtained only in the marsupial Macropus eugenii.

GENERAL CONCLUSIONS

During embryonic diapause, the blastocysts of all species investigated, enter a state of metabolic and developmental arrest in association with the

simultaneous inhibition of endometrial secretory activity.   A causal relationship between the lack of secretory material within the uterine lumen and the
inhibition of embryonic growth is suggested by in vitro experiments employing
mouse blastocysts.   If active day 3 blastocysts are cultured in media deprived
of serum, certain amino acids or glucose, they enter a state of diapause closely
resembling that observed in vivo (Gwatkin, 1966a,b; McLaren, 1973; Nilsson,
1974).   The ability of blastocysts to reduce their rate of metabolism is
presumably confined to those species exhibiting embryonic diapause, although
hard data are lacking on this point.   It would be of great interest, for
example, to determine whether blastocysts obtained from those subspecies of
spotted skunk not exhibiting delayed implantation (e.g. S.p. interrupta, S.p.
ambarvalis and S.p. putorius) become dormant when transferred to the uteri of
subspecies exhibiting the phenomenon (e.g. S.p. leucoparia, S.p. latifrons and
S.p. gracilis).

Blastocyst reactivation at the end of diapause appears to be induced by a
sudden increase in uterine secretory activity.   A cause-effect relationship
between the availability of this secretory material and the termination of
diapause is suggested by the fact that dormant mouse blastocysts can be activated in vitro by the addition of a non-specific embryotrophic substrate, such
as foetal bovine serum to the culture medium (McLaren, 1973).   The characteristics of the activation process appear to be identical, regardless of whether
it is induced by serum in vitro or oestradiol-17β in vivo.   In both instances
the blastocyst responds with an increase in the rates of nucleic acid synthesis,
carbon dioxide production and cell division, followed by the differentiation of
giant trophoblast cells (McLaren, 1973).

In addition to an increase in uterine secretory activity, the termination
of diapause may also involve a change in the surface properties of the endometrium.   This requirement is thought to apply particularly to those species
exhibiting an 'attachment reaction' at the time of implantation, such as the rat
and mouse (Nilsson, 1970).   The exact nature of these surface changes is
unknown, but an alteration in the pattern of protein synthesis exhibited by the
luminal epithelium, the biochemical composition of the glycocalyx or the electrochemical charge on the surface of the endometrium may be involved.

The endometrial events responsible for blastocyst activation and
implantation are, in turn, induced by alterations in the pattern of ovarian
steroid biosynthesis.   In those mammals not exhibiting embryonic diapause,
implantation takes place in the presence of progesterone alone.   In the roe deer,
rat, mouse and armadillo, however, implantation is delayed despite the presence
of fully active corpora lutea in the ovaries.   In these species, oestrogen as
well as progesterone is required for the induction of blastocyst activity, the
timing of implantation being determined by controlled release of oestrogen.

It must be emphasized that this oestrogen requirement is a unique feature which has evolved in association with the diapause phenomenon. For this reason, the rat and mouse are particularly unsuitable as animal models for contraceptive research involving the implantation process.

In the Carnivora and Marsupalia, embryonic diapause is associated with a period of luteal inactivity. In these species, the eventual reactivation of the blastocyst is preceded by an increase in luteal size, vascularity and progesterone production. Whether or not this group of animals also exhibits an oestrogen requirement for implantation is unknown. Blastocyst reactivation is associated with an increase in plasma oestrogen levels in the spotted skunk and fur seal, while implantation has been induced in several marsupials by the exogenous administration of either oestrogen or progesterone during diapause. Curiously, all attempts to induce premature implantation in mustelids with exogenous steroids have failed.

Very different roles are played by the pituitary gland in controlling embryonic diapause in Metatherian and Eutherian mammals. In the only Metatherian species studied, Macropus eugenii, the pituitary gland is thought to actively suppress luteal activity throughout the dormant phase since hypophysectomy at this time results in the sudden termination of diapause. The nature of the inhibitory factor has not yet been elucidated, but prolactin and oxytocin are possible candidates. In most Eutherian mammals, diapause is associated with the inhibition of pituitary gonadotrophin secretion. Hence, plasma LH levels are depressed during diapause in the spotted skunk and a paucity of gonadotrophin secreting cells has been observed in the pituitary gland of the badger, mink and long-winged bat at this time. The best studied species in this respect is the rat. Lactational delayed implantation in this species is associated with low plasma levels of LH and FSH and a high concentration of prolactin. With litter sizes large enough to induce embryonic diapause (i.e. ≥ 6 young), the suckling stimulus is directly responsible for the inhibition of LH and the stimulation of prolactin secretion, while the production of FSH is hormonally suppressed by prolactin or, more probably, prolactin-induced progesterone synthesis.

In animals exhibiting the seasonal form of delayed implantation, gonado-trophin secretion is probably regulated by environmental factors, such as temperature or photoperiod. Accordingly, manipulation of these factors during the delay phase has been found to induce premature implantation in a variety of animals, such as the mink, badger and pine marten. Since delayed implantation in the roe deer is terminated within days of the winter solstice, environmental factors may also play a role in controlling embryonic diapause in this species. However, preliminary studies have not only failed to detect any marked changes in gonadotrophin secretion during the termination of diapause but have also failed to influence the length of gestation by the experimental modification of daylength. It is, therefore, possible that neither environmental nor pituitary factors play an important part in controlling the duration of diapause in this species. This possibility illustrates one of the most striking features of embryonic diapause - the great diversity of physiological mechanisms that have been employed by different species to achieve the same end, the prolonged inhibition of embryonic growth.

REFERENCES

AITKEN, R.J. (1974a) Delayed implantation in the roe deer (Capreolus capreolus). Ph.D. Thesis, University of Cambridge.

AITKEN, R.J. (1974b) Delayed implantation in roe deer (Capreolus capreolus). J. Reprod. Fert. 39, 225-233.

AITKEN, R.J. (1974c) Calcium and zinc in the endometrium and uterine flushings of the roe deer (Capreolus capreolus) during delayed implantation. J. Reprod. Fert. 40, 333-340.

AITKEN, R.J. (1975) Ultrastructure of the blastocyst and endometrium of the roe deer (Capreolus capreolus) during delayed implantation. J. Anat. 119, 369-384.

AITKEN, R.J. (1976) Uterine secretion of fructose in the roe deer. J. Reprod. Fert. 46, 439-440.

AITKEN, R.J., BURTON, J., HAWKINS, J., KERR-WILSON, R., SHORT, R.V. & STEVEN, D.H. (1973) Histological and ultrastructural changes in the blastocyst and reproductive tract of the roe deer, Capreolus capreolus, during delayed implantation. J. Reprod. Fert. 34, 481-493.

AMENOMORI, Y., CHEN, C.L. & MEITES, J. (1970) Serum prolactin levels in rats during different reproductive states. Endocrinology 86, 506-510.

ANON (1825) Cited in: H. Prell (1938) Die Tragzeit des Rehes. Zuechtungskunde 13, 325-340.

ARTHUR, A.T. & DANIEL, J.C. Jr. (1972) Progesterone regulation of blastokinin production and maintenance of rabbit blastocysts transferred into uteri or castrate recipients. Fert. Steril. 23, 115-122.

AXELROD, J., WURTMAN, R.J. & SNYDER, S.H. (1965) Control of hydroxyindole-O-methyltransferase activity in the rat pineal gland by environmental lighting. J. Biol. Chem. 240, 949-954.

BAEVSKY, U.B. (1963) The effect of embryonic diapause on the nuclei and mitotic activity of mink and rat blastocysts. In: A.C. Enders (Ed.). Delayed Implantation. University of Chicago Press, Chicago, pp. 141-154.

BAEVSKY, U.B. (1964) Changes in the anterior lobe of the hypophysis, corpora lutea of pregnancy, and the thyroid in the mink (Mustela vison) associated with implantation of embryos. USSR Acad. Sci. 157, 522-525.

BELYAEV, D.K., PERELDIK, N.S. & PORTNOVA, N.T. (1951) Experimental reduction of the period of embryonal development in sables (Martes zibellina). J. Gen. Biol. 12, 260-265.

BERGER, P.J. & SHARMAN, G.B. (1969) Progesterone-induced development of dormant blastocysts in the tammar wallaby, Macropus eugenii Desmarest; Marsupialia. J. Reprod. Fert. 20, 201-210.

BERGSTROM, S. (1972) Delay of implantation by ovariectomy or lactation. A scanning electron microscope study. Fert. Steril. 23, 548-561.

BERGSTROM, S. & NILSSON, O. (1973) Various types of embryo-endometrial contacts during delay of implantation in the mouse. J. Reprod. Fert. 32, 531-533.

BINDON, B.M. (1970) Preimplantation changes after litter removal from suckling mice. J. Endocr. 46, 511-516.

BISCHOFF, T.L.W. (1854) Entwickelungsgeschichte des Rehes. J. Ricker'sche Buchhandling, Giessen, Germany.

BLAND, K.P. (1969) Reproduction in the female Indian gerbil (Tatera indica). J. Zool. Lond. 157, 47-61.

BOWMAN, P. & McLAREN, A. (1970) Cleavage rate of mouse ova in vivo and in vitro. J. Embryol. exp. Morph. 24, 203-207.

BRAMBELL, F.W.R. & ROWLANDS, I.W. (1936) Reproduction of the bank vole (Evotomys glareolus, Schreber) 1. The oestrous cycle of the female. Phil. Trans. R. Soc. B. 226, 71-97.

BUCHANAN, G.D. (1969) Blastocyst survival in ovariectomised rats. J. Reprod. Fert. 19, 279-284.

CANIVENC, R. (1960) L'ovo-implantation differee des animaux sauvages. In: G. Masson (Ed.), Les fonctions de nidation uterine et leurs troubles, Masson, Paris, pp. 33-68.

CANIVENC, R. & BONNIN-LAFFARGUE, M. (1963) Inventory of problems raised by the delayed ova implantation in the European badger (Meles meles L.). In: A.C. Enders (Ed.), Delayed implantation, University of Chicago Press, Chicago, pp. 115-128.

CANIVENC, R. & LAFFARGUE, M. (1958) Action de differents equilibres hormonaux sur la phase de vie libre de l'oeuf feconde chez le blaireau europeen (Meles meles L.). C.R. Seanc. Soc. Biol. (Paris) 152, 58.

CANIVENC, R., SHORT, R.V. & BONNIN-LAFFARGUE, M. (1966) Etude histologique et biochimique du corps jaune du blaireau europeen (Meles meles L.). Annls. Endocr. (Paris) 27, 401-413.

CANIVENC, R., BONNIN-LAFFARGUE, M. & LAJUS-BOUE, M. (1971) Realisation experimentale precoce de l'ovo implantation chez le Blaireau europeen (Meles meles L.) pendant la periode de latence blastocystaire. C.R. Acad. Sci. (Paris) 273, 1855-1856.

CLARK, M.J. (1966) Blastocyst of the red kangaroo during diapause. Aust. J. Zool. 14, 19-25.

CLARK, M.J. (1968) Termination of embryonic diapause in the red kangaroo, Megaleia rufa, by injection of progesterone or oestrogen. J. Reprod. Fert. 15, 347-355.

CLARK, M.J. & POOLE, W.E. (1967) The reproductive system and embryonic diapause in the female grey kangaroo, Macropus giganteus. Aust. J. Zool. 15, 441-459.

COCHRANE, R.L. & SHACKELFORD, R.M. (1962) Effects of exogenous oestrogen alone and in combination with progesterone on pregnancy in the intact mink. J. Endocr. 25, 101-106.

CUMMING, I.A., BAXTER, R. & LAWSON, R.A.S. (1974) Steroid hormone requirements for the maintenance of early pregnancy in sheep: a study using ovariectomized, adrenalectomized ewes. J. Reprod. Fert. 40, 443-446.

DANIEL, J.C. Jr. (1971) Growth of the preimplantation embryo of the northern fur seal and its correlation with changes in uterine protein. Devl. Biol. 26, 316-332.

DANIEL, J.C. Jr. (1972) Local production of protein during implantation in the rabbit. J. Reprod. Fert. 31, 303-306.

DANIEL, J.C. (1974) Circulating levels of oestradiol-17β during early pregnancy in the Alaskan fur seal showing an oestrogen surge preceding implantation. J. Reprod. Fert. 37, 425-428.

DASS, C.M., MOHLA, S. & PRASAD, M.R. (1969) Time sequence of action of estrogen on nucleic acid and protein synthesis in the uterus and blastocyst during delayed implantation in the rat. Endocrinology, 85, 528-536.

DEANESLY, R. (1935) The reproductive processes in some mammals. Part IX. Growth and reproduction in the stoat (Mustela mustela). Phil. Trans. R. Soc. B. 225, 459-492.

DEANESLY, R. (1943) Delayed implantation in the stoat (Mustela mustela). Nature 151, 365.

DEMPSEY, E.W., WISLOCKI, G.B. & AMOROSO, E.C. (1955) Electron microscopy of the pigs placenta with especial reference to the cell-membranes of the endometrium and chorion. Am. J. Anat. 96, 65-101.

DEY, S.K. & DICKMAN, Z. (1974a) Δ^5-3β -hydroxysteroid dehydrogenase activity in rat embryos on days 1 through 7 of pregnancy. Endocrinology 95, 321-322.

DEY, S.K. & DICKMAN, Z. (1974b) Estradiol-17 -hydroxysteroid dehydrogenase activity in preimplantation rat embryos. Steroids 24, 57-62.

DICKMAN, Z. (1968) Can the rat blastocyst survive in the absence of stimulation by the ovarian hormones? J. Endocrin. 42, 605-606.

DICKMAN, Z. (1973) Post coital contraceptive effects of medroxyprogesterone acetate and oestrone in rats. J. Reprod. Fert. 32, 65-69.

DICKMAN, Z. & DE FEO, V.J. (1967) The rat blastocyst during normal pregnancy and during delayed implantation including an observation on the shedding of the zona pellucida. J. Reprod. Fert. 13, 3-9.

DICKSON, A.D. (1966) Form of the mouse blastocyst. J. Anat. 100, 335-348.

DOUGLAS, C.P., GARROW, J.S. & PUGH, E.W. (1970) Sugar content of endometrial secretion. J. Obstet. Gynaec. Br. Commonw. 77, 891-894.

EALEY, E.H.M. (1963) The ecological significance of delayed implantation in a population of the hill kangaroo (Macropus robustus). In: A.C. Enders (Ed.), Delayed Implantation, University of Chicago Press, Chicago, pp. 33-48.

ENDERS, A.C. (1961) Comparative studies on the endometrium of delayed implantation. Anat. Rec. 139, 483-497.

ENDERS, A.C. (1962) The structure of the armadillo blastocyst. J. Anat. 96, 39-48.

ENDERS, A.C. (1967) Uterus in delayed implantation. In: R.M. Wynn (Ed.), Cellular Biology of the Uterus, Meredith, Amsterdam, pp. 151-190.

ENDERS, A.C. (1971) The fine structure of the blastocyst. In: R.J. Blandau (Ed.), The Biology of the Blastocyst, University of Chicago Press, Chicago, pp. 71-94.

ENDERS, A.C. & SCHLAFKE, S. (1974) Surface coats of the mouse blastocyst and uterus during the preimplantation period. Anat. Rec. 180, 31-46.

ENDERS, A.C., ENDERS, R.K. & SCHLAFKE, S.J. (1963) An electron microscope study of the gland cells of the mink endometrium. J. Cell. Biol. 18, 405-418.

ENDERS, R.K. (1952) Reproduction in the mink (Mustela vison). Proc. Am. Phil. Soc. 96, 691-755.

ENDERS, R.K. & ENDERS, A.C. (1963) Morphology of the female reproductive tract during delayed implantation in the mink. In: A.C. Enders (Ed.), Delayed Implantation, University of Chicago Press, Chicago, pp. 129-139.

FINN, C.A. (1974) The induction of implantation in mice by actinomycin D. J. Endocr. 60, 199-200.

FINN, C.A. & DOWNIE, J.M. (1975) Changes in the endometrium of mice after the induction of implantation by actinomycin D. J. Endocr. 65, 259-264.

354

FINN, C.A. & MARTIN, L. (1974) The control of implantation. J. Reprod. Fert. 39, 195-206.

FORD, J.J. & MELAMPY, R.M. (1973) Gonadotrophin levels in lactating rats: effect of ovariectomy. Endocrinology 93, 540-547.

FORD, J.J. & YOSHINAGA, K. (1975) The role of prolactin in the luteotrophic process of lactating rats. Endocrinology 96, 335-339.

FORESMAN, K.R. & MEAD, R.A. (1973) Luteinizing hormone levels and pituitary responsiveness to luteinizing hormone-releasing hormone in the western spotted skunk. Biol. Reprod. 9, 76.

FORESMAN, K.R. & MEAD, R.A. (1974) Pattern of luteinizing hormone secretion during delayed implantation in the spotted skunk (Spilogale putorius latifrons). Biol. Reprod. 11, 475-480.

FRASCHINI, F., MESS, B. & MARTINI, L. (1968) Pineal gland, melatonin and the control of luteinizing hormone secretion. Endocrinology 82, 919-924.

GOULD, L.A. (1965) Delayed implantation in the Mustelidae. M.Sc. Thesis, University of Wales.

GULYAS, B.J. & DANIEL, J.C. Jr. (1969) Incorporation of labelled nucleic acid and protein precursors by diapausing and non-diapausing blastocysts. Biol. Reprod. 1, 11-16.

GWATKIN, R.B.L. (1966a) Defined media and development of mammalian eggs in vitro. Ann. N.Y. Acad. Sci. 139, 79-90.

GWATKIN, R.B.L. (1966b) Amino acid requirements for attachment and outgrowth of the mouse blastocyst in vitro. J. cell. Physiol. 68, 335-343.

HAM, K.N., HURLEY, S.V., LOPATA, A. & RYAN, G.B. (1970) A combined isotopic and electron microscope study of the response of the rat uterus to exogenous oestradiol. J. Endocr. 46, 71-81.

HAMILTON, W.J. Jr. & EADIE, W.R. (1964) Reproduction in the otter, Lutra canadensis. J. Mammal. 45, 242-251.

HAMMOND, J. Jr. (1951) Failure of progesterone treatment to affect delayed implantation in the mink. J. Endocr. 7, 330-334.

HANSSON, A. (1947) Physiology of reproduction in the mink (Mustela vison, Schreb.) with special reference to delayed implantation. Acta zool. 28, 1-136.

HAYNES, N.B. & LAMMING, G.E. (1967) Carbohydrate content of sow uterine flushings. J. Reprod. Fert. 14, 335-337.

HEARN, J. (1973) Pituitary inhibition of pregnancy. Nature 241, 207-208.

HENSLEIGH, H.C. (1971) Dry weight and lipid content of normal and delayed implanting mouse blastocysts. Anat. Rec. 169, 338-339.

HERLANT, M. & CANIVENC, R.M. (1960) Les modifications hypophysaires chez la femelle du blaireau (Meles meles L.) au cours du cycle annuel. C.R. Acad. Sci. (Paris) 250, 606-608.

HOLCOMB, L.C. (1967) Effects of progesterone treatments on delayed implantation in the mink. Ohio J. Sci. 67, 24-31.

HOLCOMB, L.C., SCHAIBLE, P.J. & RINGER, R.K. (1962) The effects of varied lighting regimes on reproduction in the mink. Mich. Agr. Exp. Sta. Quart. Bull. 44, 666-678.

HOYES, A.D. (1972) The endometrial glands of the pregnant sheep: an ultrastructural study. J. Anat. 111, 55-67.

HUGHES, R.L. (1974) Morphological studies on implantation in marsupials. J. Reprod. Fert. 39, 173-186.

JONES, L.T., HEAP, R.B. & PERRY, J.S. (1976) Protein synthesis in vitro by pig blastocyst tissue before attachment. J. Reprod. Fert. 47, 129-131.

KEIBEL, F. (1902) Die Entwicklung des Rehes bis zur Anlage des Mesoblast. Arch. Anat. Physiol. Suppl. 24, 293-314.

KENNAWAY, D. & SEAMARK, R.F. (1976) Pineal gland changes during the period of blastocyst activation in the tammar wallaby, Macropus eugenii, J. Reprod. Fert. 46, 503.

KNIGHT, J.W., BAZER, F.W. & WALLACE, H.D. (1973) Hormonal regulation of porcine uterine protein secretion. J. Anim. Sci. 36, 546-553.

KREHBIEL, R.H. (1941) The effects of lactation on the implantation of ova of a concurrent pregnancy in the rat. Anat. Rec. 81, 43-65.

LABHSETWAR, A.P. (1967) Progesterone in the luteal and placental tissues of the armadillo. Anat. Rec. 157, 273.

LABHSETWAR, A.P. & ENDERS, A.C. (1969) Pituitary LH content during delayed and post-implantation periods in the armadillo (Dasypus novemcinctus). J. Reprod. Fert. 18, 383-389.

LATASTE, M.F. (1891) Des variations de duree de la gestation chez les mammiferes et des circonstances qui determinent ces variations: theorie de la gestation retardee. C.R. Soc. Biol. 9, 21-31.

LINCOLN, G.A. & GUINESS, F.E. (1972) Effect of altered photoperiod on delayed implantation and moulting in roe deer. J. Reprod. Fert. 31, 455-457.

LU, K.H., CHEN, H.T., HUANG, H.H., GRANDISON, L., MARSHALL, S. & MEITES, J. (1976) Relation between prolactin and gonadotrophin secretion in post-partum lactating rats. J. Endocr. 68, 241-250.

MANECKJEE, R. & MOUDGAL, N.R. (1975) Induction and inhibition of implantation in lactating rats. J. Reprod. Fert. 43, 33-40.

MANN, T. (1946) Fructose as a normal constituent of seminal plasma. Site of formation and function of fructose in semen. Biochem. J. 40, 481-491.

MARSHALL, A.J. (1967) Origin of delayed implantation in marsupials. Nature 216, 192-193.

MAYER, G. (1963) Delayed nidation in rats: a method of exploring the mechanisms of ovo implantation. In: A.C. Enders (Ed.), Delayed Implantation, University of Chicago Press, Chicago, pp. 213-231.

McLAREN, A. (1968) A study of blastocysts during delay and subsequent implantation in lactating mice. J. Endocr. 42, 453-463.

McLAREN, A. (1970) Early embryo-endometrial relationships. In: P.O. Hubinot, F. Leroy, C. Robyn and P. Leleux (Eds.), Ovo Implantation, Human Gonadotrophins and Prolactin, S. Karger, Basel, pp. 18-37.

McLAREN, A. (1971) Blastocysts in the mouse uterus: the effect of ovariectomy, progesterone and oestrogen. J. Endocr. 50, 515-526.

McLAREN, A. (1973) Blastocyst activation. In: S.J. Segal, R. Crozier, P.A. Corfman and P.G. Condliffe (Eds.), The Regulation of Mammalian Reproduction, Thomas, Springfield, Illinois, pp. 321-328.

McLAREN, A. & MENKE, T.M. (1971) CO_2 output of mouse blastocysts in vitro, in normal pregnancy and in delay. J. Reprod. Fert. Suppl. 14, 23-29.

MEAD, R.A. (1968) Reproduction in western forms of the spotted skunk (genus Spilogale). J. Mammal. 49, 373-390.

MEAD, R.A. (1971) Effects of light and blinding upon delayed implantation in the spotted skunk. Biol. Reprod. 5, 214-220.

356

MEAD, R.A. (1972) Pineal gland: its role in controlling delayed implantation in the spotted skunk. J. Reprod. Fert. 30, 147-150.

MEAD, R.A. (1975) Effects of hypophysectomy on blastocyst survival, progesterone secretion and nidation in the spotted skunk. Biol. Reprod. 12, 526-533.

MEAD, R.A. & EIK-NES, K.B. (1969a) Seasonal variation in plasma levels of progesterone in western forms of spotted skunk. J. Reprod. Fert. Suppl. 6, 397-403.

MEAD, R.A. & EIK-NES, K.B. (1969b) Oestrogen levels in peripheral blood plasma of the spotted skunk. J. Reprod. Fert. 18, 351-353.

MENKE, T.M. & McLAREN, A. (1970a) Carbon dioxide production by mouse blastocysts during lactational delay of implantation or after ovariectomy. J. Endocr. 47, 287-294.

MENKE, T.M. & McLAREN, A. (1970b) Mouse blastocysts grown in vivo and in vitro: carbon dioxide production and trophoblast outgrowth. J. Reprod. Fert. 23, 117-127.

MØLLER, O.M. (1973) The progesterone concentration in the peripheral plasma of the mink (Mustela vison) during pregnancy. J. Endocr. 56, 121-132.

MOORE, G.P.M. (1976) Reactivation of RNA synthesis in macropod embryos after diapause. J. Reprod. Fert. 46, 504.

MURPHY, B.D. (1972) Population of gonadotrophic cells in the adenohypophysis of the mink during embryonic diapause and early post-implantation. Anat. Rec. 172, 372.

MURPHY, B.D. & MEAD, R.A. (1976) Effects of antibodies to oestrogens on implantation in ferrets. J. Reprod. Fert. 46, 261-263.

MUTERE, F.A. (1965) Delayed implantation in the equatorial fruit bat. Nature 207, 780.

NALBANDOV, A.V. (1971) Endocrine control of implantation. In: R.J. Blandau (Ed.), The Biology of the Blastocyst, University of Chicago Press, Chicago, pp. 383-392.

NEAL, E.G. & HARRISON, R.J. (1958) Reproduction in the european badger (Meles meles L.). Trans. Zool. Soc. London 29, 67-130.

NILSSON, O. (1970) Some ultrastructural aspects of ovo-implantation. In: P.O. Hubinot, F. Leroy, C. Robyn and P. Leleux (Eds.), Ovo Implantation, Human Gonadotrophins and Prolactin, S. Karger, Basel, pp. 52-72.

NILSSON, O. (1974) The morphology of blastocyst implantation. J. Reprod. Fert. 39, 187-194.

NORRIS, M.L. & ADAMS, C.E. (1971) Delayed implantation in the Mongolian gerbil, Meriones unguiculatus. J. Reprod. Fert. 27, 486.

PEARSON, O.P. & ENDERS, R.F. (1944) Duration of pregnancy in some mustelids. J. Exp. Zool. 95, 21-35.

PEPLOW, V., BREED, W.G., JONES, C.M.J. & ECKSTEIN, M.D. (1973) Studies on uterine flushings in the baboon. 1. Method of collection, cellular composition and protein electrophoretic profiles in animals with and without intrauterine contraceptive devices. Am. J. Obstet. Gynec. 116, 771-779.

PERRY, J.S., HEAP, R.B. & AMOROSO, E.C. (1973) Steroid hormone production by pig blastocysts. Nature 245, 45-47.

PEYRE, A. & HERLANT, M. (1963) Correlations hypophyso-genitales chez la femelle du Minioptere (Miniopterus schreibersii B.). Gen. Comp. Endocrinol. 3, 726-727.

PFEIL, W. (1807) Wann ist die wahre Brunstzeit des Rehes? Eine Bermerkung. J. Forst-Jagd Fischeregwesen 2, 49-54.

POCKELS, D. (1836) Uber die Brunstzeit der Rehe. Arch. Anat. Physiol. 193-204.

POLLARD, R.M. & FINN, C.A. (1972) Ultrastructure of the uterine epithelium during hormonal induction of sensitivity and insensitivity to a decidual stimulus in the mouse. J. Endocr. 55, 293-298.

POLLARD, J.W., FINN, C.A. & MARTIN, L. (1976) Actinomycin D and uterine epithelial protein synthesis. J. Endocr. 69, 161-162.

POTTS, M. & PSYCHOYOS, A. (1967) L'ultrastructure des relations ovo-endometriales au cours du retard experimental de nidation chez la souris. Compt. Rend. 264, 956-958.

PRASAD, M.R.N., DASS, C.M.S. & MOHLA, S. (1968) Action of oestrogen on the blastocyst and uterus in delayed implantation. An autoradiographic study. J. Reprod. Fert. 16, 97-103.

PSYCHOYOS, A. (1973) Hormonal control of ovo implantation. Vitams. Horm. 31, 201-256.

PSYCHOYOS, A. & BITTON-CASIMIRI, V. (1969) Caption in vitro d'un precurseur d'acide ribonucleique (ARN) (uridine-5-^{3}H) par le blastocyste du rat: difference entre blastocyste normaux et blastocyste en diapause. C.R. Acad. Sci. (Paris) 268, 188-192.

RENFREE, M.B. (1972) Influence of the embryo on the marsupial uterus. Nature 240, 475-477.

RENFREE, M.B. (1973) Proteins in the uterine secretions of the marsupial Macropus eugenii. Dev. Biol. 32, 41-49.

RENFREE, M.B. & TYNDALE-BISCOE, C.H. (1973) Intrauterine development after diapause in the marsupial, Macropus eugenii. Dev. Biol. 32, 28-40.

SADLIER, R.M.F.S. (1969) The ecology of reproduction in wild and domestic mammals. Menthuen, London.

SANYAL, M.K. & MEYER, R.K. (1972) Deoxyribonucleic acid synthesis in vitro in normal and delayed nidation preimplantation blastocysts of adult rats. J. Reprod. Fert. 29, 439-442.

SCHLAFKE, S. & ENDERS, A.C. (1963) Observations on the fine structure of the rat blastocyst. J. Anat. 97, 353-360.

SEKI, K., SEKI, M. & OKUMUA, T. (1974) Serum FSH rise induced by CB154(2-Br-α-ergocryptine) in post partum women. J. Clin. Endocr. Metab. 39, 184-186.

SHARMAN, G.B. (1954) Reproduction in marsupials. Nature 173, 302.

SHARMAN, G.B. (1963) Delayed implantation in marsupials. In: A.C. Enders (Ed.), Delayed Implantation, University of Chicago, Chicago, pp. 3-14.

SHARMAN, G.B. (1970) Reproductive physiology of marsupials. Science 167, 1221-1228.

SHARMAN, G.B. & BERGER, P.J. (1969) Embryonic diapause of marsupials. In: A. McLaren (Ed.), Advances in reproductive physiology, Vol. 4, Logos Press, London, pp. 212-240.

SHARMAN, R.M. (1972) Fate of short-tailed weasel, Mustela erminea, blastocysts following ovariectomy during diapause. J. Reprod. Fert. 31, 347-352.

SHELDON, R.M. (1973) Failure of ovarian steroids to influence blastocysts of weasels (Mustela erminea) ovariectomized during delayed implantation. Endocrinology 92, 638-641.

SHERMAN, M.I. & BARLOW, P.W. (1972) Deoxyribonucleic acid content in delayed mouse blastocysts. J. Reprod. Fert. 29, 123-126.

SHORT, R.V. & HAY, M.F. (1966) Delayed implantation in the roe deer (Capreolus capreolus). In: I.W. Rowlands (Ed.), Comparative Biology of Reproduction in Mammals, Academic Press, London, pp. 173-194.

SIMPSON, A.A., SIMPSON, M.H.W., SINHA, Y.N. & SCHMIDT, G.H. (1973) Changes in concentration of prolactin and adrenal corticosteroids in rat plasma during pregnancy and lactation. J. Endocr. 58, 675-676.

SINGH, M.M., KAMBOF, V.P., CHOWDHURY, S.R., PANDE, S.K. & ROY, S.K. (1973) Histochemical and biochemical changes in rat uterus during delayed implantation. Indian J. Exp. Biol. 11, 488-493.

SINHA, A.A. & MEAD, R.A. (1975) Ultrastructural changes in granulosa lutein cells and progesterone levels during preimplantation, implantation, and early placentation in the western spotted skunk. Cell. Tiss. Res. 164, 179-192.

SMITH, M.J. & SHARMAN, G.B. (1969) Development of dormant blastocysts induced by oestrogen in the ovariectomized marsupial, Macropus eugenii. Aust. J. Biol. Sci. 22, 171-180.

STIEVE, H. (1950) Anatomische-biologische Untersuchungen uber die Fortpflanzungstatigkeit des europaischen Rehes (Capreolus capreolus L.). Z. Mikr. Anat. Forsch. 55, 427-530.

SUGA, T. & MASAKI, J. (1973) Studies on the uterine secretion of the cow. 6. Sugar and polyolconstituents in the luminal fluid of bovine uterus. Jap. J. Anim. Reprod. 18, 143-147.

SURANI, M.A.H. (1975) Hormonal regulation of proteins in the uterine secretion of ovariectomized rats and the implications for implantation and embryonic diapause. J. Reprod. Fert. 43, 411-417.

SVIHLA, A. (1932) A comparative life history study of the mice of the genus Peronmyscus. Univ. Mich. Mus. Zool. Misc. Publ. 24.

TALMAGE, R.V., BUCHANAN, G.D., KRAINTZ, F.W., LASOWASEM, E.A. & ZARROW, M.X. (1954) The presence of a functional corpus luteum during delayed implantation in the armadillo. J. Endocr. 11, 44-54.

TOMOGANE, H., OTA, K. & YOKOYAMA, A. (1969) Progesterone and 20α hydroxypregn-4-en-3-one levels in ovarian vein blood of the rat throughout lactation. J. Endocr. 44, 101-106.

TOMOGANE, H., OTA, K. & YOKOYAMA, A. (1975) Suppression of progesterone secretion in lactating rats by administration of ergocornine and the effect of prolactin replacement. J. Endocr. 65, 155-161.

TORBIT, C.A. & WEITLAUF, H.M. (1974) The effect of oestrogen and progesterone on CO_2 production by 'delayed implanting' mouse embryos. J. Reprod. Fert. 39, 379-382.

TRAMS, G., BREWITT, H., MOLIMANN, H. & MAASS, H. (1973) Effect of progesterone on RNA and protein synthesis in the rat uterus. Acta endocr., Copenh. 73, 740-750.

TYNDALE-BISCOE, C.H. (1963a) The role of the corpus luteum in the delayed implantation of marsupials. In: A.C. Enders (Ed.), Delayed Implantation, University of Chicago Press, Chicago, pp. 15-32.

TYNDALE-BISCOE, C.H. (1963b) Effects of ovariectomy in the marsupial, Setonix brachyurus. J. Reprod. Fert. 6, 25-40.

TYNDALE-BISCOE, C.H. (1968) Reproduction and post-natal development in the marsupial Bettongia lesueur (Quoy Gaimard). Aust. J. Zool. 16, 577-602.

TYNDALE-BISCOE, C.H., HEARN, J.P. & RENFREE, M.B. (1974) Control of reproduction in macropodid marsupials. J. Endocr. 63, 589-614.

URZUA, M.A., STAMBAUGH, R., FLICKINGER, G. & MASTROIANNI, L. (1970) Uterine and oviduct fluid protein patterns in the rabbit before and after ovulation. Fert. Steril. 21, 860-865.

WARREN, R.H. & ENDERS, A.C. (1964) Electron microscope study of the rat endometrium during delayed implantation. Anat. Rec. 148, 177-195.

WEICHERT, C.K. (1942) The experimental control of prolonged pregnancy in the lactating rat by means of oestrogen. Anat. Rec. 83, 1-17.

WEITLAUF, H.M. (1971) Influence of ovarian hormones on the incorporation of amino acids by blastocysts in vivo. In: R.J. Blandau (Ed.), The Biology of the Blastocyst, University of Chicago Press, Chicago, pp. 277-290.

WEITLAUF, H.M. (1973) Changes in the protein content of blastocysts from normal and delayed implanting mice. Anat. Rec. 176, 121-124.

WEITLAUF, H.M. (1976) Effect of uterine flushings on RNA synthesis by 'implanting' and 'delayed implanting' mouse blastocysts in vitro. Biol. Reprod. 14, 566-571.

WEITLAUF, H.M. & GREENWALD, G.S. (1965) Comparison of ^{35}S methionine incorporation by the blastocysts of normal and delayed implanting mice. J. Reprod. Fert. 10, 203-208.

WEITLAUF, H.M. & GREENWALD, G.S. (1968) Survival of blastocysts in the uteri of ovariectomized mice. J. Reprod. Fert. 17, 515-520.

WIMSATT, W.A. (1963) Delayed implantation in the Ursidae with particular reference to the black bear (Ursus americanus). In: A.C. Enders (Ed.), Delayed Implantation, University of Chicago, Chicago, pp. 49-76.

WOLF, D.P. & MASTROIANNI, L. (1975) Protein composition of human uterine fluid. Fert. Steril. 26, 240-247.

WRIGHT, P.L. (1942) Delayed implantation in the long-tailed weasel (Mustela frenata), the short-tailed weasel (Mustela cicognani) and the marten (Martes americana). Anat. Rec. 83, 341-353.

WRIGHT, P.L. (1963) Variations in reproductive cycles in North American mustelids. In: A.C. Enders (Ed.), Delayed Implantation, University of Chicago Press, Chicago, pp. 77-98.

WRIGHT, P.L. & RAUSCH, R. (1955) Reproduction in the wolverine (Gulo gulo). J. Mammal. 36, 346-355.

WURTMAN, R.J., AXELROD, J. & CHU, E.W. (1963a) Melatonin, a pineal substance: effect on the rat ovary. Science 141, 277.

WURTMAN, R.J., AXELROD, J. & PHILLIPS, L.S. (1963b) Melatonin synthesis in the pineal gland: control by light. Science 142, 1071-1073.

WURTMAN, R.J., AXELROD, J. & FISCHER, J.E. (1964) Melatonin synthesis in the pineal gland: effect of light mediated by the sympathetic nervous system. Science 143, 1328-1329.

WURTMAN, R.J., AXELROD, J. & KELLY, D.W. (1968) The pineal, Academic Press, New York.

YASUKAWA, J.J. & MEYER, R.K. (1966) Effect of progesterone and oestrone on the preimplantation and implantation stages of embryo development in the rat. J. Reprod. Fert. 11, 245-255.

YOSHINAGA, K. (1961) Effect of local application of ovarian hormones on the delay in implantation in lactating rats. J. Reprod. Fert. 2, 35-41.

YOSHINAGA, K. (1974) Ovarian progestin secretion in lactating rats: effect of intrabursal injection of prolactin antiserum, prolactin and LH. Endocrinology 94, 829-834.

YOSHINAGA, K. & ADAMS, C.E. (1966) Delayed implantation in the spayed, progesterone treated adult mouse. J. Reprod. Fert. 12, 593-595.

ZEIGLER, L. (1843) Beobachtungen uber die Brunst und den Embryo der Rehe, Hanover. Cited in: Prell, H. (1938) Die Tragzeit des Rehes. Zuechtungskunde 13, 325-340.

HORMONAL CONTROL OF UTEROGLOBIN BIOSYNTHESIS

Miguel Beato

Institut fur Physiologische Chemie
der Philipps-Universitat
Deutschhausstr. 1-2
3550 Marburg, F.R.G.

The mechanism of implantation of the embryo has received considerable
attention during the past few years.   One aspect of this process which deserves
particular consideration is the interaction of the maternal milieu with the
developing embryo during the period preceding implantation, i.e. before a
physical contact between the two organisms has been established.   It is becoming
clear that this interaction is under hormonal control, and that the balanced
production of ovarian hormones after fertilization, determines the development
of the uterine surroundings which are essential for a successful implantation
(Schwick, 1965; Beier, 1974).

In the rabbit, the structure of the endometrium and the composition of the
uterine secretion change dramatically during the few days after the entry of the
blastocyst into the uterine cavity.   One of the most characteristic changes
during the preimplantation phase is the appearance in the uterine secretion of
specific protein bands, among which the most prominent was called uteroglobin
by Beier (1968) and blastokinin by Krishnan and Daniel (1967).   Since the
function of this protein is still unknown, and its postulated role in blastocyst
development has not been confirmed (Maurer & Beier, 1976), the name uteroglobin
will be used in the following description, in spite of the fact that uteroglobin-
like antigens have been detected outside the uterus (see below).   This
designation seems justified, as the uterus is not only the site of maximal
uteroglobin production, but also the organ where this protein was originally
described.   Certainly other proteins of the uterine secretion, including glycoproteins and a postalbumin fraction, are also under hormonal control (Kirchner,
1969), but this discussion will be limited to the regulation of uteroglobin
synthesis and, more specifically, to the intracellular mechanisms involved.

The structure of uteroglobin will be the subject of another chapter in this
series. Suffice it to say, that it is composed of two similar subunits, each
containing some 75 aminoacid residues, held together by two disulfide bridges
and other non-covalent interactions (Nieto et al., 1976).   Its function may be
related to the hormonal interaction between the maternal organism and the

conceptus, as uteroglobin exhibits the interesting property of binding ovarian steroids, and particularly progesterone, with relative high affinity and specificity (Arthur et al., 1972; Beato, 1976; Beato & Baier, 1975; Urzua et al., 1970).

This report will comprise, first, of a brief summary of the available data on the biosynthesis of uteroglobin in vivo, and a description of experiments carried out with isolated uteri. It will then proceed to discuss more recent work on the characterization of the messenger RNA (mRNA) for uteroglobin, and its hormonal control. Finally, the recent reports on uteroglobin synthesis in organs other than the uterus will be briefly reviewed.

SYNTHESIS OF THE UTEROGLOBIN IN VIVO

In the uterus, uteroglobin is detectable in very small amounts during estrous (Bullock & Connell, 1973; Kirchner, 1972) but its concentration increases markedly during the few days preceding implantation. At the time of maximal uteroglobin production (day 5-6 post-coitum) this protein may represent up to 40-50% of the total protein content of the uterine fluid. Interestingly, administration of estradiol shortly after normal mating, results in a delayed secretion of uteroglobin (Beier et al., 1971), and estrogens might also be responsible for the termination of uteroglobin synthesis which occurs after implantation (Beier, 1976). Experiments with intact and castrated rabbits have shown that optimal induction of uteroglobin secretion is achieved by the sequential administration of estrogens and progesterone, although small amounts of uteroglobin can be detected in the uteri of castrated animals treated only with estrogens (Arthur & Daniel, 1972; Beato & Arnemann, 1975; Bullock & Willen, 1974; Mayol & Longenecker, 1974; Rahman et al., 1975). In the castrated rabbits, progesterone alone can cause an accumulation of uteroglobin in the uterine secretion, which is detectable two days after the beginning of the hormonal treatment (Arthur & Daniel, 1972). Studies with synthetic steroids, including several contraceptive compounds, did not show a good correlation between progestational activity and uteroglobin induction (Arthur & Chang, 1974), although this may be due to the oral administration of the compounds (Nishino, 1976). There appears to be no correlation either between the affinity of various steroids for uteroglobin (Beato, 1976), and their ability to induce uteroglobin synthesis.

The data reported above demonstrate an accumulation of uteroglobin in the uterine lumen of the rabbit under certain hormonal conditions, but they say nothing about the site of uteroglobin synthesis. The immunofluorescence methods, used by Kirchner (1972), demonstrate the presence of uteroglobin in the epithelial cells of the endometrium, but still they do not prove that the protein is synthesized by these cells. More convincing evidence for the uterine origin of uteroglobin was provided by Krishnan (1970) and Daniel (1971), who showed that after injection of radio-active amino acids into the uterine cavity of rabbits in different stages of pregnancy, a preferential incorporation of the radio-activity into protein fractions behaving as uteroglobin on Sephadex columns was detected.

SYNTHESIS OF UTEROGLOBIN IN ISOLATED UTERI

In order to answer the question of whether the influence of ovarian steroids on the production of uteroglobin was due to changes in the rate of protein synthesis or to the regulation of protein secretion, we have investigated the rate of uteroglobin synthesis in isolated uteri (Beato & Arnemann, 1975). Using the technique of intraluminal perfusion with medium containing radioactive amino acids, we studied the relationship between the uteroglobin content in the uterine fluid and the ability of the uterus to synthesize and secrete uteroglobin in vitro. Three groups of animals were initially selected: 1) normal estrous animals; 2) animals treated with estradiol for two days; and 3) animals treated with estradiol for two days and with progesterone for the four following days. After flushing the uteri and determining the concentration of uteroglobin in the uterine fluids by polyacrylamide gel electrophoresis, the uteri were perfused in vitro with Eagle's minimum medium containing (^{3}H)leucine and 10^{-5}M progesterone equilibrated with a mixture of O_2 and CO_2 (95:5). Under these conditions, the incorporation of radioactivity into total secreted protein was linear for at least four hours (Figure 1). Experiments carried out with cyclohexymide demonstrated that the incorporation of radioactivity into secreted proteins was the consequence of de novo protein synthesis. The radioactivity in the medium was analyzed by chromatography on Sephadex columns, together with carrier uteroglobin, and by immunoprecipitation with specific anti-uteroglobin antiserum. As can be seen from Figure 1, the rate of uteroglobin synthesis is also linear for at least three hours, and accounted for almost half of the total protein synthesized and secreted in the medium under conditions of optimal stimulation.

The results obtained with the three experimental groups of animals are depicted in Figure 2. In control animals no uteroglobin was detected in the uterine flushes nor was any detectable synthesis of uteroglobin in perfused uteri (Figures 2 a and d). In animals, treated with estradiol alone, the percentage of uteroglobin in the uterine lumen accounts for up to 5% of the total proteins (Figure 2b) and a similar percentage of radioactive leucine was incorporated into secreted proteins (Figure 2e). After repeated administration of progesterone, however, uteroglobin represents about 50% of the total protein in the uterine flushes (Figure 2c), and around 40% of the radioactive amino acids incorporated into the secreted proteins during the perfusion experiment are

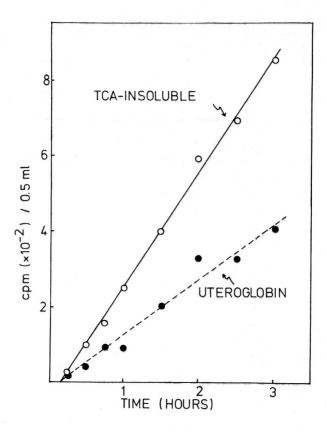

Figure 1: Rate of incorporation of (^{3}H)leucine into secreted proteins of isolated perfused uteri. Uteri from rabbits treated sequentially with estradiol-17β (2 days) and progesterone (5 days), were perfused with medium containing (^{3}H)leucine as previously described (Beato & Arnemann, 1975). At the indicated time intervals, aliquots were taken from the perfusion medium for the determination of radioactivity incorporated into total proteins (o——o), by the method of Bollum (1968). Aliquots were also used for the measurements of the radioactivity incorporated into uteroglobin, as determined by a double antibody technique (Beato & Nieto, 1976).

localized in the uteroglobin region (Figure 2f). These studies were repeated under different hormonal conditions, and a good correlation was observed between the uteroglobin content in the uterine flushes, before starting the perfusion and the percentage of the total secreted proteins accounted for by uteroglobin (Figure 3). Identification of in vitro synthesized uteroglobin was further achieved by gel electrophoresis, immunoprecipitation and tryptic digestion of the in vitro product followed by chromatography on Dowex 50 x 8 (Beato & Nieto, 1976). The peptide mapping of in vitro synthesized uteroglobin, labelled with (^{3}H)lysine as radioactive precursor, shows a perfect coincidence with that

of mature uteroglobin (Figure 4).   As expected from the amino acid composition
(Nieto et al., 1976) which yields 7 lysines and 2 arginines per chain, seven
radioactive peptides were detected in Dowex chromatography, coincident with
ninhydrin positive peaks.   The other three ninhydrin-peaks, which do not have
radio-active counterparts, probably correspond to the two arginine peptides
and the C-terminal lysine peptide.

These experiments clearly show that estradiol treatment already induces
the synthesis of uteroglobin, and that the effect of progesterone on uteroglobin content is due to an increased rate of uteroglobin synthesis.   Attempts
to measure the presence of radioactive uteroglobin in endometrial fractions of
the perfused uteri (microsomes and cytosol) were unsuccessful, suggesting that
the protein is immediately secreted after its synthesis is completed.

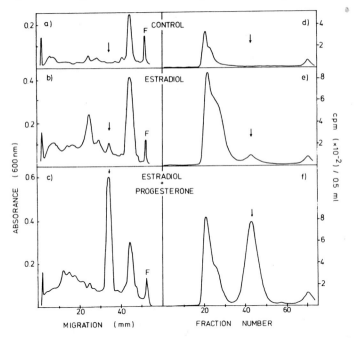

Figure 2: Comparison between the uteroglobin content of the uterine secretion
and uteroglobin synthesis in isolated uteri, under different hormonal conditions.
Uteri were extirpated from control rabbits, and from animals treated with either
estradiol alone, or estradiol and progesterone. One horn was flushed with tris-
saline and the proteins analyzed by polyacrylamide gel electrophoresis. The
densitometric scans of the three experimental groups are depicted on the left
panel. The other uterine horn was used for perfusion experiments with (3H)
leucine, and the proteins secreted in the medium were submitted to chromatography on columns of Sephadex G75. The right panel shows the corresponding
radioactive elution profiles. In both panels the arrows indicate the position
of authentic uteroglobin, added as internal marker. "F" indicates the dye
front.

TRANSLATION OF THE UTEROGLOBIN mRNA IN VITRO

The next step in the elucidation of the molecular mechanism involved in the modulation of uteroglobin biosynthesis, was the quantitation of its specific mRNA, as this will allow a decision as to whether the increased rate of utero-globin synthesis caused by progesterone was due to a more efficient translation of pre-existing mRNA or to an accumulation of specific mRNA. In order to answer this question, it was necessary to develop an assay for uteroglobin mRNA.

Endometrial polysomes obtained from rabbits sequentially treated with estradiol and progesterone were used as a source for the isolation of utero-globin mRNA. Total polysomal RNA was extracted with SDS-phenol-chloroform (Beato & Rungger, 1975) and submitted to chromatography on a column of oligo (dT)-cellulose (Beato & Nieto, 1976). The RNA which did not bind to the column was designated as flow-through RNA, and the RNA fraction which was eluted from the column with distilled water was called messenger RNA (mRNA). These two fractions were assayed in three different protein synthesizing systems: a) oocytes from Xenopus laevis (Beato & Rungger, 1975); b) post-mitochondrial supernatant of Krebs II ascites cells supplemented with reticulocyte initiation factors (Schultz et al., 1974); c) post-mitochondrial supernatant of wheat germ (Roberts & Paterson, 1973).

The products obtained in oocytes injected with endometrial mRNA and labelled with (^{35}S) methionine were subjected to polyacrylamide gel electro-phoresis in gels containing urea and SDS (Swank & Munkres, 1971), and the radioactivity determined after cutting the gels in 1 mm slices. The results are depicted in Figure 5. It is evident that one of the main polypeptide chains

Figure 3 (upper): Correlation between the uteroglobin content of the uterine secretion and uteroglobin synthesizing capacity of isolated uteri. The utero-globin content in the uterine flushes of animals treated with various com-binations of estradiol and progesterone was determined by planimetric measure-ments of polyacrylamide gel scans (see Figure 2, left panel), and is plotted in the abscissa as percentage of the total protein stain in the gel. The utero-globin synthesizing capacity of the contralateral horn was determined by a double antibody technique (Beato & Nieto, 1976), and is plotted in the ordinate as percentage of total radioactivity incorporated into secreted proteins, in a 3h perfusion experiment (Beato & Arnemann, 1975).

Figure 4 (lower): Tryptic peptides of uteroglobin synthesized in isolated uteri. The uterine horns of rabbits treated with estradiol (2 days) and progesterone (5 days), were perfused for 3 hours with medium containing (^{3}H)lysine. The radioactive uteroglobin was isolated from the medium by repeated column chroma-tography (Beato & Arnemann, 1975). After mixing with 7 mg of authentic utero-globin, the sample was oxydized with performic acid, lyophilized, and digested with trypsin (Beato & Nieto, 1976). The tryptic digest was chromatographed at 38°C in a column of Dowex AG 50W x 8, using linear gradients of pyridine/acetate of increasing molarity and pH. The open circles represent the absorbance at 570 nm after the ninhydrin reaction, and the closed circles the radioactivity of each fraction. The dotted line and the internal scale refer to the pH gradients.

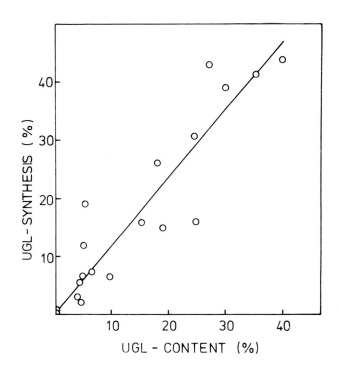

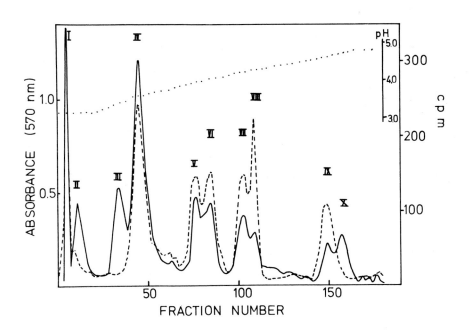

synthesized in the oocytes under the influence of endometrial mRNA comigrates with authentic uteroglobin, both before and after reduction of the disulfide bridges (Figures 5a, b).   The identification of this polypeptide chain as uteroglobin was performed by incubation of the radioactive products with a specific anti-uteroglobin antiserum, followed by addition of anti-immunoglobulin, and electrophoresis of the immunoprecipitates on urea-SDS gels.   Between 10 and 12% of the total radioactive proteins synthesized in oocytes injected with endometrial mRNA, were precipitated with the specific antiserum and comigrated

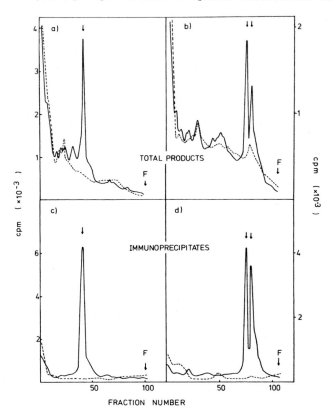

Figure 5: Electrophoresis in gels containing urea and SDS of the radioactive products obtained in Xenopus laevis oocytes injected with endometrial mRNA. Oocytes were injected with either buffer (---) or endometrial mRNA (——), and incubated with (^{35}S)methionine as described (Beato & Rungger, 1975). The upper panel shows the electrophoretic pattern of the total radioactive products in the 10,000 x g supernatants, before (a) and after (b) reduction with dithioerythritol (DTE). The lower panel shows a similar electrophoretic separation of the radioactive products after immunoprecipitation with a double antibody technique, specific for uteroglobin (Beato & Nieto, 1976). The arrows indicate the position of uteroglobin given as internal marker. The position of the dye front is labelled with "F".

with authentic uteroglobin upon gel electrophoresis (Figures 5c, d).   In control
oocytes injected with buffer, only 1-2% of the in vitro synthesized polypeptides
were found in the immunoprecipitate, and no radioactive peak was detected in the
uteroglobin band (Figures 5c, d).   The identity of the in vitro synthesized
uteroglobin with native uteroglobin was further demonstrated by comparison of
the peptide mappings obtained after tryptic digestion (data not shown).   These
findings show that uteroglobin mRNA is one of the main components of total polysomal mRNA obtained from the endometrium of induced rabbits.

    A precise quantitation of uteroglobin mRNA, however, requires the use of
cell-free translational systems, with low endogenous mRNA activity.   Two such
mRNA-dependent systems, the Krebs II ascites S-30, and the wheat germ S-30, were
optimized with purified rabbit globin mRNA and with endometrial mRNA (Beato &
Nieto, 1976).   At the optimal concentration of KC1 and MgC1$_2$, one of the main
products encoded by endometrial mRNA of induced rabbits in these cell-free
systems migrates slower than the uteroglobin subunits, but faster than the native
protein (Figures 6a, 6b, 7a, 7b).   In order to identify the nature of this cell-
free product, an immunoprecipitation with a monospecific anti-uteroglobin antiserum from guinea pig was carried out.   The cell-free products were incubated
with anti-uteroglobin immunoglobulin, and precipitated by the addition of antiserum directed against immunoglobulin from guinea pigs.   After washing the
immunoprecipitates, the samples were applied to polyacrylamide gels containing
urea and SDS (Figures 6c, 6d, 7c, 7d).   The results clearly demonstrate the
existence of polypeptide chains synthesized in the cell-free system under the
direction of endometrial mRNA which are antigenically related to uteroglobin and
migrate faster than the native protein but slower than the uteroglobin subunit.

    Estimation of the apparent molecular weight of the cell-free product,
using proteins and peptides of known size indicates a molecular weight of about
7,000 (Figure 8).   However, it is known that the determination of uteroglobin
molecular weight in this type of gel, leads to considerable underestimation of
the actual values (Bullock & Connell, 1973; Nieto et al., 1976).   Values of
11,000 and 5,500 are obtained for native uteroglobin and the uteroglobin subunit,
respectively, although we know that the actual molecular weight is considerably
larger (Nieto et al.,1976).   The important information from these experiments,
however, is that the cell-free product is 1,000 to 1,500 daltons larger than the
uteroglobin subunit, corresponding to 10-15 additional amino acids.

    The immunological identity with uteroglobin of the cell-free product was
demonstrated by competition experiments in which the radioactive polypeptide
could be displaced from the complex with the specific anti-uteroglobin by addition
of authentic uteroglobin (Beato & Nieto, 1976).

    Further identification of the immunoprecipitated cell-free products was
carried out by trypsin digestion and chromatography of the tryptic peptides on a

column of Dowex 80 x 5 (Figure 9). For these experiments, endometrial mRNA from induced animals was translated in the ascites cell-free system using (^{3}H)lysine as the radioactive amino acid, and the products were immunoprecipitated with anti-uteroglobin, and digested with trypsin together with authentic uteroglobin (Beato & Nieto, 1976). Eight radioactive peptides were detected after ion-exchange chromatography of which seven were coincident with ninhydrin positive peaks, as expected from the amino acid composition of the uteroglobin subunit (see above). In addition, another radioactive peptide was detected in the cell-free product (labelled XI) which was not present in the mature protein. This finding in conjunction with the immunological relationship and the larger size of the cell-free product strongly suggests that the endometrial mRNA from induced rabbits contains the information for a precursor of the uteroglobin subunit with 10-15 additional amino acids, presumably at the N-terminal end of the nascent polypeptide chain. In the Xenopus oocytes this precursor, that we will call pre-uteroglobin, is proteolytically cleaved and the subunits are covalently bound to form the mature uteroglobin, whereas in the cell-free translational system this processing machinery is not operative.

It is important to note that the difference in the electrophoretic migration between pre-uteroglobin and authentic uteroglobin is not observed upon electrophoresis in conventional SDS-polyacrylamide gels (Laemmli, 1970; Weber & Osborn, 1969) containing less than 10% acrylamide. Under these conditions, the resolution in the lower molecular weight range is very poor and no clear distinction is possible between native uteroglobin and its subunits (Beato & Nieto, 1976). Only at concentrations of acrylamide above 15% is a separation of the cell-free product from the uteroglobin subunit possible (unpublished observation).

Figure 6 (upper): Electrophoresis in polyacrylamide gels containing urea and SDS of the total cell-free products and the immunoprecipitation obtained in the ascites S-30 system. Standard cell-free incubation assays (50μ 1) were carried out using (^{3}H)leucine as radioactive amino acid, with or without saturating amounts of endometrial mRNA from induced animals. The total cell-free products (upper panel), and the immunoprecipitates obtained with anti-uteroglobin and anti-immunoglobulin (lower panel), were submitted to electrophoresis in gels containing 8M urea and 0.1% SDS, before (a and c) and after (b and d) reduction with dithioerythritol (Beato & Nieto, 1976). The same total amount of radioactive products (20-30,000 cpm), were applied to all gels, in the case of assays containing mRNA (continuous line). The control gels, from assays without added mRNA, contain considerably lower total radioactivity (broken line). To all gels, authentic uteroglobin was added as internal marker and its position is indicated by arrows.

Figure 7 (lower): Electrophoresis in polyacrylamide gels containing urea and SDS of the total cell-free products and the immunoprecipitates obtained in the wheat germ system. The experimental conditions and the symbols are as described in the legend to Figure 6, with the only difference, that the wheat germ system was used instead of the ascites system.

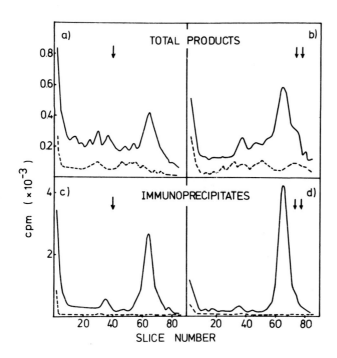

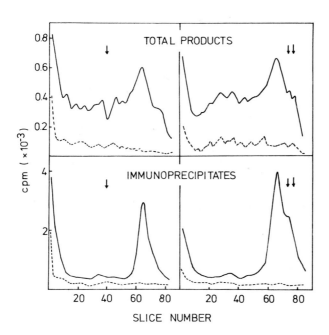

SIZE OF THE UTEROGLOBIN SYNTHESIZING POLYSOMES AND OF UTEROGLOBIN mRNA

The experiments reported above indicated that uteroglobin mRNA is one of the main messenger RNAs in endometrial polysomes. Therefore, identification of the polysomes involved in the synthesis of uteroglobin was attempted by immunological techniques. Anti-uteroglobin immunoglobulin, prepared against uteroglobin denatured with 2-mercaptoethanol and SDS, was purified by affinity chromatography, using a column of uteroglobin covalently bound to Affi-Gel 10 (Bio-Rad Laboratories) (Cuatrecasas & Parickh, 1972). The purified immunoglobulin was eluted with glycine-HCl buffer, pH 3.0, and iodinated with ^{125}I to a specific activity of 6 x 10^5 cpm/μg (Bayse et al., 1972). All preparations of immunoglobulin were freed of ribonuclease by a passage through a column of carboxymethyl and DEAE-cellulose (Palacios et al., 1972), followed by precipitation with 40% saturated ammonium sulphate. The radioactive immunoglobulins were incubated with endometrial polysomes from induced animals in the absence or in the presence of an excess of non-radioactive anti-uteroglobin immunoglobulin. The incubated polysomes were then fractionated in a linear sucrose gradient and the distribution of the radioactivity and the absorbance at 260 nm in the different fractions was determined (Figure 10). Although relatively little radioactivity was bound to the polysomes, it is obvious that monosomes, disomes and trisomes are the main polysomal fractions involved in uteroglobin biosynthesis as the radioactivity in these fractions was displaced by an excess of cold anti-uteroglobin. The fact that the radioactivity bound to these fractions is

Figure 8 (upper): Determination of the apparent molecular weight of the immunoprecipitable cell-free product in polyacrylamide gels containing urea and SDS. The products of the standard ascites cell-free system containing endometrial mRNA, were precipitated by the double antibody technique using a monospecific antiserum against uteroglobin, and electrophoresed in polyacrylamide gels containing urea and SDS, together with the following markers: 1) myoglobin; 2) complex of CNBr-peptide I and II obtained from myoglobin; 3) myoglobin CNBr-peptide I; 4) myoglobin CNBr-peptide II, and 5) myoglobin CNBr-peptide III. The position of native uteroglobin (UGL$_2$) and the uteroglobin subunit (UGL$_1$) are also indicated. The figure is a plot of the R_f values of the different polypeptide chains, against the logarithm of their molecular weight.

Figure 9 (lower): Tryptic peptides of the immunoprecipitable products encoded by endometrial mRNA of induced rabbits in the ascites cell-free system. A standard ascites cell-free system (100μl), containing (^{3}H)lysine as the radioactive amino acid, and saturating amounts of endometrial mRNA, was used to prepare the immunoprecipitable product by means of the double antibody technique (Beato & Nieto, 1976). The material was mixed with authentic uteroglobin and submitted to oxidation with performic acid and digestion with trypsin (Beato & Nieto, 1976). The tryptic peptide was chromatographed as described in the legend to Figure 4. The radioactivity (---), the intensity of the ninhydrin reaction (——), and the pH-gradient (...) are indicated.

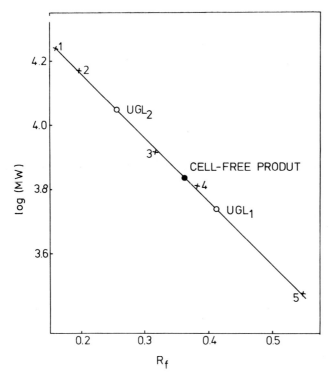

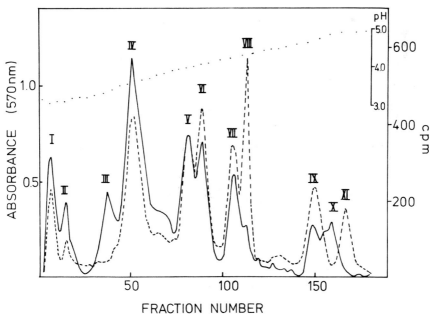

374

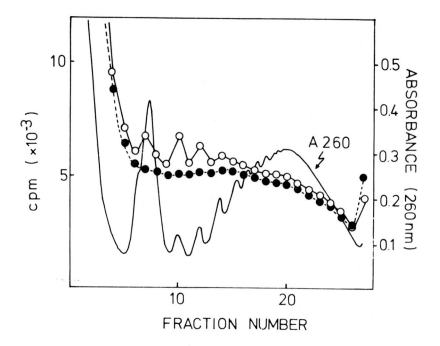

Figure 10: Binding of (^{125}I) anti-uteroglobin immunoglobulin to the endometrial polysomes of induced rabbits. Purified and iodinated immunoglobulin (25μ g) prepared against uteroglobin denatured with 0.1% SDS and 1% 2-mercaptoethanol, was incubated with endometrial polysomes of induced rabbits at 0°C for 60 mins. The mixture was centrifuged through 1 M sucrose, containing 1% Na deoxycholate and 1% Triton X-100, and the pellet applied to a linear sucrose gradient (0.5 M-1.5 M sucrose). After centrifugation in a Beckman SW 41 Rotor (40,000 rpm for 3 h), the radioactivity (o—o), and the absorbance at 260 nm (----) was determined. In a control experiment, the polysomes were preincubated at 0°C for 30 min with 200μ g of purified anti-uteroglobin, before addition of the iodinated immuno-globulin. The results of this control experiment are plotted (●—●), in the same graph. Note displacement of curve in fractions corresponding to disomes/trisomes.

relatively low, is probably due to the shortness of the pre-uteroglobin chain (85-90 amino acids). In addition, pre-uteroglobin contains 10-15 amino acids at the growing N-terminal end of the polypeptide chain, which are not present in the mature protein used as antigen, and could interfere with antibody binding. It is known that around 40 amino acids are covered by the large ribosomal subunit, and therefore only some 35 residues of what will be mature uteroglobin are available for binding of the antibodies. The relative unaccessibility of the growing polypeptide chains on the polysomes is confirmed by the finding that antibodies against the native protein, did not bind at all to the polysomes of induced animals. It was only after an antiserum to the denatured uteroglobin subunits was available, that the binding of the specific immunoglobulin to the polysomes could be demonstrated.

Further identification of the uteroglobin synthesizing polysomes was carried out by preliminary fractionation of the endometrial polysomes in linear sucrose gradients, followed by resuspension of the different fractions in 0.5 M NaC1, 0.5% SDS, and chromatography on oligo(dt)-cellulose (Krystosek et al., 1975). The poly(A)-containing RNA was tested in the wheat germ system for its ability to direct the synthesis of pre-uteroglobin (Figure 11). As can be seen, most of the mRNA coding activity for pre-uteroglobin is localized in the disome-trisome region of the gradient, indicating that during incubation with the antibodies only very little degradation of the mRNA occurs.

The size of the pre-uteroglobin mRNA was determined in sucrose gradient containing 50% formamide (Anderson et al., 1974) and was shown to be around 10 S (Figure 12). This size is considerably larger than expected of a polypeptide chain of some 85-90 amino acids, even after taking into consideration the usual length of the poly(A) fragment. This finding can be explained assuming the existence of long untranslated sequences in the mRNA. An alternative possibility, however, is that the actual primary product of the translation of this mRNA in vivo is larger than pre-uteroglobin, and is not detected in the cell-free systems. As seen in Figure 12, some pre-uteroglobin mRNA activity was also detected, in a region of the gradient corresponding to RNA larger than 18 S, indicating that aggregation is still taking place under these conditions.

QUANTITATION OF UTEROGLOBIN mRNA AFTER DIFFERENT HORMONAL TREATMENTS

Once an assay for uteroglobin mRNA was available, the quantitation of its cellular content was undertaken in order to determine whether the progesterone induced synthesis of uteroglobin is mediated through an accumulation of specific mRNA or rather results from a more efficient translation of pre-existing mRNA. Obviously, measurement of the cellular content of total uteroglobin mRNA sequences will require the use of hybridization techniques, employing DNA complementary to the purified messenger. With this procedure uteroglobin mRNA lacking poly(A) can also be measured, and we know that a considerable percentage of total uteroglobin mRNA does not bind to oligo(dT)-cellulose (Beato & Nieto, 1976). The measurements presented here are limited to determinations of the polysomal content of translatable and poly(A)-containing pre-uteroglobin mRNA.

The poly(A)-containing mRNA prepared from endometrial polysomes by chromatography on oligo(dT)-cellulose, was assayed in the ascites cell-free system, using the double antibody technique described above (Beato & Nieto, 1976). The results of experiments carried out under different hormonal conditions are depicted in Table 1. Within the limitations of our methodology, a good correlation is found between the uteroglobin content of the uterine secretion, the uteroglobin synthesis and secretion by isolated perfused uteri, and the polysomal content of pre-uteroglobin mRNA (Table 1). In animals castrated three weeks before the beginning of the experiments, and not submitted to hormonal treatment, the uteri are atrophic and very little material can be obtained for a preparation of endometrium. No data are, therefore, available from these groups of animals. In intact control animals the uteroglobin secretion in isolated uteri and the polysomal content of the specific mRNA are at the lowest detectable level (Table 1). As mentioned before, treatment with estradiol alone was sufficient to induce a certain accumulation of uteroglobin in the uterine secretion, accounting for up to 5% of the total uterine protein. Similarly, the uteroglobin production in isolated perfused uteri from animals treated with estradiol alone also resulted in the incorporation of about 4% of the radioactive amino acids into uteroglobin. In good correlation with these findings, the polysomal RNA from the endometrium of animals treated with estradiol alone contained mRNA for pre-uteroglobin, which accounted for 2 to 4% of the total mRNA population (Table 1).

After repeated treatment with progesterone, the uteroglobin content of the uterine lumina increases up to 45% of the total proteins, in good correlation with the increased ability of these uteri to incorporate radioactive amino acids into uteroglobin (Table 1). The polysomes obtained from the endometrium of these animals treated sequentially with estradiol and progesterone contain a

Figure 11 (upper): Identification of the pre-uteroglobin synthesizing polysomes. Polysomes prepared from the endometrium of induced rabbits were fractionated on linear sucrose gradients (0.5 M to 1 M sucrose; Beckman SW 27 Rotor; 27,000 rpm for 3 hrs) into seven fractions as indicated in the graph. The polysomes in each fraction were precipitated with ethanol, and submitted to chromatography on oligo(dT)-cellulose (Krystosek et al., 1975). The poly(A)-containing mRNA was translated in the wheat germ system using (^{35}S)-methionine as labelled amino acid, and the percentage of synthesized pre-uteroglobin was determined. Equal amounts of radioactivity were applied to a slab gel containing urea and SDS (Swank & Munkres, 1971), and after staining and destaining, the radioactive bands were detected by fluorography (Laskey & Mills, 1975), and quantitated with a microdensitometer. Continuous line, absorbance at 200 nm; blocks represent mRNA in each fraction.

Figure 12 (lower): Sizing of pre-uteroglobin mRNA on sucrose gradients containing 50% formamide. Polysomal RNA (130 A$_{260}$), extracted from the endometrium of induced rabbits, was treated with formamide and centrifuged through 5-20% sucrose gradients containing 50% formamide (Anderson et al., 1974). Six identical gradients (35 ml) were run in the Beckman SW 27 rotor at 27,000 rpm for 40 hrs; and after fractionation, 9 fractions were pooled with average sedimentation values of 6,8,10,12,14,16,18.5,23 and 28.5 S respectively. The RNA was precipitated with ethanol and translated in the ascites cell-free system, using (^{3}H)leucine as radioactive amino acid. The radioactivity in total cell-free products (●--●), and in pre-uteroglobin (o——o), was determined as described (Beato & Nieto, 1976). The absorbance at 260 nm is also depicted (——).

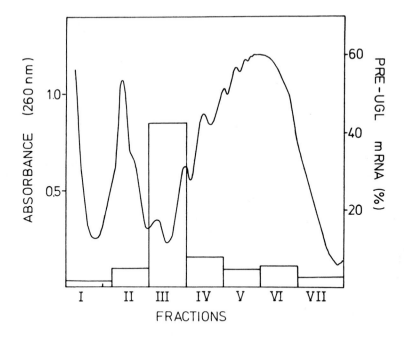

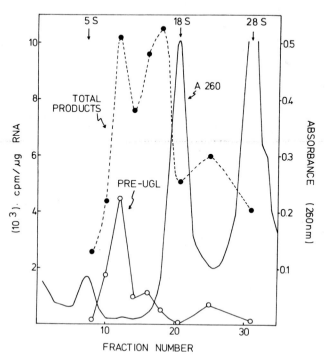

378

TABLE 1

EFFECT OF HORMONAL TREATMENT ON THE UTEROGLOBIN CONTENT OF THE
UTERINE SECRETION, ON THE SYNTHESIS OF UTEROGLOBIN IN ISOLATED UTERI,
AND ON THE POLYSOMAL CONTENT OF PRE-UTEROGLOBIN mRNA

Treatment	Uteroglobin content (%)*	Uteroglobin synthesis in isolated uteri (%)**	Content of pre-uteroglobin mRNA in endometrial polysomes (%)***
Control intact animals	0.0	0.2	0.05
Estradiol (100 ꝺg, 2 days)	3.0	2.0	2.0
Estradiol (100 ꝺg, 3 days)	4.9	3.6	4.3
Estradiol (100 ꝺg, 2 days) Progesterone (4 mg 3 days)	29.8	-	20.1
Estradiol (100 ꝺg, 2 days) Progesterone (4 mg 4 days)	48.0	44.0	-
Estradiol (100 ꝺg, 2 days) Progesterone (4 mg 5 days)	45.0	46.1	27.1
HCG, a single injection (5 days previously)	39.0	32.0	22.3

* Determined by scanning polyacrylamide gels of uterine flushes (Beato & Arnemann, 1975).

** Determined by incorporation of (^{3}H)leucine into secreted uteroglobin, isolated by immunoprecipitation (Beato & Arnemann, 1976).

*** Determined by translation in the ascites cell-free system as described (Beato & Nieto, 1976).

considerable amount of mRNA for pre-uteroglobin, varying between 20 and 30% of the total mRNA population. Similar results were obtained in animals treated with a single injection of human chorionic gonadotropin five days before the beginning of the experiments (Table 1). In this case, too, a correlation existed between the uteroglobin content of the uterine flushes, the ability of perfused uteri to synthesize and secrete uteroglobin, and the polysomal content of uteroglobin specific mRNA. The relatively low values obtained in the determination of mRNA are probably due to the fact that the endometrium used to prepare the polysomes contains a mixed population of cell, including epithelial and stromal cells, of which only the first type synthesizes uteroglobin (Kirchner, 1972).

Bullock et al. (1976) have performed titrations of the pre-uteroglobin mRNA during early pregnancy, and found that its concentration in the endometrium rises after Day 2, reaches a peak at Day 4, and then declines progressively to non-pregnant values on Day 8, a pattern very similar to that of uteroglobin secretion.

These findings suggest that the hormonal control of uteroglobin synthesis and secretion is primarily exerted by modulation of the cellular level of its specific mRNA. However, a possible control at the translational level cannot be excluded until a more precise quantitation of total and polysomal mRNA is carried out using hybridization to radioactive cDNA complementary to purified pre-uteroglobin mRNA.

UTEROGLOBIN SYNTHESIS OUTSIDE THE ENDOMETRIUM

Blastocyst

Although uteroglobin has been detected in the blastocyst fluid (Beier, 1968b; Kirchner, 1969; Hamana & Hafez, 1970; Petzoldt, 1972, 1974a; Beier & Maurer, 1975), and in the blastocyst coverings by immunofluorescence (Kirchner, 1972, 1975), it is not known whether it is synthesized by the blastocyst itself, or penetrates the blastocyst from the uterine secretion. Recent experiments indicate that blastocysts cultured in a chemically defined medium are not able to synthesize uteroglobin (Beier & Maurer, 1975), but this type of experiment is not conclusive as little is known about the conditions, hormonal and other, which could induce uteroglobin synthesis by the blastocyst in vivo (Van Blerkom & Manes, 1974). In fact, autoradiographic experiments carried out with blastocysts exposed for a brief period of time to radioactive amino acids, suggest an incorporation of radioactivity into a protein band with similar electrophoretic properties to uteroglobin (Petzoldt, 1974b). Probably, this question will remain open, until hybridization experiments with DNA complementary to the uteroglobin mRNA are carried out with RNA fractions prepared from blastocyst.

Oviduct

Feigelson and her co-workers (Kay & Feigelson, 1972; Goswami & Feigelson, 1974) have demonstrated the existence of a protein very similar to uteroglobin, that they call the "cone" protein, in the oviduct of the rabbit. Recently, the immunological identity of these two proteins has been proved unambiguously (Kirchner, 1976; Noske, 1976; Feigelson, 1976). In the oviduct, however, estrogens instead of progesterone appear to be the inducing steroid (Goswami & Feigelson, 1974). It will therefore be interesting to compare the mechanism of induction in uterus and oviduct, as this could lead to a better understanding of the role played by estrogen in the induction process.

Male genital tract and other organs

Recently (Noske, 1976; Feigelson, 1976; Beier et al., 1975; Kirchner & Schroer, 1976), a uteroglobin-like antigen with electrophoretic properties very similar to uteroglobin has been detected in male genital tract secretions. Obviously, if the synthesis of uteroglobin in this site is hormonally regulated, the mechanisms ought to be different from those operating in the uterus, and, therefore, will offer an appropriate system to study differential gene regulation.

The presence of uteroglobin in other organs, especially the lung, appears to be independent of the hormonal conditions, as it is detected in male as well as in normal and pregnant female (Noske, 1976; Feigelson, 1976). If the postulated synthesis of uteroglobin in the bronchial epithelia is confirmed, it may offer an example of hormone independent expression of the uteroglobin gene. In fact, a mRNA activity coding for a polypeptide, which is immunologically and electrophoretically related to pre-uteroglobin, has recently been detected (Bullock, 1976).

CONCLUDING REMARKS

The evidence presented in this paper does not suffice to decide the question of the onset of uteroglobin synthesis in the uterus. Experiments should be undertaken with immature and castrated rabbits, using very sensitive immunological detection methods and hybridization techniques, in order to investigate whether the uteroglobin gene is expressed at all in the absence of ovarian hormones.

In the adult animal, estradiol appears to be sufficient for the initiation of uteroglobin synthesis and secretion, but progesterone is required to reach the high levels of uteroglobin secretion which are characteristic of the pre-implantation phase. The preliminary experiments on titration of translatable uteroglobin mRNA indicate a good correlation between the uteroglobin synthesizing capacity of the endometrium and the polysomal content of uteroglobin mRNA. These data suggest that the steroid hormones control the synthesis of uteroglobin primarily by regulating the cellular levels of its specific mRNA. It remains to be decided whether the accumulation of uteroglobin mRNA is due to a more efficient transcription of the uteroglobin gene or to a reduced rate of degradation of the specific mRNA. This question, as well as the possible existence of a hormonal modulation of the translation of the uteroglobin mRNA, will only be answered after a purified messenger preparation is available, which will permit the synthesis of radioactive cDNA and a more precise titration of the uteroglobin mRNA in the different cellular compartments.

Among the many other mechanisms of regulation, I would like to mention the possibility of a hormonal control of uteroglobin synthesis at the post-translational level. As mentioned above, uteroglobin is probably synthesized by the small endometrial polysomes in the form of a precursor of the utero-globin subunit containing some 10 to 15 additional amino acids. This "pre-uteroglobin" must be proteolytically processed in the endometrium before it can be covalently dimerized to form mature uteroglobin. It is, therefore, con-ceivable that the ovarian steroids modulate the rate of this post-translational processing, eventually by their ability to bind to the protein itself (Beato & Beier, 1975), and it is this property of uteroglobin, to bind the inducing steroids, which confers special interest to the study of the molecular mechanisms involved in the regulation of its biosynthesis.

ACKNOWLEDGEMENTS

Part of the work reported in this review was carried out in collaboration with Dr. A. Nieto and J. Arnemann, who have provided skilful technical assistance throughout. I would like to thank the Schering AG, Berlin, for the gift of steroids, and the Behringwerke, Marburg, for the free supply of the rabbits used for these studies.

This work was supported in part by a grant from the Deutsche Forschungs-gemeinschaft (SFB 103 - B 1.1).

REFERENCES

ANDERSON, C.W., LEWIS, J.B., ATKINS, J.F. & GESTELAND, R.F. (1974) Cell-free synthesis of adenovirus 2 proteins programmed by fractionated mRNA: A comparison of polypeptide products and mRNA lengths. Proc. Natl. Acad. Sci. USA 71, 2756-2760.

ARTHUR, A.L. & CHANG, M.C. (1974) Induction of blastokinin by oral contra-ceptive steroids: implications for fertility control. Fertil. Steril. 25, 217-221.

ARTHUR, A.T., COWAN, B.D. & DANIEL, J.C. Jr. (1972) Steroid binding to blastokinin. Fertil. Steril. 23, 85-92.

ARTHUR, A.L. & DANIEL, J.C. Jr. (1972) Progesterone regulation of blastokinin production and maintenance of rabbit blastocysts transferred into uteri of castrated recipients. Fertil. Steril. 23, 115-122.

BAYSE, G.S., MICHAELS, A.W. & MORRISON, M. (1972) Lacto-peroxidase-catalyzed iodination of tyrosine peptides. Biochim. Biophys. Acta 284, 30-33.

BEATO, M. (1976) Binding of steroids to uteroglobin. J. Ster. Biochem. 7, 327-334.

BEATO, M. & ARNEMANN, J. (1975) Hormone-dependent synthesis and secretion of uteroglobin in isolated rabbit uterus. FEBS Letters 58, 126-129.

BEATO, M. & BAIER, R. (1975) Binding of progesterone to the proteins of the uterine luminal fluid. Identification of uteroglobin as the binding protein. Biochim. Biophys. Acta 392, 346-356.

BEATO, M. & NIETO, A. (1976) Translation of the mRNA for rabbit uteroglobin in cell-free systems. Evidence for a precursor protein. Eur. J. Biochem. 64, 15-25.

BEATO, M. & RUNGGER, D. (1975) Translation of the mRNA for rabbit uteroglobin in Xenopus oocytes. FEBS Letters 59, 305-309.

BEIER, H.M. (1968a) Uteroglobin: a hormone-sensitive endometrial protein involved in blastocyst development. Biochim. Biophys. Acta 160, 289-291.

BEIER, H.M. (1968b) Biochemisch-entwicklungsphysiologische Untersuchungen am Proteinmilieu fur die Blastocystenentwicklung des Kaninchens (Oryctogalus cuniculus). Zool. Jb. Abt. Anat. u. Ontog. 85, 72-190.

BEIER, H.M. (1974) Hormonal regulation of embryonic development before nidation. Adv. Biosci. 13, 199-219.

BEIER, H.M. (1976) Uteroglobin and related biochemical changes in the reproductive tract during early pregnancy. J. Reprod. Fert. (Suppl.), in press.

BEIER, H.M., OHN, H. & MULLER, W. (1975) Uteroglobinlike antigen in the male genital tract secretions. Cell Tiss. Res. 165, 1-11.

BEIER, H.M., KUHNEL, W. & PETRY, C. (1971) Uterine secretion proteins as extrinsic factors in preimplantation development. Adv. Biosci. 6, 165-189.

BEIER, H.M. & MAURER, R.R. (1975) Uteroglobin and other proteins in rabbit blastocyst fluid after development in vivo and in vitro. Cell. Tiss. Res. 159, 1-10.

BOLLUM, F.J. (1968) Filter paper disk techniques for assaying radioactive macromolecules. Methods Enzymol. 12b, 169-173.

BULLOCK, D.W. (1976) Progesterone induction of mRNA and protein synthesis in rabbit uterus. Annals N.Y. Acad. Sci., in press.

BULLOCK, D.W. & CONNELL, K.M. (1973) Occurrence and molecular weight of rabbit uterine "blastokinin". Biol. Reprod. 9, 125-132.

BULLOCK, D.W. & WILLEN, G.F. (1974) Regulation of a specific uterine protein by estrogen and progesterone in ovariectomized rabbits. Proc. Soc. exp. Biol. Med. 146, 294-298.

BULLOCK, D.W., WOO, S.L.C., & O'MALLEY, B.W. (1976) Uteroglobin mRNA translation in vitro. Biol. Reprod., in press.

CUATRECASAS, P. & PARIKH, I. (1972) Adsorbents for affinity chromatography. Use of N-hydroxysuccinimide esters of agarose. Biochemistry 11, 2291-2298.

DANIEL, J.C. Jr. (1971) Uterine proteins and embryonic development. Adv. Biosci. 6, 191-203.

FEIGELSON, M. (1976) Reproductive tract fluid proteins. XXIV Ann. Coll. "Protides of the Biological Fluids". Brugge, Belgium. Abstr. 60.

GOSWAMI, A. & FEIGELSON, M. (1974) Differential regulation of a low-molecular weight protein in oviductal and uterine fluids by ovarian hormones. Endocrinology 95, 669-775.

HAMANA, K. & HAFEZ, E.S.E. (1970) Disc electrophoretic patterns of uteroglobin and serum proteins in rabbit blastocoelic fluid. J. Reprod. Fert. 21, 557-560.

KAY, E. & FEIGELSON, M. (1972) An estrogen modulated protein in rabbit oviducal fluid. Biochim. Biophys. Acta 271, 436-441.

KIRCHNER, C. (1969) Untersuchungen an uterusspecifischen Glycoproteinen wahrend der fruhen Graviditat des Kaninchens (Oryctogalus cuniculus). Wilhelm Roux Archiv Entw. Mec. Org. 164, 97-133.

KIRCHNER, C. (1972) Immune histologic studies on the synthesis of a uterine-specific protein in the rabbit and its passage through the blastocyst coverings. Fertil. Steril. 23, 131-136.

KIRCHNER, C. (1976) Uteroglobin in the rabbit: I. Intracellular localization in the oviduct, uterus and the preimplantation blastocyst. Cell. Tiss. Res., in press.

KIRCHNER, C. & SCHROER, H.G. (1976) Uterine secretionlike proteins in the seminal plasma of the rabbit. J. Reprod. Fert. 47, 325-330.

KRISHNAN, R.S. (1970) Incorporation in vivo of ^{14}C-labeled precursors into the proteins of rabbit uterine secretions. Arch. Biochem. Biophys. 141, 764-765.

KRISHNAN, R.S. & DANIEL, J.C. Jr. (1967) "Blastokinin": Inducer and regulator of blastocyst development in the rabbit uterus. Science 158, 490-492.

KRYSTOSEK, A., CAWTHON, M.L. & KABAT, D. (1975) Improved methods for purification and assay of eukaryotic mRNAs and ribosomes. Quantitative analysis of their interaction in a fractionated reticulocyte cell-free system. J. Biol. Chem. 250, 6077-6084.

LAEMMLI, U.K. (1970) Cleavage of structural proteins during the assembly of the head of bacteriophage T 4. Nature 227, 680-685.

LASKEY, R.A. & MILLS, A.D. (1975) Quantitative film detection of ^{3}H and ^{14}C in polyacrylamide gels by fluorography. Eur. J. Biochem. 56, 335-341.

MAURER, R.R. & BEIER, H.M. (1976) Uterine proteins and development in vitro of preimplantation embryos of the rabbit. J. Reprod. Fert., in press.

MAYOL, R.F. & LONGENECKER, D.E. (1974) Development of a radio-immunoassay for blastokinin. Endocrinology 95, 1534-1542.

NIETO, A., PONSTINGL, H. & BEATO, M. (1976) Purification and quaternary structure of the hormonally induced protein uteroglobin. Arch. Biochem. Biophys. in press.

NISHINO, Y. (1976) Uteroglobin as a sensitive indicator for the biological activity of progestagens. Symposium "Proteins and Steroids in Early Mammalian Development", Aachen; F.R.G., July 18-24.

NOSKE, I.G. (1976) Rabbit genital tract sections: special reference to a uteroglobin-like protein. Symposium "Proteins and Steroids in Early Mammalian Development", Aachen; F.R.G., July 18-24.

PALACIOS, R., PALMITER, R.D. & SCHIMKE, R.T. (1972) Identification and isolation of ovalbumin-synthesizing polysomes. I. Specific binding of ^{125}I-anti-ovalbumin to polysomes. J. Biol. Chem. 247, 2316-2321.

PETZOLDT, U. (1972) Protein patterns of rabbit blastocyst tissues. Cytobiologie 6, 473-475.

PETZOLDT, U. (1974a) Micro-disc electrophoresis of soluble proteins in rabbit blastocysts. J. Embryol. exp. Morphol. 31, 479-487.

PETZOLDT, U. (1974b) Autoradiographic studies on the origin of rabbit blastocyst fluid proteins. Cytobiologie 9, 401-406.

RAHMAN, S.S.U., BILLIAR, R.B. & LITTLE, B. (1975) Induction of uteroglobin in rabbits by progestagens, estradiol-17β and ACTH. Biol. Reprod. 12, 305-314.

ROBERTS, B.E. & PATERSON, B.M. (1973) Efficient translation of TMV RNA and rabbit globin 9S RNA in a cell-free system from commercial wheat germ. Proc. Natl. Acad. Sci. USA 70, 2330-2334.

SCHUTZ, G., BEATO, M. & FEIGELSON, P. (1974) Isolation on cellulose of ovalbumin and globin mRNA and their translation in an ascites cell-free system. Methods Enzymol. 30, 701-708.

SCHWICK, H.G. (1965) Chemisch-entwicklungsphysiologische Beziehungen von Uterus zu Blastocyste des Kaninchens. Wilhelm Roux Arch. Entw. Mech. Org. 156, 283-343.

SWANK, R.T. & MUNKRES, K.D. (1971) Molecular weight analysis of oligopeptides by electrophoresis in polyacrylamide gel with sodium dodecyl sulfate. Anal. Biochem. 39, 462-477.

URZUA, M.A., STAMBAUGH, R., FLICKINGER, G. & MASTROIANNI, L. Jr. (1970) Uterine and oviduct fluid protein patterns in the rabbit before and after ovulation. Fert. Steril. 21, 860-865.

VAN BLERKOM, J. & MANES, C. (1974) Developmental of preimplantation rabbit embryos in vivo and in vitro. II. A comparison of qualitative aspects of protein synthesis. Devel. Biol. 40, 40-51.

WEBER, K. & OSBORN, M. (1969) Reliability of molecular weight determinations by dodecyl sulfate-polyacrylamide gel electrophoresis. J. Biol. Chem. 244, 4406-4412.

SUBJECT INDEX